ELECTROMAGNETO-MECHANICS OF MATERIAL SYSTEMS AND STRUCTURES

ELECTROMAGNETO-MECHANICS OF MATERIAL SYSTEMS AND STRUCTURES

Yasuhide Shindo

Tohoku University, Japan

Registered office
John Wiley & Sons Singapore Pte. Ltd., 1 Fusionopolis Walk, #07-01 Solaris South Tower, Singapore 138628.

For details of our global editorial offices, for customer services and for information about how to apply for permission to reuse the copyright material in this book please see our website at www.wiley.com.

Library of Congress Cataloging-in-Publication Data applied for.

ISBN: 9781118837962

Set in 10/12pt, TimesLtStd by SPi Global, Chennai, India.
Printed and bound in Singapore by Markono Print Media Pte Ltd

1   2015

Contents

About the Author   ix

Preface   xi

Acknowledgments   xiii

1   **Introduction**   1
        References   2

2   **Conducting Material Systems and Structures**   5
2.1   Basic Equations of Dynamic Magnetoelasticity   5
2.2   Magnetoelastic Plate Vibrations and Waves   7
        2.2.1   *Classical Plate Bending Theory*   9
        2.2.2   *Mindlin's Theory of Plate Bending*   13
        2.2.3   *Classical Plate Bending Solutions*   16
        2.2.4   *Mindlin Plate Bending Solutions*   23
        2.2.5   *Plane Strain Plate Solutions*   26
2.3   Dynamic Magnetoelastic Crack Mechanics   32
2.4   Cracked Materials Under Electromagnetic Force   40
2.5   Summary   45
        References   45

3   **Dielectric/Ferroelectric Material Systems and Structures**   47

Part 3.1   Dielectrics   47
3.1   Basic Equations of Electroelasticity   48
3.2   Static Electroelastic Crack Mechanics   49
        3.2.1   *Infinite Dielectric Materials*   49
        3.2.2   *Dielectric Strip*   57
3.3   Electroelastic Vibrations and Waves   60
3.4   Dynamic Electroelastic Crack Mechanics   68
3.5   Summary   72

Part 3.2   Piezoelectricity   **72**
3.6   Piezomechanics and Basic Equations   73
          3.6.1   Linear Theory   73
          3.6.2   Model of Polarization Switching   77
          3.6.3   Model of Domain Wall Motion   80
          3.6.4   Classical Lamination Theory   82
3.7   Bending of Piezoelectric Laminates   90
          3.7.1   Bimorphs   90
          3.7.2   Functionally Graded Bimorphs   100
          3.7.3   Laminated Plates   111
3.8   Electromechanical Field Concentrations   113
          3.8.1   Laminates   113
          3.8.2   Disk Composites   123
          3.8.3   Fiber Composites   126
          3.8.4   MEMS Mirrors   136
3.9   Cryogenic and High-Temperature Electromechanical Responses   140
          3.9.1   Cryogenic Electromechanical Response   140
          3.9.2   High-Temperature Electromechanical Response   147
3.10   Electric Fracture and Fatigue   149
          3.10.1   Fracture Mechanics Parameters   150
          3.10.2   Cracked Rectangular Piezoelectric Material   173
          3.10.3   Indentation Fracture Test   185
          3.10.4   Modified Small Punch Test   189
          3.10.5   Single-Edge Precracked Beam Test   193
          3.10.6   Double Torsion Test   201
          3.10.7   Fatigue of SEPB Specimens   203
3.11   Summary   212
          References   213

4   Ferromagnetic Material Systems and Structures   **219**

Part 4.1   Ferromagnetics   **219**
4.1   Basic Equations of Magnetoelasticity   220
          4.1.1   Soft Ferromagnetic Materials   220
          4.1.2   Magnetically Saturated Materials   221
          4.1.3   Electromagnetic Materials   222
4.2   Magnetoelastic Instability   224
          4.2.1   Buckling of Soft Ferromagnetic Material   225
          4.2.2   Buckling of Magnetically Saturated Material   228
          4.2.3   Bending of Soft Ferromagnetic Material   231
4.3   Magnetoelastic Vibrations and Waves   233
          4.3.1   Vibrations and Waves of Soft Ferromagnetic Material   233
          4.3.2   Vibrations and Waves of Magnetically Saturated Material   243
4.4   Magnetic Moment Intensity Factor   250
          4.4.1   Simply Supported Plate Under Static Bending   251
          4.4.2   Fixed-End Plate Under Static Bending   252
          4.4.3   Infinite Plate Under Dynamic Bending   255

4.5   Tensile Fracture and Fatigue   256
        4.5.1   Cracked Rectangular Soft Ferromagnetic Material   257
        4.5.2   Fracture Test   261
        4.5.3   Fatigue Crack Growth Test   263
4.6   Summary   265

Part 4.2   Magnetostriction   **265**
4.7   Basic Equations of Magnetostriction   265
4.8   Nonlinear Magneto-Mechanical Response   267
        4.8.1   Terfenol-D/Metal Laminates   267
        4.8.2   Terfenol-D/PZT Laminates   270
4.9   Magnetoelectric Response   272
4.10   Summary   273
        References   273

Index   **277**

About the Author

Dr. Shindo received his Doctorate of Engineering from Tohoku University in 1977. He is currently a professor in the Department of Materials Processing in the Graduate School of Engineering at Tohoku University. Dr. Shindo also served on the Board of Editors of the *International Journal of Solids and Structures* and is currently serving on the editorial board of *Journal of Mechanics of Materials and Structures*, the Advisory Board of *Acta Mechanica*, the International Editorial Board of *AES Technical Reviews International Journal (Part A: International Journal of Nano and Advanced Engineering Materials (IJNAEM), Part B: International Journal of Advances in Mechanics and Applications of Industrial Materials (IJAMAIM), Part C: International Journal of Advances and Trends in Engineering Materials and their Applications (IJATEMA), Part D: International Journal of Reliability and Safety of Engineering Systems and Structures (IJRSESS))*, the Editorial Advisory Board of *The Open Civil Engineering Journal/The Open Textile Journal/The Open Conference Proceeding Journal/The Open Physics Journal* (formerly *The Open Mechanics Journal*), the Editorial Board of *Advances in Theoretical and Applied Mechanics, Chemical Engineering and Process Techniques, International Scholarly Research Notices (Mechanical Engineering), Journal of Applied Mathematics*, the editorial board of *Strength, Fracture and Complexity, An International Journal*, and the Editor-in-Chief of *The Open Mechanical Engineering Journal* and *International Journal of Metallurgical & Materials Engineering*. His primary research interests are in the areas of mesomechanics of material systems and structures, electromagnetic fracture and damage mechanics, dynamics and cryomechanics of advanced composite materials/structural alloys, and reliability and durability of micro-/nanocomponents and devices.

Preface

The science of electromagneto-mechanics, which is concerned with the interaction of electromagnetic fields and deformation in material systems and structures, has developed because of the possibility of its practical applications in various fields such as electronic and electromechanical devices. As the area of science and technology expands, it becomes important that newly acquired knowledge and expertise are communicated effectively to those who can gain most by applying them in practice. This book covers a very wide and varied range of subject areas that fall under its subject and all aspects (theoretical, experimental, computational studies, and/or industrial applications) of electromagneto-mechanics from state-of-the-art fundamental research to applied research and applications in emerging technologies.

YASUHIDE SHINDO
Sendai, Japan
September, 2014

Acknowledgments

I am indebted to many authors whose writings are classics in the field of electromagneto-mechanics. It is also a pleasure to acknowledge the help received from my students and colleagues. Special thanks go to Professor Fumio Narita of Tohoku University, who read the entire manuscript and gave me many valuable suggestions for improvement. Finally, I would like to thank the publisher John Wiley & Sons for their continuous support for this project.

1

Introduction

The electromagneto-mechanics of material systems and structures has been developing rapidly with extensive applications in, for example, electronic industry, magnetic fusion engineering, superconducting devices, and smart materials and microelectromechanical systems (MEMS). Researchers in this interdisciplinary field are with diverse background and motivation. This book reflects a cross section of recent activities in the electromagneto-mechanics of conducting materials, dielectric materials, piezoelectric materials and devices, ferromagnetic materials, magnetostrictive material systems, and so on.

Chapter 2 deals with the magneto-mechanics of conducting material systems and structures. Here, the theory of dynamic magnetoelasticity is presented. Vibrations and waves of conducting plates are then considered, and the effect of the magnetic field on the flexural waves is examined. The theory is also applied to various problems for cracked conducting plates, and the influence of the magnetic field on the dynamic singular stresses is displayed graphically and discussed. In addition, the results for the cracked plates under large electric current and strong magnetic field are presented, and the effect of the electromagnetic force on the mechanical behavior is shown.

Chapter 3 provides the electromechanical interactions of dielectric/ferroelectric material systems and structures. In Part 3.1, we present the theory of dielectrics. Basic equations of electroelasticity are given. Applications are then made to static electroelastic crack mechanics, electroelastic vibrations and waves, and dynamic electroelastic crack mechanics of dielectric materials. Part 3.2 is devoted to the discussion of linear and nonlinear piezoelectricity. For a literature on this topic, we refer readers to Tiersten [1]. Piezomechanics and basic equations are presented. Theory is then applied to various problems, including bending behavior, electromechanical field concentrations, and cryogenic electromechanical response. Experimental data are also shown to validate the theoretical model. Furthermore, the theoretical and experimental results on the electric field dependence of fracture and fatigue of piezoelectric material systems are presented.

In Chapter 4, we deal with the magneto-mechanics of ferromagnetic material systems and structures. Part 4.1 presents the theory and test of ferromagnetics. Reference on this topic may be made to Brown [2]. Basic equations of magnetoelasticity are developed. Theory is then applied to various problems, including magnetoelastic instability, magnetoelastic vibrations, and waves of soft ferromagnetic and magnetically saturated materials under magnetic

---

Electromagneto-Mechanics of Material Systems and Structures, First Edition. Yasuhide Shindo.
© 2015 John Wiley & Sons Singapore Pte Ltd. Published 2015 by John Wiley & Sons Singapore Pte Ltd.

fields, and some experiments are performed to validate the theoretical predictions. The magnetoelastic analysis and experimental evidence are also presented for cracked plates under bending, and the effect of magnetic fields on the moment intensity factor is shown. Moreover, the tensile fracture and fatigue of soft ferromagnetic materials under magnetic fields are dedicated. Part 4.2 is concerned with a discussion of magnetostriction. Works on the subject are found to be in du Tremolet de Lacheisserie [3]. Basic equations of magnetostriction are given. Theoretical and experimental treatments of the nonlinear magneto-mechanical response in magnetostrictive material systems are then presented. Here, the material systems consist of the magnetostrictive and elastic layers, and later, we consider the magnetostrictive layer bonded to the piezoelectric layer. In addition, the piezomagnetoelectric effect of particle-reinforced composites is discussed.

There are extensive literatures on this subject. Some books are listed as follows. That is, Moon [4] organized the existing literatures on magneto-solid mechanics and gave a presentation of the basic principles and some useful method of analysis. Parton and Kudryavtsev [5] analyzed the behavior of piezoelectric materials and considered strength and failure problems for piezoelectric and electrically conducting materials. In addition, Eringen and Maugin [6, 7] presented a unified approach to the nonlinear continuum theory of deformable and fluent materials subjected to electromagnetic and thermal loads. Also, there are the following conference proceedings books of IUTAM symposium: Maugin [8], Yamamoto and Miya [9], and Hsieh [10], and of other mini-symposiums: Lee et al. [11], Yang and Maugin [12], and Shindo [13]. Moreover, the following monographs present a good discussion of this subject: Paria [14], Parkus [15, 16], Alblas [17], Moon [18], Hutter and van de Ven [19], Pao [20], Hsieh [21], Ambartsumian [22], and the set of chapters edited by Parkus [23]. In the above-listed literatures, references to other papers can be found.

References

[1] H. F. Tiersten, *Linear Piezoelectric Plate Vibration*, Plenum Press, New York, 1969.

[2] W. F. Brown, Jr., *Magnetoelastic Interactions*, Springer-Verlag, Berlin, 1966.

[3] E. du Tremolet de Lacheisserie, *Magnetostriction: Theory and Applications of Magnetoelasticity*, CRC Press, Boca Raton, FL, 1993.

[4] F. C. Moon, *Magneto-Solid Mechanics*, John Wiley & Sons, Inc., New York, 1984.

[5] V. Z. Parton and B. A. Kudryavtsev, *Electromagnetoelasticity*, Gordon and Breach Science Publishers, New York, 1988.

[6] A. C. Eringen and G. A. Maugin, *Electrodynamics of Continua I*, Springer-Verlag, New York, 1989.

[7] A. C. Eringen and G. A. Maugin, *Electrodynamics of Continua II*, Springer-Verlag, New York, 1990.

[8] G. A. Maugin (ed.), *Proceedings of the IUTAM/IUPAP Symposium on the Mechanical Behavior of Electromagnetic Solid Continua*, North-Holland, Amsterdam, 1984.

[9] Y. Yamamoto and K. Miya (eds.), *Proceedings of the IUTAM Symposium on Electromagnetomechanical Interactions in Deformable Solids and Structures*, North-Holland, Amsterdam, 1987.

[10] R. K. T. Hsieh (ed.), *Proceedings of the IUTAM Symposium on Mechanical Modeling of New Electromagnetic Materials*, Elsevier, Amsterdam, 1990.

[11] J. S. Lee, G. A. Maugin and Y. Shindo (eds.), *Mechanics of Electromagnetic Materials and Structures*, AMD-Vol. 161, MD-Vol. 42, ASME, New York, 1993.

[12] J. S. Yang and G. A. Maugin (eds.), *Mechanics of Electromagnetic Materials and Structures*, IOS Press, Amsterdam, 2000.

[13] Y. Shindo (ed.), *Mechanics of Electromagnetic Material Systems and Structures*, WIT Press, Southampton, 2003.

[14] G. Paria, "Magneto-elasticity and magneto-thermo-elasticity," *Adv. Appl. Mech.* **10**, 73 (1967).

[15] H. Parkus, "Variational principles in thermo- and magneto-elasticity," *CISM Courses and Lectures* Vol. **58**, Springer-Verlag, Wien, 1970.

[16] H. Parkus, "Magneto-thermoelasticity," *CISM Courses and Lectures* Vol. **118**, Springer-Verlag, Wien, 1972.

[17] J. B. Alblas, "Electro-magneto-elasticity," *Topics in Applied Continuum Mechanics*, J. L. Zeman and F. Ziegler (eds.), Springer-Verlag, Wien, p. 71 (1974).

[18] F. C. Moon, "Problems in magneto-solid mechanics," *Mech. Today* **4**, 307 (1978).

[19] K. Hutter and A. A. F. van de Ven, "Field matter interactions in thermoelastic solids: a unification of existing theories of electro-magneto-mechanical interactions," *Lecture Notes in Physics* Vol. **88**, Springer-Verlag, Berlin, 1978.

[20] Y.-H. Pao, "Electromagnetic forces in deformable continua," *Mech. Today* **4**, 209 (1978).

[21] R. K. T. Hsieh, "Micropolarized and magnetized media," *Mechanics of Micropolar Media*, O. Brulin and R. K. T. Hsieh (eds.), World Scientific, Singapore, p. 187 (1982).

[22] S. A. Ambartsumian, "Magneto-elasticity of thin plates and shells," *Appl. Mech. Rev.* **35**(1), 1 (1982).

[23] H. Parkus (ed.), "Electromagnetic interactions in elastic solids," *CISM Courses and Lectures* Vol. **257**, Springer-Verlag, Wien, 1979.

2

Conducting Material Systems and Structures

If electrically conducting materials are used in strong magnetic field, we must consider the effect of induced current. Figure 2.1 shows the dynamic magnetoelastic interactions of conducting materials. In this chapter, first, magnetoelastic vibrations and waves of conducting materials are discussed. Next, the influence of magnetic field on the dynamic singular stresses in cracked conducting materials is described.

The components of the superconducting structures are most often used in environments with large electric currents and strong magnetic fields. The singular stresses in cracked conducting materials under electromagnetic force are also examined in this chapter.

2.1   Basic Equations of Dynamic Magnetoelasticity

Let us now consider the rectangular Cartesian coordinates $x_i(O\text{-}x_1, x_2, x_3)$. Electrically conducting materials are permeated by a static uniform magnetic field $\mathbf{H}_0$. We consider small perturbations characterized by the displacement vector $\mathbf{u}$ produced in the material.

The magnetic and electric fields may be expressed in the form

$$\mathbf{H} = \mathbf{H}_0 + \mathbf{h}, \quad \mathbf{E} = \mathbf{0} + \mathbf{e} \tag{2.1}$$

where $\mathbf{H}$ and $\mathbf{E}$ are magnetic and electric field intensity vectors, and $\mathbf{h}$ and $\mathbf{e}$ are the fluctuating fields and are assumed to be of the same order of magnitude as the particle displacement $\mathbf{u}$. The extension of Maxwell's theory from materials at rest to those in motion was performed by Minkowski in 1908 [1].

The linearized field equations are listed as follows [2]:

$$\sigma_{ji,j} + \varepsilon_{ijk}J_j B_{0k} = \rho u_{i,tt} \tag{2.2}$$

$$\varepsilon_{ijk}H_{0k,j} = 0, \quad B_{0i,i} = 0 \tag{2.3}$$

$$\varepsilon_{ijk}h_{k,j} - d_{i,t} = j_i, \quad \varepsilon_{ijk}e_{k,j} + b_{i,t} = 0, \quad b_{i,i} = 0, \quad d_{i,i} = \rho_e \tag{2.4}$$

Electromagneto-Mechanics of Material Systems and Structures, First Edition. Yasuhide Shindo.
© 2015 John Wiley & Sons Singapore Pte Ltd. Published 2015 by John Wiley & Sons Singapore Pte Ltd.

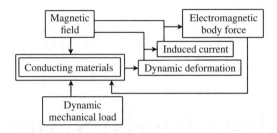

Figure 2.1 Dynamic magnetoelastic interactions of conducting materials

where u_i is the displacement vector component, σ_{ij} is the stress tensor component, H_{0i}, B_{0i}, h_i, e_i, b_i, d_i, j_i are the components of $\mathbf{H}_0$, magnetic induction vector $\mathbf{B}_0$, $\mathbf{h}$, $\mathbf{e}$, magnetic induction vector $\mathbf{b}$, electric displacement vector $\mathbf{d}$, current density vector $\mathbf{j}$, respectively, ρ is the mass density, ρ_e is the free electric charge density, a comma followed by an index denotes partial differentiation with respect to the space coordinate x_i or the time t, and the summation convention over repeated indices is used. The permutation symbol ε_{ijk} is defined by

$$\varepsilon_{ijk} = \begin{array}{l} +1 \ \text{ if } ijk \text{ is a cyclic permutation of 1, 2, 3} \\ \ \ \ 0 \ \text{ if any two indices are equal} \\ -1 \ \text{ if } ijk \text{ is an anticyclic permutation} \end{array} \tag{2.5}$$

The equation of conservation of charge follows from the first and fourth of Eqs. (2.4)

$$\rho_{e,t} + j_{i,i} = 0 \tag{2.6}$$

The linearized constitutive equations can be written as

$$\sigma_{ij} = \lambda u_{k,k}\delta_{ij} + \mu(u_{i,j} + u_{j,i}) \tag{2.7}$$

$$\sigma_{ij}^M = H_{0j}b_i + h_j B_{0i} - \frac{1}{2}(H_{0k}b_k + h_k B_{0k})\delta_{ij} \tag{2.8}$$

$$B_{0i} = \kappa H_{0i}, \quad b_i = \kappa h_i, \tag{2.9}$$

$$d_i = \epsilon e_i + (\epsilon\kappa - \epsilon_0\kappa_0)\varepsilon_{ijk}u_{j,t}H_{0k} \tag{2.10}$$

$$j_i = \sigma(e_i + \varepsilon_{ijk}u_{j,t}B_{0k}) \tag{2.11}$$

where σ_{ij}^M is the Maxwell stress tensor component, $\lambda = 2Gv/(1-2v)$ and $\mu = G$ are the Lamé constants, $G = E/2(1+v)$ is the shear modulus, E and v are Young's modulus and Poisson's ratio, respectively, κ is the magnetic permeability, ϵ is the permittivity, $\kappa_0 = 1.26 \times 10^{-6}$ H/m is the magnetic permeability of free space, $\epsilon_0 = 8.85 \times 10^{-12}$ C/Vm is the permittivity of free space, and σ is electric conductivity. The Kronecker delta δ_{ij} is defined by

$$\delta_{ij} = \begin{array}{l} 1 \ \text{ if } i = j \\ 0 \ \text{ if } i \neq j \end{array} \tag{2.12}$$

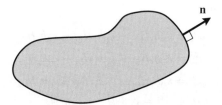

Figure 2.2 An arbitrary material volume element

The linearized boundary conditions are obtained as

$$[\![\sigma_{ji} + \sigma_{ji}^M]\!]n_j = 0 \tag{2.13}$$

$$\varepsilon_{ijk}n_j[\![h_k]\!] = j_i^s, \quad \varepsilon_{ijk}n_j[\![e_k + u_{k,t}B_{0k}]\!] = 0$$
$$[\![b_i]\!]n_i = 0, \quad [\![d_i]\!]n_i = \rho_e^s \tag{2.14}$$

where ρ_e^s is the surface charge density, j_i^s is the component of surface current density vector $\mathbf{j}^s$, n_i is the component of outer unit vector $\mathbf{n}$ normal to an undeformed material as shown in Fig. 2.2, and $[\![f_i]\!]$ means the jump in any field quantity f_i across the boundary; that is, $[\![f_i]\!] = f_i^e - f_i$. The superscript e denotes the quantity outside the material. The conservation of charge applied to the region gives the following normal boundary condition:

$$[\![j_i]\!]n_i = -\rho_{e,t}^s \tag{2.15}$$

2.2 Magnetoelastic Plate Vibrations and Waves

In this section, the magnetoelastic vibrations and waves of a conducting material are discussed. Consider an electrically conducting elastic plate with thickness $2h$ in a rectangular Cartesian coordinate system (x, y, z). The coordinate axes x and y are in the middle plane of the plate, and the z-axis is normal to this plane. It is assumed that the plate has the permittivity $\epsilon = \epsilon_0$ and magnetic permeability $\kappa = \kappa_0$, respectively. A uniform magnetic field $\mathbf{H}_0$ is applied.

Using the first of Eqs. (2.9), Eqs. (2.3) can be expressed as

$$H_{0z,y}^e - H_{0y,z}^e = 0, \quad H_{0x,z}^e - H_{0z,x}^e = 0, \quad H_{0y,x}^e - H_{0x,y}^e = 0,$$
$$H_{0x,x}^e + H_{0y,y}^e + H_{0z,z}^e = 0$$
$$H_{0z,y} - H_{0y,z} = 0, \quad H_{0x,z} - H_{0z,x} = 0, \quad H_{0y,x} - H_{0x,y} = 0,$$
$$H_{0x,x} + H_{0y,y} + H_{0z,z} = 0 \tag{2.16}$$

The mechanical constitutive equations are taken to be the usual Hooke's law

$$\sigma_{xx} = \lambda(u_{x,x} + u_{y,y} + u_{z,z}) + 2\mu u_{x,x}$$
$$\sigma_{yy} = \lambda(u_{x,x} + u_{y,y} + u_{z,z}) + 2\mu u_{y,y}$$
$$\sigma_{zz} = \lambda(u_{x,x} + u_{y,y} + u_{z,z}) + 2\mu u_{z,z}$$
$$\sigma_{xy} = \sigma_{yx} = \mu(u_{x,y} + u_{y,x}) \tag{2.17}$$
$$\sigma_{yz} = \sigma_{zy} = \mu(u_{y,z} + u_{z,y})$$
$$\sigma_{xz} = \sigma_{zx} = \mu(u_{z,x} + u_{x,z})$$

and the Maxwell stresses are

$$\sigma_{xx}^M = \kappa_0 h_x H_{0x} - \kappa_0 h_y H_{0y} - \kappa_0 h_z H_{0z}$$
$$\sigma_{yy}^M = \kappa_0 h_y H_{0y} - \kappa_0 h_z H_{0z} - \kappa_0 h_x H_{0x}$$
$$\sigma_{zz}^M = \kappa_0 h_z H_{0z} - \kappa_0 h_x H_{0x} - \kappa_0 h_y H_{0y}$$
$$\sigma_{xy}^M = \sigma_{yx}^M = \kappa_0 h_x H_{0y} + \kappa_0 h_y H_{0x}$$
$$\sigma_{yz}^M = \sigma_{zy}^M = \kappa_0 h_y H_{0z} + \kappa_0 h_z H_{0y}$$
$$\sigma_{zx}^M = \sigma_{xz}^M = \kappa_0 h_z H_{0x} + \kappa_0 h_x H_{0z}$$

(2.18)

The currents are determined by Ohm's law, Eq. (2.11), and they are

$$j_x = \sigma\{e_x + \kappa_0(u_{y,t}H_{0z} - u_{z,t}H_{0y})\}$$
$$j_y = \sigma\{e_y + \kappa_0(u_{z,t}H_{0x} - u_{x,t}H_{0z})\}$$
$$j_z = \sigma\{e_z + \kappa_0(u_{x,t}H_{0y} - u_{y,t}H_{0x})\}$$

(2.19)

The stress equations of motion, Eq. (2.2), are given by

$$\sigma_{xx,x} + \sigma_{yx,y} + \sigma_{zx,z} = \rho u_{x,tt} - \kappa_0(j_y H_{0z} - j_z H_{0y})$$
$$\sigma_{xy,x} + \sigma_{yy,y} + \sigma_{zy,z} = \rho u_{y,tt} - \kappa_0(j_z H_{0x} - j_x H_{0z})$$
$$\sigma_{xz,x} + \sigma_{yz,y} + \sigma_{zz,z} = \rho u_{z,tt} - \kappa_0(j_x H_{0y} - j_y H_{0x})$$

(2.20)

Neglecting displacement currents compared to the conduction currents, and using the second of Eqs. (2.9) and Eq. (2.10), the Maxwell's equations, Eqs. (2.4), are

$$h_{z,y}^e - h_{y,z}^e = 0, \quad h_{x,z}^e - h_{z,x}^e = 0, \quad h_{y,x}^e - h_{x,y}^e = 0$$
$$h_{z,y} - h_{y,z} = j_x, \quad h_{x,z} - h_{z,x} = j_y, \quad h_{y,x} - h_{x,y} = j_z$$

(2.21)

$$e_{z,y}^e - e_{y,z}^e = -\kappa_0 h_{x,t}^e, \quad e_{x,z}^e - e_{z,x}^e = -\kappa_0 h_{y,t}^e, \quad e_{y,x}^e - e_{x,y}^e = -\kappa_0 h_{z,t}^e$$
$$e_{z,y} - e_{y,z} = -\kappa_0 h_{x,t}, \quad e_{x,z} - e_{z,x} = -\kappa_0 h_{y,t}, \quad e_{y,x} - e_{x,y} = -\kappa_0 h_{z,t}$$

(2.22)

$$h_{x,x}^e + h_{y,y}^e + h_{z,z}^e = 0$$
$$h_{x,x} + h_{y,y} + h_{z,z} = 0$$

(2.23)

$$e_{x,x}^e + e_{y,y}^e + e_{z,z}^e = 0$$
$$\epsilon_0(e_{x,x} + e_{y,y} + e_{z,z}) = \rho_e$$

(2.24)

From Eqs. (2.13) – (2.15), we obtain the linearized boundary conditions

$$\sigma_{zz}^{Me}(x, y, \pm h, t) - \{\sigma_{zz}(x, y, \pm h, t) + \sigma_{zz}^M(x, y, \pm h, t)\} = 0$$
$$\sigma_{zy}^{Me}(x, y, \pm h, t) - \{\sigma_{zy}(x, y, \pm h, t) + \sigma_{zy}^M(x, y, \pm h, t)\} = 0$$
$$\sigma_{zx}^{Me}(x, y, \pm h, t) - \{\sigma_{zx}(x, y, \pm h, t) + \sigma_{zx}^M(x, y, \pm h, t)\} = 0$$

(2.25)

$$h_x^e(x, y, \pm h, t) - h_x(x, y, \pm h, t) = j_y^s$$
$$h_y^e(x, y, \pm h, t) - h_y(x, y, \pm h, t) = -j_x^s \tag{2.26}$$

$$e_x^e(x, y, \pm h, t) - e_x(x, y, \pm h, t) = 0$$
$$e_y^e(x, y, \pm h, t) - e_y(x, y, \pm h, t) = 0 \tag{2.27}$$

$$h_z^e(x, y, \pm h, t) - h_z(x, y, \pm h, t) = 0 \tag{2.28}$$

$$\epsilon_0\{e_z^e(x, y, \pm h, t) - e_z(x, y, \pm h, t)\} = \rho_e^s \tag{2.29}$$

$$j_z(x, y, \pm z, t) = 0 \tag{2.30}$$

2.2.1 Classical Plate Bending Theory

Classical plate bending theory for magnetoelastic interactions in a conducting material is applied. By using magnetoelastic thin plate theory [3, 4], the rectangular displacement, magnetic field and electric field components can be expressed as follows:

$$u_x = -zw_{,x}, \quad u_y = -zw_{,y}, \quad u_z = w(x, y, t) \tag{2.31}$$

$$h_z = f(x, y, t), \quad e_x = \varphi(x, y, t), \quad e_y = \psi(x, y, t) \tag{2.32}$$

where $w(x, y, t)$ represents the deflection of the middle plane of the plate, f, φ, ψ are the functions of x, y, t. Substituting from Eqs. (2.31) and (2.32) into the fourth and fifth of Eqs. (2.21) and the second of Eqs. (2.24), using Eqs. (2.19), and assuming $\rho_e = 0$, we obtain

$$h_{x,z} = f_{,x} + \sigma\{\psi + \kappa_0(H_{0x}w_{,t} + H_{0z}zw_{,xt})\}$$
$$h_{y,z} = f_{,y} - \sigma\{\varphi - \kappa_0(H_{0y}w_{,t} + H_{0z}zw_{,yt})\} \tag{2.33}$$
$$e_{z,z} = -\varphi_{,x} - \psi_{,y}$$

For conductors with finite electric conductivity, j_x^s and j_y^s can be neglected so that, from Eqs. (2.26)–(2.30), we get

$$h_x(x, y, \pm h, t) = h_x^e(x, y, \pm h, t)$$
$$h_y(x, y, \pm h, t) = h_y^e(x, y, \pm h, t) \tag{2.34}$$

$$\varphi(x, y, \pm h, t) = e_x^e(x, y, \pm h, t)$$
$$\psi(x, y, \pm h, t) = e_y^e(x, y, \pm h, t) \tag{2.35}$$

$$f(x, y, \pm h, t) = h_z^e(x, y, \pm h, t) \tag{2.36}$$

$$e_z(x, y, \pm h, t) + \kappa_0\{H_{0y}u_{x,t}(x, y, \pm h, t) - H_{0x}u_{y,t}(x, y, \pm h, t)\} = 0 \tag{2.37}$$

Integrating the representations (2.33) with respect to z, we obtain the remaining electromagnetic field components as

$$h_x = \frac{1}{2}\{h_x(x, y, h, t) + h_x(x, y, -h, t)\} + z\{f_{,x} + \sigma(\psi + \kappa_0 H_{0x} w_{,t})\}$$
$$+ \sigma \kappa_0 H_{0z} \frac{z^2 - h^2}{2} w_{,xt}$$

$$h_y = \frac{1}{2}\{h_y(x, y, h, t) + h_y(x, y, -h, t)\} + z\{f_{,y} - \sigma(\varphi - \kappa_0 H_{0y} w_{,t})\} \qquad (2.38)$$
$$+ \sigma \kappa_0 H_{0z} \frac{z^2 - h^2}{2} w_{,yt}$$

$$e_z = \frac{1}{2}\{e_z(x, y, h, t) + e_z(x, y, -h, t)\} - z(\varphi_{,x} + \psi_{,y})$$

Therefore, all the electromagnetic field components are represented by means of the four desired functions w, f, φ, ψ. Integrating the first and second of Eqs. (2.33) and the sixth of Eqs. (2.22) with respect to z from $-h$ to h, we have

$$f_{,x} + \sigma(\psi + \kappa_0 H_{0x} w_{,t}) = \frac{1}{2h}\{h_x(x, y, h, t) - h_x(x, y, -h, t)\}$$

$$f_{,y} - \sigma(\varphi - \kappa_0 H_{0y} w_{,t}) = \frac{1}{2h}\{h_y(x, y, h, t) - h_y(x, y, -h, t)\} \qquad (2.39)$$

$$\psi_{,x} - \varphi_{,y} = -\kappa_0 f_{,t}$$

The stress boundary conditions on the plate surfaces are

$$\sigma_{zx} = \sigma_{zy} = \sigma_{zz} = 0 \quad (z = \pm h) \qquad (2.40)$$

The bending and twisting moments per unit length ($M_{xx}, M_{yy}, M_{xy} = M_{yx}$) and the vertical shear forces per unit length (Q_x, Q_y) can be expressed in terms of w as

$$M_{xx} = \int_{-h}^{h} \sigma_{xx} z \, dz = -D(w_{,xx} + v w_{,yy})$$

$$M_{yy} = \int_{-h}^{h} \sigma_{yy} z \, dz = -D(w_{,yy} + v w_{,xx}) \qquad (2.41)$$

$$M_{xy} = M_{yx} = \int_{-h}^{h} \sigma_{xy} z \, dz = -D(1 - v) w_{,xy}$$

$$Q_x = \int_{-h}^{h} \sigma_{zx} \, dz = -D(w_{,xx} + w_{,yy})_{,x}$$

$$Q_y = \int_{-h}^{h} \sigma_{zy} \, dz = -D(w_{,xx} + w_{,yy})_{,y} \qquad (2.42)$$

where $D = 4\mu h^3/3(1-v)$ is the flexural rigidity of the plate. Now, if we substitute Eqs. (2.19) into Eqs. (2.20), multiply the first and second of Eqs. (2.20) by $z\,dz$, and integrate from $-h$ to h, taking into account the boundary condition in Eq. (2.40), we obtain the results

$$M_{xx,x} + M_{yx,y} - Q_x = -\frac{2}{3}\rho h^3 w_{,xtt} - m_{xx}^C$$

$$M_{xy,x} + M_{yy,y} - Q_y = -\frac{2}{3}\rho h^3 w_{,ytt} - m_{yy}^C$$

(2.43)

The moments m_{xx}^C and m_{yy}^C are derived as

$$m_{xx}^C = \int_{-h}^{h} (j_y B_z - j_z B_y) z\,dz$$

$$= \frac{2}{3}\sigma\kappa_0 h^3 \{\kappa_0(H_{0y}^2 + H_{0z}^2)w_{,xt} - \kappa_0 H_{0x}H_{0y}w_{,yt} + H_{0y}\varphi_{,x} + H_{0y}\psi_{,y}\}$$

$$m_{yy}^C = \int_{-h}^{h} (j_z B_x - j_x B_z) z\,dz$$

$$= -\frac{2}{3}\sigma\kappa_0 h^3 \{\kappa_0 H_{0x}H_{0y}w_{,xt} - \kappa_0(H_{0x}^2 + H_{0z}^2)w_{,yt} + H_{0x}\varphi_{,x} + H_{0x}\psi_{,y}\}$$

(2.44)

If the third of Eqs. (2.20) is multiplied by dz and integrated from $-h$ to h, taking into account the boundary condition (2.40), we obtain

$$Q_{x,x} + Q_{y,y} = 2h\rho w_{,tt} - q^C$$

(2.45)

The load q^C applied to the plate is derived as

$$q^C = \int_{-h}^{h} (j_x B_y - j_y B_x)dz = 2h\sigma\kappa_0\{H_{0y}\varphi - H_{0x}\psi - \kappa_0(H_{0x}^2 + H_{0y}^2)w_{,t}\}$$

(2.46)

Eliminating Q_x, Q_y from Eqs. (2.43) and Eq. (2.45) and taking into account Eq. (2.41), we have the equation of motion for a thin plate under the influence of magnetic field

$$D(w_{,xxxx} + 2w_{,xxyy} + w_{,yyyy}) - \frac{2}{3}\rho h^3(w_{,xx} + w_{,yy})_{,tt} + 2\rho hw_{,tt} - m_{xx,x}^C - m_{yy,y}^C - q^C = 0 \quad (2.47)$$

Equations (2.39) and (2.47) are the basic equations of linear bending theory for conducting thin plates.

If we consider a perfectly conducting plate ($\sigma \to \infty$), we get from Eqs. (2.19)

$$e_x + \kappa_0(u_{y,t}H_{0z} - u_{z,t}H_{0y}) = 0$$
$$e_y + \kappa_0(u_{z,t}H_{0x} - u_{x,t}H_{0z}) = 0$$
$$e_z + \kappa_0(u_{x,t}H_{0y} - u_{y,t}H_{0x}) = 0$$

(2.48)

Hence, from the fourth, fifth, and sixth of Eqs. (2.22) and Eqs. (2.48), with Eqs. (2.31), we obtain the magnetic field intensity components as

$$h_x = H_{0z}w_{,x} + \frac{v}{1-v}H_{0x}zw_{,xx} - H_{0y}zw_{,xy} + \frac{1}{1-v}H_{0x}zw_{,yy}$$

$$h_y = -H_{0z}w_{,y} + \frac{1-2v}{1-v}H_{0y}zw_{,xx} - H_{0x}zw_{,xy} - \frac{v}{1-v}H_{0y}zw_{,yy} \tag{2.49}$$

$$h_z = -H_{0x}w_{,x} + H_{0y}w_{,y} - H_{0z}zw_{,xx} + H_{0z}zw_{,yy}$$

From the fourth, fifth, and sixth of Eqs. (2.21), with Eqs. (2.49), we also have

$$j_x = -\frac{1-2v}{1-v}H_{0y}w_{,xx} - 2H_{0x}w_{,xy} - \frac{1}{1-v}H_{0y}w_{,yy}$$

$$j_y = \left(\frac{1}{1-v}H_{0x} + H_{0z}\right)w_{,xx} - 2H_{0y}w_{,xy} + \left(\frac{1}{1-v}H_{0x} - H_{0z}\right)w_{,yy} \tag{2.50}$$

$$j_z = -2H_{0z}w_{,xy} + \frac{1-2v}{1-v}H_{0y}zw_{,xxx} - \frac{1}{1-v}H_{0x}zw_{,xxy} + \frac{1-2v}{1-v}H_{0y}zw_{,xyy}$$

$$-\frac{1}{1-v}H_{0x}zw_{,yyy}$$

Moreover, from Eqs. (2.20), with Eqs. (2.50), we have

$$M_{xx,x} + M_{yx,y} - Q_x = -\frac{2}{3}\rho h^3 w_{,xtt} - m_{xx}^{CP}$$

$$M_{xy,x} + M_{yy,y} - Q_y = -\frac{2}{3}\rho h^3 w_{,ytt} - m_{yy}^{CP} \tag{2.51}$$

$$Q_{x,x} + Q_{y,y} = 2h\rho w_{,tt} - q^{CP} \tag{2.52}$$

where

$$m_{xx}^{CP} = h\{\sigma_{zx}(x,y,h,t) - \sigma_{zx}(x,y,-h,t)\}$$

$$+\frac{\kappa_0}{1-v}\int_{-h}^{h}[\{(1-v)H_{0z}^2 + H_{0z}H_{0x}\}w_{,xx} + \{-(1-v)H_{0z}^2 + H_{0z}H_{0x}\}w_{,yy}$$

$$-(1-2v)H_{0y}^2 zw_{,xxx} + H_{0x}H_{0y}zw_{,xxy} - (1-2v)H_{0y}^2 zw_{,xyy}$$

$$+H_{0x}H_{0y}zw_{,yyy}]z\,dz$$

$$m_{yy}^{CP} = h\{\sigma_{zy}(x,y,h,t) - \sigma_{zy}(x,y,-h,t)\}$$

$$+\frac{\kappa_0}{1-v}\int_{-h}^{h}\{-(1-2v)H_{0y}H_{0z}w_{,xx} - H_{0y}H_{0z}w_{,yy} - (1-2v)H_{0x}H_{0y}zw_{,xxx}$$

$$+H_{0x}^2 zw_{,xxy} - (1-2v)H_{0x}H_{0y}zw_{,xyy} + H_{0x}^2 zw_{,yyy}\}z\,dz$$

$$\tag{2.53}$$

$$q^{CP} = \sigma_{zz}(x,y,h,t) - \sigma_{zz}(x,y,-h,t)$$

$$-\frac{\kappa_0}{1-v}\int_{-h}^{h}\left[\{H_{0x}^2 + (1-2v)H_{0y}^2 + (1-v)H_{0z}H_{0x}\}w_{,xx}\right.$$

$$+\{H_{0x}^2 - H_{0y}^2 + (1-v)H_{0z}H_{0x}\}w_{,yy}\right]dz \tag{2.54}$$

From Eqs. (2.51) and (2.52), with Eqs. (2.41), we obtain

$$D(w_{,xxxx} + 2w_{,xxyy} + w_{,yyyy}) - \frac{2}{3}\rho h^3(w_{,xx} + w_{,yy})_{,tt}$$

$$+2\rho h w_{,tt} - m_{xx,x}^{CP} - m_{yy,y}^{CP} - q^{CP} = 0 \tag{2.55}$$

2.2.2 Mindlin's Theory of Plate Bending

Mindlin's theory of plate bending [5], which accounts for the rotatory inertia and shear effects, is applied for magnetoelastic interactions in a conducting material. The rectangular displacement components may assume the forms

$$u_x = z\Psi_x(x, y, t), \quad u_y = z\Psi_y(x, y, t), \quad u_z = \Psi_z(x, y, t) \tag{2.56}$$

where Ψ_z represents the normal displacement of the plate, and Ψ_x and Ψ_y denote the rotations of the normals about the x- and y-axes. The magnetic and electric field components are expressed by Eqs. (2.32). Substituting from Eqs. (2.56) and (2.32) into the fourth and fifth of Eqs. (2.21) and the second of Eqs. (2.24), using Eqs. (2.19), and assuming $\rho_e = 0$, we obtain

$$h_{x,z} = f_{,x} + \sigma\{\psi + \kappa_0(H_{0x}\Psi_{z,t} - H_{0z}z\Psi_{x,t})\}$$

$$h_{y,z} = f_{,y} - \sigma\{\varphi - \kappa_0(H_{0y}\Psi_{z,t} - H_{0z}z\Psi_{y,t})\} \tag{2.57}$$

$$e_{z,z} = -\varphi_{,x} - \psi_{,y}$$

Integrating the representations (2.57) with respect to z, we obtain

$$h_x = \frac{1}{2}\{h_x(x, y, h, t) + h_x(x, y, -h, t)\} + z\{f_{,x} + \sigma(\psi + \kappa_0 H_{0x}\Psi_{z,t})\}$$
$$- \sigma\kappa_0 H_{0z}\frac{z^2 - h^2}{2}\Psi_{x,t}$$

$$h_y = \frac{1}{2}\{h_y(x, y, h, t) + h_y(x, y, -h, t)\} + z\{f_{,y} - \sigma(\varphi - \kappa_0 H_{0y}\Psi_{z,t})\} \tag{2.58}$$
$$- \sigma\kappa_0 H_{0z}\frac{z^2 - h^2}{2}\Psi_{y,t}$$

$$e_z = \frac{1}{2}\{e_z(x, y, h, t) + e_z(x, y, -h, t)\} - z(\varphi_{,x} + \psi_{,y})$$

Therefore, all of the electromagnetic field components are represented by means of the six desired functions $\Psi_x, \Psi_y, \Psi_z, f, \varphi, \psi$. Integrating the first and second of Eqs. (2.33) and the sixth of Eqs. (2.22) with respect to z from $-h$ to h, we have

$$f_{,x} + \sigma(\psi + \kappa_0 H_{0x}\Psi_{z,t}) = \frac{1}{2h}\{h_x(x, y, h, t) - h_x(x, y, -h, t)\}$$

$$f_{,y} - \sigma(\varphi - \kappa_0 H_{0y}\Psi_{z,t}) = \frac{1}{2h}\{h_y(x, y, h, t) - h_y(x, y, -h, t)\} \tag{2.59}$$

$$\psi_{,x} - \varphi_{,y} = -\kappa_0 f_{,t}$$

The bending and twisting moments per unit length ($M_{xx}, M_{yy}, M_{xy} = M_{yx}$) and the vertical shear forces per unit length (Q_x, Q_y) can be expressed in terms of Ψ_x, Ψ_y, and Ψ_z as

$$M_{xx} = \int_{-h}^{h} \sigma_{xx} z \, dz = D(\Psi_{x,x} + v\Psi_{y,y})$$

$$M_{yy} = \int_{-h}^{h} \sigma_{yy} z \, dz = D(\Psi_{y,y} + v\Psi_{x,x}) \tag{2.60}$$

$$M_{xy} = M_{yx} = \int_{-h}^{h} \sigma_{xy} z \, dz = \frac{1-v}{2} D(\Psi_{y,x} + \Psi_{x,y})$$

$$Q_x = \int_{-h}^{h} \sigma_{zx} \, dz = \frac{\pi^2}{6} \mu h(\Psi_{z,x} + \Psi_x)$$

$$Q_y = \int_{-h}^{h} \sigma_{zy} \, dz = \frac{\pi^2}{6} \mu h(\Psi_{z,y} + \Psi_y) \tag{2.61}$$

Now, substituting Eqs. (2.19) into Eqs. (2.20), multiplying the first and second of Eqs. (2.20) by $z \, dz$, and integrating from $-h$ to h, with the boundary condition (2.40), we find

$$M_{xx,x} + M_{yx,y} - Q_x = \frac{2}{3} \rho h^3 \Psi_{x,tt} - m_{xx}^M$$

$$M_{xy,x} + M_{yy,y} - Q_y = \frac{2}{3} \rho h^3 \Psi_{y,tt} - m_{yy}^M \tag{2.62}$$

where

$$m_{xx}^M = \frac{2}{3} \sigma \kappa_0 h^3 [\kappa_0 (H_{0y}^2 + H_{0z}^2)\Psi_{x,t} - \kappa_0 H_{0x} H_{0y} \Psi_{y,t} - H_{0y}\varphi_{,x} - H_{0y}\psi_{,y}]$$

$$m_{yy}^M = \frac{2}{3} \sigma \kappa_0 h^3 [\kappa_0 H_{0x} H_{0y} \Psi_{x,t} - \kappa_0 (H_{0x}^2 + H_{0z}^2)\Psi_{y,t} - H_{0x}\varphi_{,x} - H_{0x}\psi_{,y}] \tag{2.63}$$

Multiplying the third of Eqs. (2.20) by dz and integrating from $-h$ to h, with the boundary condition (2.40), we obtain

$$Q_{x,x} + Q_{y,y} = 2h\rho\Psi_{z,tt} - q^M \tag{2.64}$$

where

$$q^M = 2h\sigma\kappa_0 \{H_{0y}\varphi - H_{0x}\psi - \kappa_0(H_{0x}^2 + H_{0y}^2)\Psi_{z,t}\} \tag{2.65}$$

Substituting Eqs. (2.60) and (2.61) into Eqs. (2.62) and Eq. (2.64), we have

$$\frac{S}{2}\{(1-v)(\Psi_{x,xx} + \Psi_{x,yy}) + (1+v)\Phi_{,x}\} - \Psi_x - \Psi_{z,x} = \frac{4h^2\rho}{\pi^2\mu}\Psi_{x,tt} - \frac{6}{\pi^2\mu h}m_{xx}^M$$

$$\frac{S}{2}\{(1-v)(\Psi_{y,xx} + \Psi_{y,yy}) + (1+v)\Phi_{,y}\} - \Psi_y - \Psi_{z,y} = \frac{4h^2\rho}{\pi^2\mu}\Psi_{y,tt} - \frac{6}{\pi^2\mu h}m_{yy}^M \tag{2.66}$$

$$\Psi_{z,xx} + \Psi_{z,yy} + \Phi = \frac{4h^2\rho}{\pi^2\mu}\frac{1}{R}\Psi_{z,tt} - \frac{6}{\pi^2\mu h}q^M$$

where

$$\Phi = \Psi_{x,x} + \Psi_{y,y} \tag{2.67}$$

$$R = \frac{h^2}{3}, \quad S = \frac{6D}{\pi^2\mu h} \tag{2.68}$$

Equations (2.59) and (2.66) are the basic equations of linear bending theory for conducting Mindlin plates.

If we consider a perfectly conducting Mindlin plate ($\sigma \to \infty$), we obtain the magnetic field intensity components, from the fourth, fifth, and sixth of Eqs. (2.22) and Eqs. (2.48), with Eqs. (2.56), as

$$h_x = -H_{0z}\Psi_x + \frac{v}{1-v}H_{0x}z\Psi_{x,x} + H_{0y}z\Psi_{x,y} + \frac{1}{1-v}H_{0x}z\Psi_{y,y}$$

$$h_y = -\frac{1-2v}{1-v}H_{0y}z\Psi_{x,x} + H_{0z}\Psi_y + H_{0x}z\Psi_{y,x} + \frac{v}{1-v}H_{0y}z\Psi_{y,y} \qquad (2.69)$$

$$h_z = H_{0z}z\Psi_{x,x} - H_{0z}z\Psi_{y,y} - H_{0x}\Psi_{z,x} + H_{0y}\Psi_{z,y}$$

From the fourth, fifth, and sixth of Eqs. (2.21), with Eqs. (2.69), we also have

$$j_x = \frac{1-2v}{1-v}H_{0y}\Psi_{x,x} + H_{0z}z\Psi_{x,xy} - H_{0x}\Psi_{y,x} - \frac{v}{1-v}H_{0y}\Psi_{y,y} - H_{0z}z\Psi_{y,yy}$$
$$\quad -H_{0x}\Psi_{z,xy} + H_{0y}\Psi_{z,yy}$$

$$j_y = \frac{v}{1-v}H_{0x}\Psi_{x,x} + H_{0y}\Psi_{x,y} + H_{0z}z\Psi_{x,xx} + \frac{1}{1-v}H_{0x}\Psi_{y,y} - H_{0z}z\Psi_{y,xy} \qquad (2.70)$$
$$\quad +H_{0x}\Psi_{z,xx} - H_{0y}\Psi_{z,xy}$$

$$j_z = H_{0z}\Psi_{x,y} - \frac{1-2v}{1-v}H_{0y}z\Psi_{x,xx} - \frac{v}{1-v}H_{0x}z\Psi_{x,xy} - H_{0y}z\Psi_{x,yy}$$
$$\quad +H_{0z}\Psi_{y,x} + H_{0x}z\Psi_{y,xx} + \frac{v}{1-v}H_{0y}z\Psi_{y,xy} - \frac{1}{1-v}H_{0x}z\Psi_{y,yy}$$

Moreover, from Eqs. (2.20), with Eqs. (2.70), we have

$$M_{xx,x} + M_{yx,y} - Q_x = \frac{2}{3}\rho h^3 \Psi_{x,tt} - m_{xx}^{MP}$$
$$\qquad (2.71)$$
$$M_{xy,x} + M_{yy,y} - Q_y = \frac{2}{3}\rho h^3 \Psi_{y,tt} - m_{yy}^{MP}$$

$$Q_{x,x} + Q_{y,y} = 2h\rho\Psi_{z,tt} - q^{MP} \qquad (2.72)$$

where

$$m_{xx}^{MP} = h\{\sigma_{zx}(x,y,h,t) - \sigma_{zx}(x,y,-h,t)\}$$
$$+\frac{\kappa_0}{1-v}\int_{-h}^{h}[H_{0z}H_{0x}v\Psi_{x,x} + \{(1-2v)H_{0y}^2 + (1-v)H_{0z}^2\}z\Psi_{x,xx}$$
$$+vH_{0x}H_{0y}z\Psi_{x,xy} + (1-v)H_{0y}^2 z\Psi_{x,yy} - (1-v)H_{0y}H_{0z}\Psi_{y,x}$$
$$-H_{0z}H_{0x}\Psi_{y,y} - (1-v)H_{0x}H_{0y}z\Psi_{y,xx} - \{vH_{0y}^2 + (1-v)H_{0z}^2\}z\Psi_{y,xy}$$
$$-H_{0x}H_{0y}z\Psi_{y,yy} + (1-v)H_{0z}H_{0x}\Psi_{z,xx} - (1-v)H_{0y}H_{0z}\Psi_{z,xy}]dz$$

$$m_{yy}^{MP} = h\{\sigma_{zy}(x,y,h,t) - \sigma_{zy}(x,y,-h,t)\}$$
$$+\frac{\kappa_0}{1-v}\int_{-h}^{h}[(1-2v)H_{0y}H_{0z}\Psi_{x,x} - (1-v)H_{0z}H_{0x}\Psi_{x,y}$$
$$+(1-2v)H_{0x}H_{0y}z\Psi_{x,xx} + \{vH_{0x}^2 + (1-v)H_{0z}^2\}z\Psi_{x,xy}$$
$$+(1-v)H_{0x}H_{0y}z\Psi_{x,yy} - 2(1-v)H_{0z}H_{0x}\Psi_{y,x} - vH_{0y}H_{0z}\Psi_{y,y}$$
$$-(1-v)H_{0x}^2 z\Psi_{y,xx} - vH_{0x}H_{0y}z\Psi_{y,xy} - \{H_{0x}^2 + (1-v)H_{0z}^2\}z\Psi_{y,yy}$$
$$-(1-v)H_{0z}H_{0x}\Psi_{z,xy} + (1-v)H_{0y}H_{0z}\Psi_{z,yy}]dz \qquad (2.73)$$

$$q^{MP} = \sigma_{zz}(x, y, h, t) - \sigma_{zz}(x, y, -h, t)$$

$$+ \frac{\kappa_0}{1 - v} \int_{-h}^{h} [\{-vH_{0x}^2 + (1 - 2v)H_{0y}^2\}\Psi_{x,x} - (1 - v)H_{0x}H_{0y}\Psi_{x,y}$$

$$-(1 - v)H_{0z}H_{0x}z\Psi_{x,xx} + (1 - v)H_{0y}H_{0z}z\Psi_{x,xy}$$

$$-(1 - v)H_{0x}H_{0y}\Psi_{y,x} + (H_{0x}^2 - vH_{0y}^2)\Psi_{y,y}$$

$$-(1 - v)(H_{0y}H_{0z} - H_{0z}H_{0x})z\Psi_{y,xy}$$

$$-(1 - v)H_{0x}^2\Psi_{z,xx} + (1 - v)H_{0y}^2\Psi_{z,yy}]dz \tag{2.74}$$

Substituting Eqs. (2.60) and (2.61) into Eqs. (2.71) and (2.72), we obtain

$$\frac{S}{2}\{(1 - v)(\Psi_{x,xx} + \Psi_{x,yy}) + (1 + v)\Phi_{,x}\} - \Psi_x - \Psi_{z,x}$$

$$= \frac{4h^2\rho}{\pi^2\mu}\Psi_{x,tt} - \frac{6}{\pi^2\mu h}m_{xx}^{MP}$$

$$\frac{S}{2}\{(1 - v)(\Psi_{y,xx} + \Psi_{y,yy}) + (1 + v)\Phi_{,y}\} - \Psi_y - \Psi_{z,y}$$

$$= \frac{4h^2\rho}{\pi^2\mu}\Psi_{y,tt} - \frac{6}{\pi^2\mu h}m_{yy}^{MP} \tag{2.75}$$

$$\Psi_{z,xx} + \Psi_{z,yy} + \Phi = \frac{4h^2\rho}{\pi^2\mu}\frac{1}{R}\Psi_{z,tt} - \frac{6}{\pi^2\mu h}q^{MP} \tag{2.76}$$

2.2.3   *Classical Plate Bending Solutions*

Consider an electrically conducting elastic plate with thickness $2h$ in a rectangular Cartesian coordinate system (x, y, z) as shown in Fig. 2.3. The coordinate axes x and y are in the middle plane of the plate, and the z-axis is normal to this plane. Let magnetic flexural waves be traveled in the y-direction. A uniform magnetic field of magnetic induction $B_0 = \kappa_0 H_0$ is also applied. We consider three possible cases of magnetic field direction. The first case is $H_{0x} = H_0, H_{0y} = H_{0z} = 0$ (Case I); the second and third cases are $H_{0y} = H_0, H_{0x} = H_{0z} = 0$ (Case II) and $H_{0z} = H_0, H_{0x} = H_{0y} = 0$ (Case III), respectively.

First, we consider the applied magnetic field in the y-direction (Case II). Basic equations (2.39) and (2.47) take the form

$$\sigma\psi = \frac{1}{2h}\{h_x(y, h, t) - h_x(y, -h, t)\}$$

$$f_{,y} - \sigma(\varphi - \kappa_0 H_0 w_{,t}) = \frac{1}{2h}\{h_y(y, h, t) - h_y(y, -h, t)\} \tag{2.77}$$

$$\varphi_{,y} = \kappa_0 f_{,t}$$

$$Dw_{,yyyy} + 2\rho h w_{,tt} - 2h\sigma\kappa_0 H_0(\varphi - \kappa_0 H_0 w_{,t}) = 0 \tag{2.78}$$

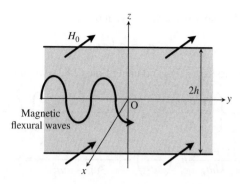

Figure 2.3   An electrically conducting elastic plate and magnetic flexural waves

We can write the solutions for w, f, φ, and ψ in the form

$$
\begin{aligned}
w &= w_0 \exp\{-i(ky + \omega t)\} \\
f &= f_0 \exp\{-i(ky + \omega t)\} \\
\varphi &= \varphi_0 \exp\{-i(ky + \omega t)\} \\
\psi &= \psi_0 \exp\{-i(ky + \omega t)\}
\end{aligned}
\tag{2.79}
$$

where w_0, f_0, φ_0, and ψ_0 are the amplitudes of the time harmonic waves, k is the wave number, and ω is the angular frequency. From the first, second, and third of Eqs. (2.21) and the first of Eqs. (2.23), we can obtain

$$
\left.
\begin{aligned}
h_x^e &= 0 \\
h_y^e &= iA_1 \exp\{-kz - i(ky + \omega t)\} \quad (z \geq h) \\
&= -iA_2 \exp\{kz - i(ky + \omega t)\} \quad (z \leq -h)
\end{aligned}
\right\}
$$
$$
\left.
\begin{aligned}
h_z^e &= A_1 \exp\{-kz - i(ky + \omega t)\} \quad (z \geq h) \\
&= A_2 \exp\{kz - i(ky + \omega t)\} \quad (z \leq -h)
\end{aligned}
\right\}
\tag{2.80}
$$

where A_1 and A_2 are undetermined constants. Making use of Eqs. (2.80) and Eqs. (2.34), Eq. (2.36) renders the x, y-magnetic intensity components h_x, h_y and the amplitude f_0

$$
\begin{aligned}
h_x(y, \pm h, t) &= 0 \\
h_y(y, \pm h, t) &= \pm i f_0 \exp\{-i(ky + \omega t)\} \\
f_0 &= A_1 \exp(-kh) = A_2 \exp(-kh)
\end{aligned}
\tag{2.81}
$$

Substituting Eqs. (2.79) and (2.81) into Eqs. (2.77) yields

$$
\begin{aligned}
\psi_0 &= 0 \\
\varphi_0 &= -\frac{c_2 \kappa_0 H_0 \sigma_h (\omega/kc_2)^2}{(1 + kh) - i\sigma_h(\omega/kc_2)} kw_0 \\
k\varphi_0 &= \omega \kappa_0 f_0
\end{aligned}
\tag{2.82}
$$

where $c_2 = (\mu/\rho)^{1/2}$ is the shear wave velocity, and

$$\sigma_h = c_2 h \sigma \kappa_0 \tag{2.83}$$

Substituting Eqs. (2.79) into Eq. (2.78) and using Eqs. (2.82), we obtain the dispersion relation

$$\left(\frac{\omega}{kc_2}\right)^4 + \left\{\frac{(1+kh)^2}{\sigma_h^2} - \frac{2(kh)^2}{3(1-v)} - \frac{1+kh}{kh}h_c\right\}\left(\frac{\omega}{kc_2}\right)^2$$

$$+i\frac{(1+kh)^2}{kh\sigma_h}h_c\left(\frac{\omega}{kc_2}\right) - \frac{2(kh)^2(1+kh)^2}{3(1-v)\sigma_h^2} = 0 \quad \text{(Case II)} \tag{2.84}$$

where

$$h_c = \frac{\kappa_0}{\mu}H_0^2 \tag{2.85}$$

If we assume a quasistatic electromagnetic state (small conductivity, high frequency, $|k|^2 \gg \sigma\kappa_0\omega$), we obtain

$$\left(\frac{\omega}{kc_2}\right)^2 + i\frac{\sigma_h}{kh}h_c\left(\frac{\omega}{kc_2}\right) - \frac{2(kh)^2}{3(1-v)} = 0 \quad \text{(Case II)} \tag{2.86}$$

Mathematically, this is accomplished by dropping $b_{i,t}$ from Maxwell's equations, Eqs. (2.4), while retaining all other time derivatives. In this case, the effect of magnetic field on the elastic waves is solely a damping [2, 6]. If we let $\sigma \to \infty$, Eq. (2.84) can be written in the form

$$\left(\frac{\omega}{kc_2}\right)^2 = \frac{2(kh)^2}{3(1-v)} + \frac{1+kh}{kh}h_c \quad \text{(Case II)} \tag{2.87}$$

To examine the effect of magnetic field $H_{0y} = H_0$ on the flexural waves, we consider two examples: graphite and aluminum. The material properties are as follows [2]:
Graphite:

$$\rho = 2250 \text{ kg/m}^3$$
$$\sigma = 1.25 \times 10^5 \text{ mho/m} \tag{2.88}$$
$$\mu = 1.96 \times 10^9 \text{ N/m}^2$$

Aluminum:

$$\rho = 2700 \text{ kg/m}^3$$
$$\sigma = 3.54 \times 10^7 \text{ mho/m} \tag{2.89}$$
$$\mu = 2.37 \times 10^{10} \text{ N/m}^2$$

Poisson's ratios of graphite and aluminum are assumed to be $v = 0.3$. Figure 2.4 shows the variation of the phase velocity $\text{Re}(\omega/kc_2)$ with the wave number kh of the plate for $\sigma_h = 1.0$ under the normalized magnetic field $h_c = 4.06 \times 10^{-2}$. $h_c = 4.06 \times 10^{-2}$ corresponds to the magnetic induction of $B_0 = 10$ T for the graphite plate. For the aluminum plate, $h_c = 4.06 \times 10^{-2}$ corresponds to $B_0 = 35$ T. This is an extremely unrealistic value. Smaller values will probably not produce any noticeable effects, and the results are of no practical importance. The dashed line represents the result of $h_c = 0$, and the dot-dashed line refers to a quasistatic electromagnetic field. We note that under $h_c = 0$, the result agrees with the solution

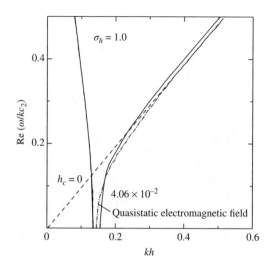

Figure 2.4   Phase velocity versus wave number ($\sigma_h = 1.0$)

of elementary theory, and $\mathrm{Re}(\omega/kc_2) = \{2/3(1-v)\}^{1/2}kh$. Figure 2.5 also shows the variation of the attenuation $-\mathrm{Im}(\omega/kc_2)$ with the wave number kh for $\sigma_h = 1.0$ under $h_c = 4.06 \times 10^{-2}$. Figures 2.6 and 2.7 show the results for $\sigma_h = 0.5$ corresponding to Figs. 2.4 and 2.5, respectively. The effect of the magnetic field on the phase velocity and attenuation is observed at low wave number. The results justify the assumption of a quasistatic electromagnetic field. In particular, this assumption is more valid for $\sigma_h = 0.5$ rather than for $\sigma_h = 1.0$. Figure 2.8 shows the variation of the phase velocity $\mathrm{Re}(\omega/kc_2)$ with the wave number kh of the plate for $\sigma_h = 10, 1000$ under the normalized magnetic field $h_c = 3.36 \times 10^{-3}$. $h_c = 3.36 \times 10^{-3}$

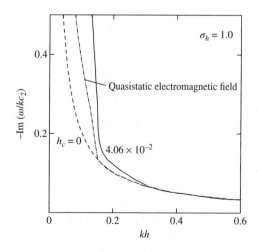

Figure 2.5   Attenuation versus wave number ($\sigma_h = 1.0$)

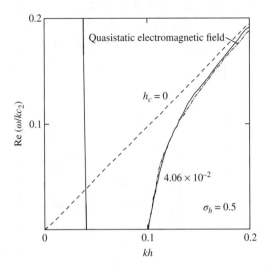

Figure 2.6 Phase velocity versus wave number ($\sigma_h = 0.5$)

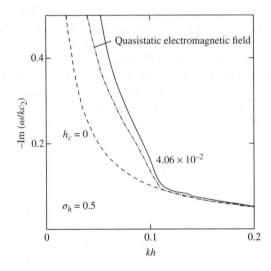

Figure 2.7 Attenuation versus wave number ($\sigma_h = 0.5$)

corresponds to the magnetic induction of $B_0 = 3$ T for the graphite plate and $B_0 = 10$ T for the aluminum plate. The dot-dashed line refers to the case of $h_c = 0$. The curve obtained for the case of a perfect conductivity, that is, $\sigma \rightarrow \infty$, coincides with the case of $\sigma_h = 1000$. The results show validity of the assumption of the perfect conductivity. As σ_h increases, $\text{Im}(\omega/kc_2)$ decreases, and $\text{Im}(\omega/kc_2)$ becomes zero in the case of $\sigma \rightarrow \infty$.

Next, we consider the magnetic fields of $H_{0x} = H_0, H_{0y} = H_{0z} = 0$ (Case I) and $H_{0z} = H_0, H_{0x} = H_{0y} = 0$ (Case III). The solution procedure is the same as that of Case II. If

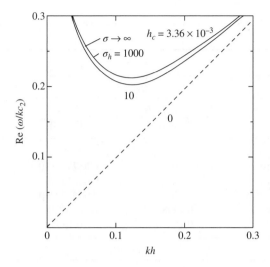

Figure 2.8 Phase velocity versus wave number ($\sigma_h = 10, 1000$)

we assume a quasistatic electromagnetic field, we obtain the dispersion relations

$$\left(\frac{\omega}{kc_2}\right)^2 + i\frac{2kh\sigma_h}{3}h_c\left(\frac{\omega}{kc_2}\right) - \frac{2(kh)^2}{3(1-v)} = 0 \quad \text{(Case I)} \tag{2.90}$$

$$\left(\frac{\omega}{kc_2}\right)^2 + i\frac{kh\sigma_h}{3}h_c\left(\frac{\omega}{kc_2}\right) - \frac{2(kh)^2}{3(1-v)} = 0 \quad \text{(Case III)} \tag{2.91}$$

If we let $\sigma \to \infty$, we obtain

$$\left(\frac{\omega}{kc_2}\right)^2 = \frac{2(kh)^2}{3(1-v)} + \frac{(1-2v)(kh)^2}{3(1-v)}h_c \quad \text{(Case I)} \tag{2.92}$$

$$\left(\frac{\omega}{kc_2}\right)^2 = \frac{2(kh)^2}{3(1-v)} + \left\{\frac{(kh)^2}{3(1-v)} + kh + 1\right\}h_c \quad \text{(Case III)} \tag{2.93}$$

Let us examine the effect of magnetic field direction on the flexural waves. Table 2.1 lists the variation of the phase velocity $\text{Re}(\omega/kc_2)$ and attenuation $\text{Im}(\omega/kc_2)$ with the wave number kh for Cases I, II, and III. A quasistatic electromagnetic field is assumed, and $\sigma_h = 0.5$ and normalized magnetic field $h_c = 4.06 \times 10^{-2}$ are chosen. Also shown are the results of $h_c = 0$. The effect of the y-direction magnetic field (Case II) on flexural waves is more pronounced than those of x- and z-direction magnetic fields (Cases I and III). Figure 2.9 shows the variation of the phase velocity $\text{Re}(\omega/kc_2)$ with the wave number kh for Cases I, II, and III. Perfect conductivity ($\sigma \to \infty$) is assumed, and $h_c = 0.003$ is chosen. The curve obtained for the x-direction magnetic field (Case I) almost coincides with that of the purely elastic case. The effect of the y-direction magnetic field (Case II) on flexural waves is pronounced, similar to the case under a quasistatic electromagnetic field.

Table 2.1   Phase velocity and attenuation versus wave number for Cases I, II, and III (quasistatic electromagnetic field)

kh	$\sigma_h = 0.5$			
	$\text{Re}(\omega/kc_2)$			
	$h_c = 4.06 \times 10^{-2}$			$h_c = 0$
	Case I	Case II	Case III	
0.0	0	–	0	0
0.05	0.48794×10^{-1}	0	0.48795×10^{-1}	0.48795×10^{-1}
0.07	0.68311×10^{-1}	0	0.68313×10^{-1}	0.68313×10^{-1}
0.10	0.97588×10^{-1}	0	0.97589×10^{-1}	0.97590×10^{-1}
0.15	0.14638	0.12981	0.14638	0.14639
0.20	0.19518	0.18846	0.19518	0.19518
kh	$\text{Im}(\omega/kc_2)$			
	$h_c = 4.06 \times 10^{-2}$			$h_c = 0$
	Case I	Case II	Case III	
0.0	0	–	0	0
0.05	-0.33834×10^{-3}	-0.40006	-0.16917×10^{-3}	0
0.07	-0.47368×10^{-3}	-0.27291	-0.23684×10^{-3}	0
0.10	-0.67668×10^{-3}	-0.12941	-0.33834×10^{-3}	0
0.15	-0.10150×10^{-2}	-0.67668×10^{-1}	-0.50751×10^{-3}	0
0.20	-0.13534×10^{-2}	-0.50751×10^{-1}	-0.67668×10^{-3}	0

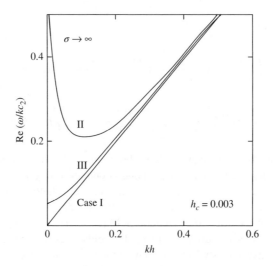

Figure 2.9   Phase velocity versus wave number for Cases I, II, and III (perfect conductivity)

2.2.4 Mindlin Plate Bending Solutions

Consider an electrically conducting Mindlin plate with thickness $2h$ in a rectangular Cartesian coordinate system (x, y, z) as shown in Fig. 2.3. The coordinate axes x and y are in the middle plane of the plate, and the z-axis is normal to this plane. Let magnetic flexural waves be traveled in the y-direction. A uniform magnetic field of magnetic induction $B_0 = \kappa_0 H_0$ is also applied.

First, we consider the applied magnetic field in the y-direction (Case II). Equations (2.59) and the second and third of Eqs. (2.66) take the form

$$
\sigma\psi = \frac{1}{2h}\{h_x(y, h, t) - h_x(y, -h, t)\}
$$
$$
f_{,y} - \sigma(\varphi - \kappa_0 H_0 \Psi_{z,t}) = \frac{1}{2h}\{h_y(y, h, t) - h_y(y, -h, t)\}
$$
$$
\varphi_{,y} = \kappa_0 f_{,t}
$$
$$(2.94)$$

$$
S\Psi_{y,yy} - \Psi_y - \Psi_{z,y} = \frac{4h^2\rho}{\pi^2\mu}\Psi_{y,tt}
$$
$$
\Psi_{z,tt} + \Psi_{y,y} = \frac{4h^2\rho}{\pi^2\mu}\frac{1}{R}\Psi_{z,tt} - \frac{12\sigma\kappa_0 H_0}{\pi^2\mu}(\varphi - \kappa_0 H_0 \Psi_{z,t})
$$
$$(2.95)$$

We can write the solutions for Ψ_x, Ψ_y, and Ψ_z in the form

$$
\Psi_x = 0
$$
$$
\Psi_y = \Psi_{y0}\exp\{-i(ky + \omega t)\}
$$
$$
\Psi_z = \Psi_{z0}\exp\{-i(ky + \omega t)\}
$$
$$(2.96)$$

where Ψ_{y0} and Ψ_{z0} are the amplitudes of the time harmonic waves. The solutions for f, φ, and ψ are given by Eqs. (2.79). Substituting Eqs. (2.96), (2.79), and (2.81) into Eqs. (2.94) yields

$$
\psi_0 = 0
$$
$$
\varphi_0 = -\frac{c_2\kappa_0 H_0\sigma_h(\omega/kc_2)^2}{(1 + kh) - i\sigma_h(\omega/kc_2)}k\Psi_{z0}
$$
$$
k\varphi_0 = \omega\kappa_0 f_0
$$
$$(2.97)$$

Substituting Eqs. (2.96) and (2.79) into Eq. (2.95) and using Eqs. (2.97), we obtain

$$
\frac{4(kh)^3\sigma_h}{3}\left(\frac{\omega}{kc_2}\right)^5 + i\frac{4(kh)^3(1 + kh)}{3}\left(\frac{\omega}{kc_2}\right)^4
$$
$$
-\left[\left\{\frac{8}{3(1-v)} + \frac{\pi^2}{9}\right\}(kh)^3\sigma_h + \frac{\pi^2(kh)\sigma_h}{3} + \frac{4(kh)^2(1+kh)}{3}\Gamma_h\right]\left(\frac{\omega}{kc_2}\right)^3
$$
$$
-i\left[\left\{\frac{8}{3(1-v)} + \frac{\pi^2}{9}\right\}(kh)^3 + \frac{\pi^2(kh)}{3}\right](1 + kh)\left(\frac{\omega}{kc_2}\right)^2
$$
$$
+\left[\frac{2\pi^2(kh)^3\sigma_h}{9(1-v)} + \left\{\frac{8(kh)^2}{3(1-v)} + \frac{\pi^2}{3}\right\}(1+kh)\Gamma_h\right]\left(\frac{\omega}{kc_2}\right)
$$
$$
+i\frac{2\pi^2(kh)^3(1+kh)}{9(1-v)} = 0 \quad \text{(Case II)}
$$
$$(2.98)$$

where

$$\Gamma_h = h_c \sigma_h \tag{2.99}$$

The relation between Ψ_{y0} and Ψ_{z0} is

$$i\frac{\pi^2(kh)}{6}\Psi_{y0} = -\left[\frac{\pi^2(kh)}{6} - 2(kh)\left(\frac{\omega}{kc_2}\right)^2\right.$$

$$\left. - \left\{2i\left(\frac{\omega}{kc_2}\right) - \frac{\sigma_h(\omega/kc_2)^2}{(1+kh)-i\sigma_h(\omega/kc_2)}\right\}\Gamma_h\right]k\Psi_{z0} \tag{2.100}$$

If we assume a quasistatic electromagnetic state, we obtain

$$\frac{4(kh)^3}{3}\left(\frac{\omega}{kc_2}\right)^4 + i\frac{4(kh)^2}{3}\Gamma_h\left(\frac{\omega}{kc_2}\right)^3$$

$$- \left[\left\{\frac{8}{3(1-v)} + \frac{\pi^2}{9}\right\}(kh)^3 + \frac{\pi^2(kh)}{3}\right]\left(\frac{\omega}{kc_2}\right)^2$$

$$-i\left\{\frac{8(kh)^2}{3(1-v)} + \frac{\pi^2}{3}\right\}\Gamma_h\left(\frac{\omega}{kc_2}\right) + \frac{2\pi^2(kh)^3}{9(1-v)} = 0 \quad \text{(Case II)} \tag{2.101}$$

$$i\frac{\pi^2(kh)}{6}\Psi_{y0} = -\left\{\frac{\pi^2(kh)}{6} - 2(kh)\left(\frac{\omega}{kc_2}\right)^2 - 2i\left(\frac{\omega}{kc_2}\right)\Gamma_h\right\}k\Psi_{z0} \tag{2.102}$$

If we let $\sigma \to \infty$, we obtain

$$\frac{4(kh)^3}{3}\left(\frac{\omega}{c_2 k}\right)^4$$

$$- \left[\left\{\frac{8}{3(1-v)} + \frac{\pi^2}{9}\right\}(kh)^3 + \frac{\pi^2(kh)}{3} + \frac{4(kh)^2(1+kh)}{3}h_c\right]\left(\frac{\omega}{c_2 k}\right)^2$$

$$+ \frac{2\pi^2(kh)^3}{9(1-v)} + \left\{\frac{8(kh)^2}{3(1-v)} + \frac{\pi^2}{3}\right\}(1+kh)h_c = 0 \quad \text{(Case II)} \tag{2.103}$$

$$i\frac{\pi^2}{6}\Psi_{y0} = \left\{\frac{\pi^2(kh)}{6} - 2(kh)\left(\frac{\omega}{c_2 k}\right)^2 + 2(1+kh)h_c\right\}\Psi_{z0} \tag{2.104}$$

Figure 2.10 shows the variation of the phase velocity $\text{Re}(\omega/kc_2)$ with the wave number kh of the Mindlin plate for $\sigma_h = 0.5$ under the normalized magnetic field $h_c = 4.06 \times 10^{-2}$. The dashed line represents the result of $h_c = 0$. Figure 2.11 also shows the variation of the phase velocity $\text{Re}(\omega/kc_2)$ with the wave number kh of the Mindlin plate for perfect conductivity ($\sigma \to \infty$) under $h_c = 0.003$. The effect of the magnetic field on the phase velocity is observed at low wave number, similar to the case for the classical plate.

Next, we consider the magnetic fields of $H_{0x} = H_0, H_{0y} = H_{0z} = 0$ (Case I) and $H_{0z} = H_0, H_{0x} = H_{0y} = 0$ (Case III). The solution procedure is the same as that of Case II. If

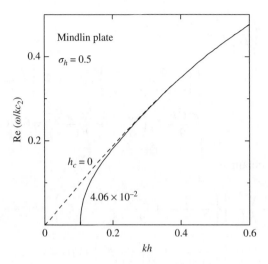

Figure 2.10   Phase velocity versus wave number (Mindlin plate, $\sigma_h = 0.5$)

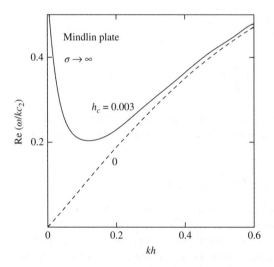

Figure 2.11   Phase velocity versus wave number (Mindlin plate, perfect conductivity)

we assume a quasistatic electromagnetic field, we obtain

$$\frac{4(kh)^2}{3}\left(\frac{\omega}{kc_2}\right)^4 + i\frac{4(kh)}{3}\Gamma_h\left(\frac{\omega}{kc_2}\right)^3$$

$$-\left[\left\{\frac{8}{3(1-v)} + \frac{\pi^2}{9}\right\}(kh)^2 + \frac{\pi^2}{3}\right]\left(\frac{\omega}{kc_2}\right)^2$$

$$-i\frac{2\pi^2(kh)}{9}\Gamma_h\left(\frac{\omega}{kc_2}\right) + \frac{2\pi^2(kh)^2}{9(1-v)} = 0 \quad \text{(Case I)} \qquad (2.105)$$

$$\frac{4(kh)^2}{3}\left(\frac{\omega}{kc_2}\right)^4 + i\frac{4(kh)}{3}\Gamma_h\left(\frac{\omega}{kc_2}\right)^3$$

$$-\left[\left\{\frac{8}{3(1-v)} + \frac{\pi^2}{9}\right\}(kh)^2 + \frac{\pi^2}{3}\right]\left(\frac{\omega}{kc_2}\right)^2$$

$$-i\frac{\pi^2(kh)}{9}\Gamma_h\left(\frac{\omega}{kc_2}\right) + \frac{2\pi^2(kh)^2}{9(1-v)} = 0 \quad \text{(Case III)} \qquad (2.106)$$

If we let $\sigma \to \infty$, we obtain

$$\frac{4(kh)^2}{3}\left(\frac{\omega}{kc_2}\right)^4 - \left[\left\{\frac{8}{3(1-v)} + \frac{\pi^2}{9} + \frac{4}{3(1-v)}h_c\right\}(kh)^2 + \frac{\pi^2}{3}\right]\left(\frac{\omega}{kc_2}\right)^2$$

$$+\left\{\frac{2\pi^2}{9(1-v)} + \frac{\pi^2}{8(1-v)}h_c\right\}(kh)^2 = 0 \quad \text{(Case I)} \qquad (2.107)$$

$$\frac{4(kh)^2}{3}\left(\frac{\omega}{kc_2}\right)^4 - \left[\left\{\frac{8}{3(1-v)} + \frac{\pi^2}{9}\right\}(kh)^2 + 4(kh)\left(\frac{kh}{3}+1\right)h_c + \frac{\pi^2}{3}\right]\left(\frac{\omega}{kc_2}\right)^2$$

$$+\frac{2\pi^2(kh)^2}{9(1-v)} + \frac{\pi^2(kh)}{3}\left(\frac{kh}{3}+1\right)h_c = 0 \quad \text{(Case III)} \qquad (2.108)$$

It is expected that the effect of the y-direction magnetic field (Case II) on flexural waves is more pronounced than those of x- and z-direction magnetic fields (Cases I and III).

2.2.5   Plane Strain Plate Solutions

Consider an electrically conducting plate with thickness $2h$ in a rectangular Cartesian coordinate system (x, y, z) as shown in Fig. 2.3. The coordinate axes x and y are in the middle plane of the plate, and the z-axis is normal to this plane. Let magnetoelastic waves be traveled in the y-direction. A uniform magnetic field of magnetic induction $B_0 = \kappa_0 H_0$ in the y-direction ($H_{0y} = H_0, H_{0x} = H_{0z} = 0$) is also applied. In addition, we assume plane strain normal to the x-axis.

The relevant components of the stress tensor follow from Hooke's law as

$$\begin{aligned}
\sigma_{yy} &= \lambda(u_{y,y} + u_{z,z}) + 2\mu u_{y,y} \\
\sigma_{zz} &= \lambda(u_{y,y} + u_{z,z}) + 2\mu u_{z,z} \\
\sigma_{yz} &= \mu(u_{y,z} + u_{z,y})
\end{aligned} \qquad (2.109)$$

The stress equations of motion, Eqs. (2.20), become

$$\begin{aligned}
\sigma_{yy,y} + \sigma_{zy,z} &= \rho u_{y,tt} \\
\sigma_{yz,y} + \sigma_{zz,z} &= \rho u_{z,tt} - \kappa_0 j_x H_0
\end{aligned} \qquad (2.110)$$

First, we assume a quasistatic electromagnetic field. The current density component j_x is obtained by the first of Eqs. (2.19) as

$$j_x = -\sigma \kappa_0 u_{z,t} H_0 \tag{2.111}$$

When the constitutive equations (2.109) and (2.111) are substituted in the stress equations of motion, Eqs. (2.110), we obtain the displacement equations of motion

$$\{2\mu u_{y,y} + \lambda(u_{y,y} + u_{z,z})\}_{,y} + \mu(u_{y,z} + u_{z,y})_{,z} = \rho u_{y,tt}$$
$$\mu(u_{y,z} + u_{z,y})_{,y} + \{2\mu u_{z,z} + \lambda(u_{y,y} + u_{z,z})\}_{,z} - \sigma \kappa_0^2 H_0^2 u_{z,t} = \rho u_{z,tt} \tag{2.112}$$

If the expressions for the stresses are also substituted in the stress boundary conditions on the plate surfaces, Eq. (2.40), we obtain

$$\lambda u_{y,y}(y, \pm h, t) + (2\mu + \lambda)u_{z,z}(y, \pm h, t) = 0$$
$$u_{y,z}(y, \pm h, t) + u_{z,y}(y, \pm h, t) = 0 \tag{2.113}$$

To investigate wave motion in the conducting plane strain plate, we consider solutions of Eqs. (2.112) of the form

$$u_y = u_{y0} \exp\{pz - i(ky + \omega t)\}$$
$$u_z = u_{z0} \exp\{pz - i(ky + \omega t)\} \tag{2.114}$$

where u_{y0}, u_{z0} are the amplitudes of the time harmonic waves, and p is the modification factor of the wave amplitude with respect to the thickness. Substitution of Eqs. (2.114) into Eqs. (2.112) yields the following algebraic equations:

$$\begin{bmatrix} a_{11} & a_{12} \\ a_{21} & a_{22} \end{bmatrix} \begin{bmatrix} u_y \\ u_z \end{bmatrix} = 0 \tag{2.115}$$

where

$$a_{11} = \mu p^2 - (\lambda + 2\mu)k^2 + \rho\omega^2$$
$$a_{12} = -i(\lambda + \mu)kp$$
$$a_{21} = -i(\lambda + \mu)kp \tag{2.116}$$
$$a_{22} = (\lambda + 2\mu)p^2 - \mu k^2 + \rho\omega^2 + i\omega\sigma\kappa_0^2 H_0^2$$

A nontrivial solution of Eq. (2.115) will exist when p is related to k and ω such that the determinant of the coefficient matrix is zero, that is, when

$$\frac{c_1^2}{c_2^2}\left(\frac{p}{k}\right)^4 + \left\{\left(1 + \frac{c_1^2}{c_2^2}\right)\left(\frac{\omega}{kc_2}\right)^2 - 2\frac{c_1^2}{c_2^2} + i\frac{1}{kh}\Gamma_h\left(\frac{\omega}{kc_2}\right)\right\}\left(\frac{p}{k}\right)^2$$

$$+ \left(\frac{\omega}{kc_2}\right)^4 - \left(1 + \frac{c_1^2}{c_2^2}\right)\left(\frac{\omega}{kc_2}\right)^2 + \frac{c_1^2}{c_2^2}$$

$$+ i\frac{1}{kh}\Gamma_h\left(\frac{\omega}{kc_2}\right)^3 - i\frac{c_1^2}{c_2^2}\frac{1}{kh}\Gamma_h\left(\frac{\omega}{kc_2}\right) = 0 \tag{2.117}$$

where $c_1 = \{(\lambda + 2\mu)/\rho\}^{1/2}$ is the longitudinal wave velocity. The solutions take the form

$$
\begin{bmatrix} u_y \\ u_z \end{bmatrix} = \begin{bmatrix} 1 & 1 & 1 & 1 \\ -b_{12} & -b_{22} & b_{12} & b_{22} \end{bmatrix} \begin{bmatrix} A_{11} \exp(\lambda_{11}kz) \exp\{-i(ky + \omega t)\} \\ A_{12} \exp(\lambda_{12}kz) \exp\{-i(ky + \omega t)\} \\ B_{11} \exp(-\lambda_{11}kz) \exp\{-i(ky + \omega t)\} \\ B_{12} \exp(-\lambda_{12}kz) \exp\{-i(ky + \omega t)\} \end{bmatrix}
$$

(2.118)

where A_{11}, A_{12}, B_{11}, and B_{12} are arbitrary, $\pm\lambda_{11}, \pm\lambda_{12}$ are the roots of Eq. (2.117), and

$$
b_{j2} = i \frac{\lambda_{1j}^2 - c_1^2/c_2^2 + (\omega/kc_2)^2}{\lambda_{1j}(c_1^2/c_2^2 - 1)} \quad (j = 1, 2)
$$

(2.119)

For convenience, in Eq. (2.118) we set

$$
A_{1j} = \frac{1}{2}(B_j + C_j), \quad B_{1j} = \frac{1}{2}(B_j - C_j) \quad (j = 1, 2)
$$

(2.120)

Substituting Eqs. (2.120) into Eq. (2.118), the solutions are obtained as

$$
\begin{aligned}
u_y &= \{B_1 \cosh(\lambda_{11}kz) + B_2 \cosh(\lambda_{12}kz) + C_1 \sinh(\lambda_{11}kz) \\
&\quad + C_2 \sinh(\lambda_{12}kz)\} \exp\{-i(ky + \omega t)\} \\
u_z &= -\{B_1 b_{12} \sinh(\lambda_{11}kz) + B_2 b_{22} \sinh(\lambda_{12}kz) + C_1 b_{12} \cosh(\lambda_{11}kz) \\
&\quad + C_2 b_{22} \cosh(\lambda_{12}kz)\} \exp\{-i(ky + \omega t)\}
\end{aligned}
$$

(2.121)

Upon substituting Eqs. (2.121) into Eq. (2.113) and assuming the antisymmetric motion of the plate relative to the middle surface, we obtain

$$
\begin{bmatrix} \alpha_{11} & \alpha_{12} \\ \alpha_{21} \tanh(\lambda_{11}kh) & \alpha_{22} \tanh(\lambda_{12}kh) \end{bmatrix} \begin{bmatrix} C_1 \cosh(\lambda_{11}kh) \\ C_2 \cosh(\lambda_{12}kh) \end{bmatrix} = 0
$$

(2.122)

where

$$
\alpha_{1j} = \lambda_{1j} + ib_{j2}, \quad \alpha_{2j} = \frac{c_1^2}{c_2^2} b_{j2}\lambda_{1j} + i\left(\frac{c_1^2}{c_2^2} - 2\right) \quad (j = 1, 2)
$$

(2.123)

The frequency equation is obtained by setting the determinant of the coefficients of Eq. (2.122) equal to zero. Thus, for the antisymmetric modes, we find

$$
\alpha_{11}\alpha_{22} \tanh(\lambda_{12}kh) - \alpha_{12}\alpha_{21} \tanh(\lambda_{11}kh) = 0
$$

(2.124)

Note that in the case of the symmetric motion of the plate relative to the middle surface, the influence of the magnetic field on the elastic waves will be negligible for moderate values of H_0.

Next, we assume a perfect conductivity. From the fourth, fifth, and sixth of Eqs. (2.22) and Eqs. (2.48), we obtain the magnetic field intensity components as

$$
h_y = -H_0 u_{z,z}, \quad h_z = H_0 u_{z,y}
$$

(2.125)

Hence, the current density component j_x is obtained by the fourth of Eqs. (2.21) as

$$j_x = H_0(u_{z,yy} + u_{z,zz}) \tag{2.126}$$

Inserting the components of stress tensor and current density vector of Eqs. (2.109) and (2.126) in Eqs. (2.110), we find

$$\{2\mu u_{y,y} + \lambda(u_{y,y} + u_{z,z})\}_{,y} + \mu(u_{y,z} + u_{z,y})_{,z} = \rho u_{y,tt}$$
$$\mu(u_{y,z} + u_{z,y})_{,y} + \{2\mu u_{z,z} + \lambda(u_{y,y} + u_{z,z})\}_{,z} + \kappa_0 H_0^2(u_{z,yy} + u_{z,zz}) = \rho u_{z,tt} \tag{2.127}$$

Substitution of the relevant components of stress and Maxwell stress tensors into the stress boundary conditions in Eqs. (2.25) and the use of Eq. (2.28) lead to

$$\kappa_0 H_0\{h_y(y, \pm h, t) - h_y^e(y, \pm h, t)\} - \{\lambda u_{y,y}(y, \pm h, t) + (2\mu + \lambda)u_{z,z}(y, \pm h, t)\} = 0$$
$$u_{y,z}(y, \pm h, t) + u_{z,y}(y, \pm h, t) = 0 \tag{2.128}$$

Substituting Eqs. (2.114) into Eqs. (2.127), we find

$$\begin{bmatrix} f_{11} & f_{12} \\ f_{21} & f_{22} \end{bmatrix} \begin{bmatrix} u_y \\ u_z \end{bmatrix} = 0 \tag{2.129}$$

where

$$f_{11} = \mu p^2 - (\lambda + 2\mu)k^2 + \rho\omega^2$$
$$f_{12} = -i(\lambda + \mu)kp$$
$$f_{21} = -i(\lambda + \mu)kp$$
$$f_{22} = (\lambda + 2\mu)p^2 - \mu k^2 + \rho\omega^2 - \kappa_0(k^2 - p^2)H_0^2 \tag{2.130}$$

The determinant of the coefficient matrix must vanish, which yields the following equation:

$$\left(\frac{p}{k}\right)^4 + \left\{\left(1 + \frac{1}{c_1^2/c_2^2 + h_c}\right)\left(\frac{\omega}{kc_2}\right)^2 - \left(1 + \frac{c_1^2}{c_2^2}\frac{1 + h_c}{c_1^2/c_2^2 + h_c}\right)\right\}\left(\frac{p}{k}\right)^2$$

$$+ \left\{\left(\frac{\omega}{kc_2}\right)^2 - \frac{c_1^2}{c_2^2}\right\}\left\{\left(\frac{\omega}{kc_2}\right)^2 - 1 - h_c\right\}\frac{1}{c_1^2/c_2^2 + h_c} = 0 \tag{2.131}$$

The solutions take the form

$$\begin{bmatrix} u_y \\ u_z \end{bmatrix} = \begin{bmatrix} 1 & 1 & 1 & 1 \\ -g_{12} & -g_{22} & g_{12} & g_{22} \end{bmatrix} \begin{bmatrix} A_{21}\exp(\lambda_{21}kz)\exp\{-i(ky + \omega t)\} \\ A_{22}\exp(\lambda_{22}kz)\exp\{-i(ky + \omega t)\} \\ B_{21}\exp(-\lambda_{21}kz)\exp\{-i(ky + \omega t)\} \\ B_{22}\exp(-\lambda_{22}kz)\exp\{-i(ky + \omega t)\} \end{bmatrix}$$

$$\tag{2.132}$$

where A_{21}, A_{22}, B_{21}, and B_{22} are arbitrary, $\pm\lambda_{21}, \pm\lambda_{22}$ are the roots of Eq. (2.131), and

$$g_{j2} = i\frac{\lambda_{2j}^2 - c_1^2/c_2^2 + (\omega/kc_2)^2}{\lambda_{2j}(c_1^2/c_2^2 - 1)} \quad (j = 1, 2) \tag{2.133}$$

For convenience, in Eq. (2.132) we set

$$A_{2j} = \frac{1}{2}(F_j + G_j), \quad B_{2j} = \frac{1}{2}(F_j - G_j) \quad (j = 1, 2) \tag{2.134}$$

Substitution of Eqs. (2.134) into Eq. (2.132) leads to

$$
\begin{aligned}
u_y &= \{F_1 \cosh(\lambda_{21}kz) + F_2 \cosh(\lambda_{22}kz) + G_1 \sinh(\lambda_{21}kz) \\
&\quad + G_2 \sinh(\lambda_{22}kz)\} \exp\{-i(ky + \omega t)\} \\
u_z &= -\{F_1 g_{12} \sinh(\lambda_{21}kz) + F_2 g_{22} \sinh(\lambda_{22}kz) \\
&\quad + G_1 g_{12} \cosh(\lambda_{21}kz) + G_2 g_{22} \cosh(\lambda_{22}kz)\} \exp\{-i(ky + \omega t)\}
\end{aligned}
\tag{2.135}
$$

Substituting from the third of Eqs. (2.80) and the second of Eqs. (2.125) into the boundary condition (2.28), with the aid of the second of Eqs. (2.135), we obtain

$$
\begin{aligned}
A_1 = A_2 = -ikH_0\{F_1 g_{12} \sinh(\lambda_{21}kh) + F_2 g_{22} \sinh(\lambda_{22}kh) \\
+ G_1 g_{12} \cosh(\lambda_{21}kh) + G_2 g_{22} \cosh(\lambda_{22}kh)\} \exp(kh)
\end{aligned}
\tag{2.136}
$$

Hence

$$
\left.
\begin{aligned}
h_y^e &= kH_0\{F_1 g_{12} \sinh(\lambda_{21}kh) + F_2 g_{22} \sinh(\lambda_{22}kh) \\
&\quad + G_1 g_{12} \cosh(\lambda_{21}kh) + G_2 g_{22} \cosh(\lambda_{22}kh)\} \\
&\quad \times \exp\{k(h - z) - i(ky + \omega t)\} \qquad (z \geq -h) \\
&= -kH_0\{F_1 g_{12} \sinh(\lambda_{21}kh) + F_2 g_{22} \sinh(\lambda_{22}kh) \\
&\quad + G_1 g_{12} \cosh(\lambda_{21}kh) + G_2 g_{22} \cosh(\lambda_{22}kh)\} \\
&\quad \times \exp\{k(h + z) - i(ky + \omega t)\} \qquad (z \leq -h)
\end{aligned}
\right\}
\tag{2.137}
$$

Substituting the first of Eqs. (2.125), (2.135), and (2.137) into Eqs. (2.128) and assuming the antisymmetric motion of the plate relative to the middle surface yield

$$
\begin{bmatrix}
\lambda_{21} + ig_{12} & \lambda_{22} + ig_{22} \\
\beta_{11} \tanh(\lambda_{21}kh) + \beta_{21} & \beta_{12} \tanh(\lambda_{22}kh) + \beta_{22}
\end{bmatrix}
\begin{bmatrix}
G_1 \cosh(\lambda_{21}kh) \\
G_2 \cosh(\lambda_{22}kh)
\end{bmatrix} = 0
\tag{2.138}
$$

where

$$\beta_{1j} = \left(\frac{c_1^2}{c_2^2} + h_c\right) g_{j2}\lambda_{2j} + i\left(\frac{c_1^2}{c_2^2} - 2\right), \quad \beta_{2j} = g_{j2}h_c \qquad (j = 1, 2) \tag{2.139}$$

The frequency equations for the antisymmetric modes become

$$
\begin{aligned}
(\lambda_{21} + ig_{12})\{\beta_{12} \tanh(\lambda_{22}kh) + \beta_{22}\} \\
-(\lambda_{22} + ig_{22})\{\beta_{11} \tanh(\lambda_{21}kh) + \beta_{21}\} = 0
\end{aligned}
\tag{2.140}
$$

Figure 2.12 shows the variation of the phase velocity $\mathrm{Re}(\omega/kc_2)$ with the wave number kh of the classical, Mindlin, and plane strain plates under the normalized magnetic field $h_c = 0$. The result for the Mindlin plate agrees with that for the plane strain plate. The result for the classical plate is also in agreement with that for the plane strain plate at low wave number. Figure 2.13 shows the similar results for $\sigma_h = 0.5$ under $h_c = 4.06 \times 10^{-2}$. The result for the Mindlin plate agrees with that for the plane strain plate, and the effect of the magnetic field on the phase velocity is observed at low wave number. Figure 2.14 shows the variation of the

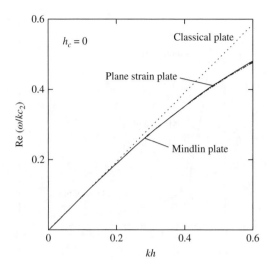

Figure 2.12 Phase velocity versus wave number of the classical, Mindlin, and plane strain plates ($h_c = 0$)

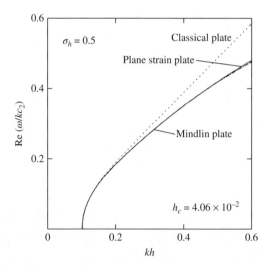

Figure 2.13 Phase velocity versus wave number of the classical, Mindlin, and plane strain plates ($\sigma_h = 0.5$)

attenuation $-\text{Im}(\omega/kc_2)$ with the wave number kh of the classical, Mindlin, and plane strain plates under the same conditions. The results for the classical and Mindlin plates agree with that for the plane strain plate.

Figure 2.15 shows the variation of the phase velocity $\text{Re}(\omega/kc_2)$ with the wave number kh of the classical, Mindlin, and plane strain plates for perfect conductivity ($\sigma \rightarrow \infty$) under $h_c = 0.003$. The result for the Mindlin plate agrees with that for the plane strain plate. The result for the classical plate is also in agreement with that for the plane strain plate at low wave number.

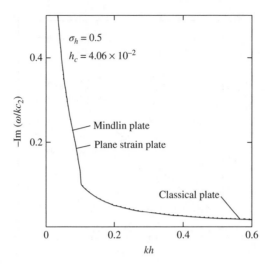

Figure 2.14 Attenuation versus wave number of the classical, Mindlin, and plane strain plates ($\sigma_h = 0.5$)

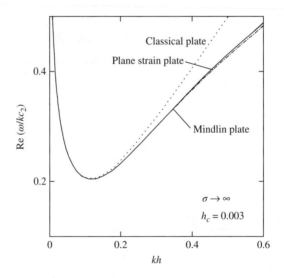

Figure 2.15 Phase velocity versus wave number of the classical, Mindlin, and plane strain plates (perfect conductivity)

2.3   Dynamic Magnetoelastic Crack Mechanics

Magnetic fusion reactor components are frequently subjected to strong magnetic fields that can have an undesirable effect on the integrity of the structures [7, 8]. From the requirement for the fusion experimental reactor, the maximum magnetic field is 12 T. The dynamic behavior of the electrically conducting elastic plates is sufficiently affected by the presence of the magnetic fields. So design and development of superconducting structures require basic research on electromagnetic fracture mechanics. The stress intensity factor approach of linear elastic fracture mechanics has proved to be very successful in predicting the unstable fracture of brittle materials [9–11]. When cracked conducting materials are subjected to strong magnetic fields, the same approach is expected to apply. Shindo et al. discussed the mechanical behavior of an electrically conducting classical plate with a through crack under the uniform magnetic field normal to the crack surface. Two special cases, quasistatic electromagnetic state [12, 13] and perfect conductivity [14], were considered. In this section, the scattering of time harmonic flexural waves by a through crack in conducting Mindlin plates under a uniform magnetic field is dealt with, and the dynamic moment intensity factor is discussed.

A frequently encountered crack shape is the surface crack. The solution of the through crack problem may be useful in studying the surface crack problem by the application of the plate theory-line spring method [15]. The results for the case with a partial crack can be estimated by the formulation of the conducting plate with a through crack under arbitrary membrane and bending loads and the solution of the corresponding plane strain problem for the conducting material with an edge crack.

Consider an electrically conducting Mindlin plate of thickness $2h$ having a through crack of length $2a$ as shown in Fig. 2.16. The coordinate axes x and y are in the middle plane of the plate, and the z-axis is normal to this plane. The crack is located on the line $y = 0, |x| < a$, and the cracked plate is permeated by the uniform magnetic field $(H_{0y} = H_0, H_{0x} = H_{0z} = 0)$ of magnetic induction $B_0 = \kappa_0 H_0$ normal to the crack surface. Incident waves giving rise to moments symmetric about the crack plane $y = 0$ are applied in an arbitrary direction.

First, we assume a quasistatic electromagnetic field. Assuming a quasistatic electromagnetic state, which is realized by cancelling the term $f_{,t}$ and retaining other time derivatives in Eqs. (2.59), and neglecting the moments m_{xx}^M and m_{yy}^M in Eqs. (2.66), we obtain

$$
\begin{aligned}
f_{,x} + \sigma \psi &= \frac{1}{2h}\{h_x(x, y, h, t) - h_x(x, y, -h, t)\} \\
f_{,y} - \sigma(\varphi - \kappa_0 H_0 \Psi_{,t}) &= \frac{1}{2h}\{h_y(x, y, h, t) - h_y(x, y, -h, t)\} \\
\psi_{,x} - \varphi_{,y} &= 0
\end{aligned}
\qquad (2.141)
$$

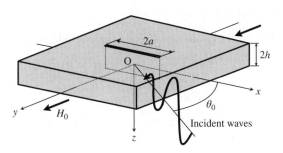

Figure 2.16   A conducting plate with a through crack and incident waves

$$\frac{S}{2}[(1-v)(\Psi_{x,xx} + \Psi_{x,yy}) + (1+v)\Phi_{,x}] - \Psi_x - \Psi_{z,x} = \frac{4h^2\rho}{\pi^2\mu}\Psi_{x,tt}$$

$$\frac{S}{2}[(1-v)(\Psi_{y,xx} + \Psi_{y,yy}) + (1+v)\Phi_{,y}] - \Psi_y - \Psi_{z,y} = \frac{4h^2\rho}{\pi^2\mu}\Psi_{y,tt} \qquad (2.142)$$

$$\Psi_{z,xx} + \Psi_{z,yy} + \Phi = \frac{4h^2\rho}{\pi^2\mu}\frac{1}{R}\Psi_{z,tt} - \frac{12\sigma\kappa_0 H_0}{\pi^2\mu}\varphi + \frac{12\sigma(\kappa_0 H_0)^2}{\pi^2\mu}\Psi_{z,t}$$

Let incident flexural waves be directed at an angle θ_0 with the x-axis so that

$$\Psi_x^i = \Psi_{x0}\exp[-i\{k(x\cos\theta_0 + y\sin\theta_0) + \omega t\}]$$
$$\Psi_y^i = \Psi_{y0}\exp[-i\{k(x\cos\theta_0 + y\sin\theta_0) + \omega t\}] \qquad (2.143)$$
$$\Psi_z^i = \Psi_{z0}\exp[-i\{k(x\cos\theta_0 + y\sin\theta_0) + \omega t\}]$$

$$f^i = f_0\exp[-i\{k(x\cos\theta_0 + y\sin\theta_0) + \omega t\}]$$
$$\varphi^i = \varphi_0\exp[-i\{k(x\cos\theta_0 + y\sin\theta_0) + \omega t\}] \qquad (2.144)$$
$$\psi^i = \psi_0\exp[-i\{k(x\cos\theta_0 + y\sin\theta_0) + \omega t\}]$$

where $\Psi_{x0}, \Psi_{y0}, \Psi_{z0}, f_0, \varphi_0, \psi_0$ are the amplitudes of the incident waves, and the superscript i stands for the incident component. Outside the plate, the solutions that vanish at $z = \pm\infty$ and have the wave factor $\exp[-i\{k(x\cos\theta_0 + y\sin\theta_0) + \omega t\}]$ are

$$h_x^{ei} = i\cos\theta_0 A_1 \exp[-kz - i\{k(x\cos\theta_0 + y\sin\theta_0) + \omega t\}] \ (z \geq h) \left.\right\}$$
$$= -i\cos\theta_0 A_2 \exp[kz - i\{k(x\cos\theta_0 + y\sin\theta_0) + \omega t\}] \ (z \leq -h)$$

$$h_y^{ei} = i\sin\theta_0 A_1 \exp[-kz - i\{k(x\cos\theta_0 + y\sin\theta_0) + \omega t\}] \ (z \geq h) \left.\right\}$$
$$= -i\sin\theta_0 A_2 \exp[kz - i\{k(x\cos\theta_0 + y\sin\theta_0) + \omega t\}] \ (z \leq -h) \qquad (2.145)$$

$$h_z^{ei} = A_1 \exp[-kz - i\{k(x\cos\theta_0 + y\sin\theta_0) + \omega t\}] \ (z \geq h) \left.\right\}$$
$$= A_2 \exp[kz - i\{k(x\cos\theta_0 + y\sin\theta_0) + \omega t\}] \quad (z \leq -h)$$

where A_1, A_2 are undetermined constants. Making use of Eqs. (2.145) and Eqs. (2.34), Eq. (2.36) renders the x, y-magnetic intensity components h_x^i, h_y^i and the amplitude f_0

$$h_x^i(x, y, \pm h, t) = \pm i\cos\theta_0 f_0 \exp[-i\{k(x\cos\theta_0 + y\sin\theta_0) + \omega t\}]$$
$$h_y^i(x, y, \pm h, t) = \pm i\sin\theta_0 f_0 \exp[-i\{k(x\cos\theta_0 + y\sin\theta_0) + \omega t\}] \qquad (2.146)$$
$$f_0 = A_1\exp(-kh) = A_2\exp(-kh)$$

Substituting Eqs. (2.143), (2.144), and (2.146) into Eqs. (2.141), we obtain

$$(kh + 1)\cos\theta_0 f_0 + i\sigma h\psi_0 = 0$$
$$\varphi_0 = -i\omega\kappa_0 H_0\cos\theta_0^2\Psi_{z0} \qquad (2.147)$$
$$\sin\theta_0\varphi_0 = \cos\theta_0\psi_0$$

Substituting Eqs. (2.143) and (2.144) into Eqs. (2.142) yields the frequency equation and the relations among the amplitudes

$$\begin{vmatrix} h_{11} & h_{12} & h_{13} \\ h_{21} & h_{22} & h_{23} \\ h_{31} & h_{32} & h_{33} \end{vmatrix} = 0 \qquad (2.148)$$

$$i\frac{\pi^2(kh)}{6}\Psi_{x0} = -\cos\theta_0 \left\{ \frac{\pi^2(kh)}{6} - 2kh\left(\frac{\omega}{kc_2}\right)^2 - 2i\left(\frac{\omega}{kc_2}\right)\Gamma_h \sin^2\theta_0 \right\} k\Psi_{z0}$$

$$i\frac{\pi^2(kh)}{6}\Psi_{y0} = -\sin\theta_0 \left\{ \frac{\pi^2(kh)}{6} - 2kh\left(\frac{\omega}{kc_2}\right)^2 - 2i\left(\frac{\omega}{kc_2}\right)\Gamma_h \sin^2\theta_0 \right\} k\Psi_{z0}$$

(2.149)

where

$$h_{11} = \frac{4(kh)^2}{3(1-v)}\left(\frac{1-v}{2}\sin^2\theta_0 + \cos^2\theta_0\right) + \frac{\pi^2}{6} - \frac{2(kh)^2}{3}\left(\frac{\omega}{kc_2}\right)^2$$

$$h_{12} = h_{21} = \frac{2(1+v)(kh)^2}{3(1-v)}\sin\theta_0\cos\theta_0$$

$$h_{13} = -h_{31} = -i\frac{\pi^2(kh)}{6}\cos\theta_0$$

$$h_{22} = \frac{4(kh)^2}{3(1-v)}\left(\sin^2\theta_0 + \frac{1-v}{2}\cos^2\theta_0\right) + \frac{\pi^2}{6} - \frac{2(kh)^2}{3}\left(\frac{\omega}{kc_2}\right)^2$$

(2.150)

$$h_{23} = -h_{32} = -i\frac{\pi^2(kh)}{6}\sin\theta_0$$

$$h_{33} = \frac{\pi^2(kh)^2}{6} - 2(kh)^2\left(\frac{\omega}{kc_2}\right)^2 - 2ikh\left(\frac{\omega}{kc_2}\right)\Gamma_h \sin^2\theta_0$$

The complete solution of the waves as diffracted by the through crack is obtained by adding the incident and scattered waves. Likewise, the plate displacements, moments, and shears can also be found by superposing the incident and scattered parts. For a traction-free crack, the quantities M_{yy}, M_{yx}, Q_y must each vanish for $|x| < a, y = 0$. The conditions to be specified on the crack for the scattered field become

$$M_{yx}^s(x, 0, t) = 0 \qquad (0 \le |x| < \infty) \tag{2.151}$$

$$Q_y^s(x, 0, t) = 0 \qquad (0 \le |x| < \infty) \tag{2.152}$$

$$e_y^s(x, 0, t) = 0 \qquad (0 \le |x| < \infty) \tag{2.153}$$

$$\begin{cases} M_{yy}^s(x, 0, t) = -M_{yy}^i & (0 \le |x| < a) \\ \Psi_y^s(x, 0, t) = 0 & (a \le |x| < \infty) \end{cases} \tag{2.154}$$

where the superscript s stands for the scattered component, and

$$M_{yy}^i = -ikD\Psi_0(\sin^2\theta_0 + v\cos^2\theta_0)\exp\{-i(kx\cos\theta_0 + \omega t)\} \tag{2.155}$$

$$\Psi_0 = \frac{1}{\cos\theta_0}\Psi_{x0} = \frac{1}{\sin\theta_0}\Psi_{y0} \tag{2.156}$$

In what follows, the time factor $\exp(-i\omega t)$ will be omitted, as it always appears with the quantity $ikD\Psi_0(\sin^2\theta_0 + v\cos^2\theta_0)\exp(-ikx\cos\theta_0)$ as indicated in Eq. (2.155).

By using Fourier transforms [16], we reduce the magnetoelastic problem to solving a pair of dual integral equations. The solution of the dual integral equations is then expressed in terms of a Fredholm integral equation of the second kind. The moment intensity factor is defined by

$$K_I = \lim_{x \to a^+} \{2\pi(x-a)\}^{1/2} M_{yy}(x,0,t) \qquad (2.157)$$

As demonstrated by Shindo and Tohyama [17], the effect of magnetic field on the dynamic bending moment intensity factor for $\theta_0 = \pi/2$ is more pronounced than that for $\theta_0 \neq \pi/2$. Here, we show some numerical examples for $\theta_0 = \pi/2$. Figure 2.17 shows a plot of the normalized moment intensity factor $|K_I/M_0(\pi a)^{1/2}|$ versus frequency ω normalized against the cut-off frequency $\omega_0 = \pi c_2/2h$ for flexural waves at normal incidence, that is, $\theta_0 = \pi/2$, crack length to plate thickness ratio $a/h = 10$, and normalized magnetic fields $\Gamma_h = 0, 0.01, 0.02$. M_0 is given by $ik_1 D\Psi_0$. The dashed curve obtained for the case of $\Gamma_h = 0$ coincides with that of the purely elastic case. The normalized magnetic field of $\Gamma_h = 0.01$ and 0.02 ($\sigma_h = 0.5$) corresponds, respectively, to about magnetic induction $B_0 = \kappa_0 H_0 = 7$ and 10 T for the graphite plate. The quantity $|K_I/M_0(\pi a)^{1/2}|$ for $\Gamma_h = 0$ is found to be smaller than that of the static case and decreases in magnitude as the frequency is increased. As the two curves for $\Gamma_h \neq 0$ possess higher amplitude than that for $\Gamma_h = 0$, the magnetic field is seen to increase the local moment with increasing Γ_h. Such an effect dies out at high frequency. A typical set of parametric curves for $a/h = 5$ is given in Fig. 2.18 to illustrate the variation of $|K_I/M_0(\pi a)^{1/2}|$ with Γ_h. Note that the moment intensity factor tends to increase with increasing Γ_h. Figure 2.19 shows the similar results for $a/h = 10$. For $\omega/\omega_0 = 0.005$, it is observed that the magnetic field effect of $\Gamma_h = 0.01$ can increase $|K_I/M_0(\pi a)^{1/2}|$ by approximately 7%. Although the data are not shown here, the magnetic field effect of $\Gamma_h = 0.01$ can increase $|K_I/M_0(\pi a)^{1/2}|$ of the classical plate by approximately 21% [12]. The classical plate theory predictions tend to overestimate the effect of magnetic field on the moment intensity factor compared to the Mindlin plate theory predictions.

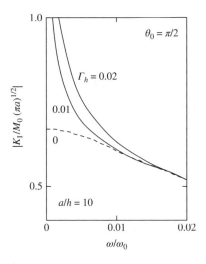

Figure 2.17 Dynamic bending moment intensity factor versus frequency of the Mindlin plate ($a/h = 10$)

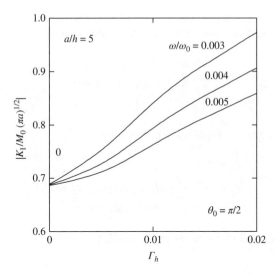

Figure 2.18 Dynamic bending moment intensity factor versus magnetic field of the Mindlin plate $(a/h = 5)$

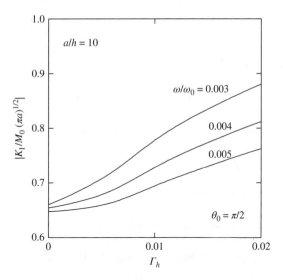

Figure 2.19 Dynamic bending moment intensity factor versus magnetic field of the Mindlin plate $(a/h = 10)$

Next, we assume a perfect conductivity. Similar procedure can be applied for the perfectly conducting Mindlin plate [18]. From Eqs. (2.69), we obtain

$$h_x = H_0 z \Psi_{x,y}$$
$$h_y = -\frac{1 - 2v}{1 - v} H_0 z \Psi_{x,x} + \frac{v}{1 - v} H_0 z \Psi_{y,y} \qquad (2.158)$$
$$h_z = H_0 \Psi_{z,y}$$

We now consider flexural waves directed at normal incidence with the x-axis, that is, $\theta_0 = \pi/2$. The following incident waves, Eqs. (2.96), are thus applied:

$$
\begin{aligned}
\Psi_x^i &= 0 \\
\Psi_y^i &= \Psi_{y0} \exp\{-i(ky + \omega t)\} \\
\Psi_z^i &= \Psi_{z0} \exp\{-i(ky + \omega t)\}
\end{aligned}
\tag{2.159}
$$

Substituting Eqs. (2.159) into Eqs. (2.158), we obtain

$$
\begin{aligned}
h_x^i &= 0 \\
h_y^i &= -ik\frac{v}{1-v}H_0 z \Psi_{y0} \exp\{-i(ky + \omega t)\} \\
h_z^i &= -ikH_0 \Psi_{z0} \exp\{-i(ky + \omega t)\}
\end{aligned}
\tag{2.160}
$$

Outside the plate, the external fields are in place of Eqs. (2.80)

$$
\begin{aligned}
h_x^{ei} &= 0 \\
h_y^{ei} &= iA_1 \exp\{-kz - i(ky + \omega t)\} \quad (z \geq h) \\
&= -iA_2 \exp\{kz - i(ky + \omega t)\} \quad (z \leq -h) \\
h_z^{ei} &= A_1 \exp\{-kz - i(ky + \omega t)\} \quad (z \geq h) \\
&= A_2 \exp\{kz - i(ky + \omega t)\} \quad (z \leq -h)
\end{aligned}
\tag{2.161}
$$

Boundary condition in Eq. (2.28) leads to the determination of A_1 and A_2 as

$$
A_1 = A_2 = -ikH_0\Psi_{z0} \exp(kh)
\tag{2.162}
$$

Making use of Eqs. (2.160) – (2.162) and Eq. (2.25) renders

$$
\sigma_{zz}^i(x,y,\pm h,t) = \left(\mp \kappa_0 k H_0^2 \Psi_{z0} \mp \kappa_0 \frac{v}{1-v} khH_0^2 \Psi_{y0} \right) \exp\{-i(ky + \omega t)\}
\tag{2.163}
$$

$$
\sigma_{zy}^i(x,y,\pm h,t) = \sigma_{zx}^i(x,y,\pm h,t) = 0
$$

From Eqs. (2.73) and (2.74), with Eqs. (2.163), we obtain

$$
m_{xx}^{MPi} = 0, \quad m_{yy}^{MPi} = 0
\tag{2.164}
$$

$$
q^{MPi} = -2\kappa_0 H_0^2 k(1 + kh)\Psi_{z0} \exp\{-i(ky + \omega t)\}
\tag{2.165}
$$

Substituting Eqs. (2.159), (2.164), and (2.165) into the second of Eqs. (2.75) and Eq. (2.76) yields the frequency equation, Eq. (2.103), and the relation between the amplitudes, Eq. (2.104).

The conditions to be specified on the crack for the scattered field become

$$
M_{xy}^s(x,0,t) = 0 \qquad (0 \leq |x| < \infty)
\tag{2.166}
$$

$$
Q_y^s(x,0,t) = 0 \qquad (0 \leq |x| < \infty)
\tag{2.167}
$$

$$\begin{cases} M_{yy}^s(x,0,t) = -M_{yy}^i & (0 \le |x| < a) \\ \Psi_y^s(x,0,t) = 0 & (a \le |x| < \infty) \end{cases} \tag{2.168}$$

where

$$M_{yy}^i = -ikD\Psi_{y0} \exp(-i\omega t) \tag{2.169}$$

In what follows, the time factor $\exp(-i\omega t)$ is suppressed.

Fourier transforms are used to reduce the magnetoelastic problem to a pair of dual integral equations, which can be further reduced to a Fredholm integral equation of the second kind. The moment intensity factor is defined by Eq. (2.157).

Figure 2.20 shows a plot of the normalized moment intensity factor $|K_{\mathrm{I}}/M_0(\pi a)^{1/2}|$ versus normalized frequency ω/ω_0 for crack length to plate thickness ratio $a/h = 10$ and normalized magnetic fields $h_c = 0, 0.003$. M_0 is given by $ik_1 D\Psi_{y0}$. The dashed curve obtained for the case of $h_c = 0$ coincides with that of the purely elastic case. For comparison, the results of the perfectly conducting classical plate [14] are also shown (dot-dashed lines). The normalized magnetic field of $h_c = 0.003$ corresponds to about magnetic induction $B_0 = 9.5$ T for the aluminum plate. The quantity $|K_{\mathrm{I}}/M_0(\pi a)^{1/2}|$ for $h_c = 0$ decays as the frequency increases. As the two curves for $h_c = 0.003$ possess lower amplitude than those for $h_c = 0$, the magnetic field is seen to decrease the local moment with increasing h_c. Such an effect dies out at high frequency.

Thus, it was shown that the dynamic moment intensity factor decreases with the magnetic field, depending on the frequency of the incident wave, crack length to plate thickness ratio, and so on. Significant increase and decrease in the local moment intensity factor occurred for the quasistatic electromagnetic and perfect conductivity states, respectively, at low wave frequency.

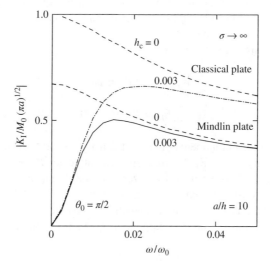

Figure 2.20   Dynamic bending moment intensity factor versus frequency ($\sigma \to \infty$)

The scattering problem of flexural waves by a through crack in conducting plates is considered in Section 2.3. Some works treated the scattering of elastic waves by a crack in conducting materials (e.g., [19, 20]). Note that in these cases, the influence of the magnetic field is contained in a factor of the order $1 + h_c$, where $h_c = \kappa_0 H_0^2 / \mu$. Therefore, the influence of the magnetic field on the elastic waves will be negligible for moderate values of H_0. If numerical results are based on $h_c = 0.25$, this corresponds, for $\mu = 10^{11}$ N/m^2, to a field intensity of over $H_0 = 10^8$ A/m and, hence, to an induction of over $B_0 = 100$ T. This is an extremely unrealistic value. Smaller values will probably not produce any noticeable effects.

2.4   Cracked Materials Under Electromagnetic Force

The components of a magnetic confinement fusion reactor, such as the Tokamak type, work under large electric currents and strong magnetic fields, and electromagnetic forces are caused by the interaction between the electric currents and magnetic fields. Therefore, inquiry into the influence of electromagnetic forces on the fracture of materials and structures is relevant. Shindo and Takeuchi [21] studied the electromagnetoelastic behavior of a conducting circular cylinder with a penny-shaped crack under a uniform axial current flow and a constant axial magnetic field and discussed the stress intensity factor. They also investigated the electromagnetoelastic problem of a conducting finite strip with a crack under a uniform current flow and a constant magnetic field [22]. Shindo and Tamura [23] examined the electromagnetoelastic behavior of a conducting classical plate with a through crack under a uniform current flow and a constant magnetic field. Moreover, Shindo et al. [24] analyzed the transient response of a conductor with a penny-shaped crack under a uniform current flow and a constant magnetic field to show the effect of the impulsive electromagnetic force on the behavior of the cracked materials. In this section, the electromagnetoelastic problems in conducting Mindlin plates with a through crack under a uniform current flow and a constant magnetic field are dealt with, and the twisting moment intensity factor and shear force intensity factor are discussed.

Let us now consider the rectangular Cartesian coordinates $x_i(O\text{-}x_1, x_2, x_3)$. An electrically conducting material is subjected to a large electric current and a strong magnetic field. We assume a quasistatic electromagnetic state.

The field equations are listed as follows:

$$\sigma_{ji,j} + \varepsilon_{ijk} J_j B_k = \rho u_{i,tt} \tag{2.170}$$

$$\varepsilon_{ijk} E_{k,j} = 0, \quad D_{i,i} = 0 \tag{2.171}$$

where J_i, B_i, E_i, D_i are the components of current density vector **J**, magnetic induction vector **B**, electric field intensity vector **E**, electric displacement vector **D**, respectively. Note that the free electric charge density is assumed to be zero in the interior of the conducting materials [25]. The mechanical constitutive equations can be written as Eq. (2.7). If we assume that the materials have the permittivity $\epsilon = \epsilon_0$ and magnetic permeability $\kappa = \kappa_0$, the component B_i is represented as a function of H_i and the component D_i as a function of E_i

$$B_i = \kappa_0 H_i \tag{2.172}$$

$$D_i = \epsilon_0 E_i \tag{2.173}$$

Ohm's law is

$$J_i = \sigma E_i \tag{2.174}$$

Consider an electrically conducting Mindlin plate of thickness $2h$ and width $2l$ with fixed ends having a through crack of length $2a$ as shown in Fig. 2.21. The coordinate axes x and y are in the middle plane of the plate, and the z-axis is normal to this plane. The crack is located on the line $y = 0$, $|x| < a$, and the cracked plate is permeated by the constant magnetic field H_0 of magnetic induction $B_0 = \kappa_0 H_0$ normal to the crack surface. Also, a uniform electric field E_0 is applied normal to the crack surface. A steady undisturbed electric current flow $J_0 = \sigma E_0$ (where J_0 is a constant with dimension of current density) passes through the plate and is uniform and normal to the crack surface.

The current flow is disturbed by the presence of the crack, and the twisting moment is caused by the interaction between the disturbed currents and magnetic fields. From Eq. (2.170), the three-dimensional equations of equilibrium, without inertia, become

$$\begin{aligned}
\sigma_{xx,x} + \sigma_{yx,y} + \sigma_{zx,z} &= 0 \\
\sigma_{xy,x} + \sigma_{yy,y} + \sigma_{zy,z} &= 0 \\
\sigma_{xz,x} + \sigma_{yz,y} + \sigma_{zz,z} &= B_0 J_x
\end{aligned} \tag{2.175}$$

The electric field equations, Eqs. (2.171), are

$$E_{x,y} - E_{y,x} = 0, \quad D_{x,x} + D_{y,y} = 0 \tag{2.176}$$

The x, y-components of the electric displacement vector are given by

$$D_x = \epsilon_0 E_x, \quad D_y = \epsilon_0 E_y \tag{2.177}$$

The x, y-components of the current density vector are

$$J_x = \sigma E_x, \quad J_y = \sigma E_y \tag{2.178}$$

By substituting Eqs. (2.177) into the second of Eqs. (2.176), the electric field equations can be expressed as

$$E_{x,x} + E_{y,y} = 0 \tag{2.179}$$

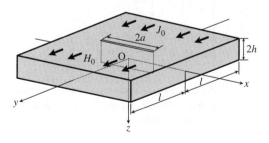

Figure 2.21   A conducting plate with a through crack

The first of Eqs. (2.176) and Eq. (2.179) are satisfied by introducing an electric potential ϕ such that

$$E_x = -\phi_{,x}, \quad E_y = -\phi_{,y}$$
$$\phi_{,xx} + \phi_{,yy} = 0 \tag{2.180}$$

Substituting the first and second of Eqs. (2.180) into Eqs. (2.178), we obtain the components of the current density vector

$$J_x = -\sigma\phi_{,x}, \quad J_y = -\sigma\phi_{,y} \tag{2.181}$$

The problem is solved for the case of electrically insulated crack and plate surfaces. The electric current density at the ends $y = \pm l$ is assumed to be $J_y = J_0$. So the boundary conditions are

$$\begin{cases} \phi_{,y}(x,0) = 0 & (0 \leq |x| < a) \\ \phi(x,0) = 0 & (a \leq |x| < \infty) \end{cases} \tag{2.182}$$

$$\phi_{,y}(x,l) = -\frac{J_0}{\sigma} \quad (0 \leq |x| < \infty) \tag{2.183}$$

Application of the Fourier transforms reduces the problem of the current density field to the solution of a pair of dual integral equations. These equations are solved by using an integral transform technique, and the result is expressed in terms of a Fredholm integral equation of the second kind [26]. For $l \to \infty$, the integral equations are solved exactly, and components of the current density vector are

$$\begin{cases} J_x = J_0 \dfrac{r}{(r_1 r_2)^2} \sin\left(\theta - \dfrac{\theta_1}{2} - \dfrac{\theta_2}{2}\right) \\ J_y = J_0 \dfrac{r}{(r_1 r_2)^2} \cos\left(\theta - \dfrac{\theta_1}{2} - \dfrac{\theta_2}{2}\right) \end{cases} \tag{2.184}$$

where

$$r = (x^2 + y^2)^{1/2}, \qquad \theta = \tan^{-1}\left(\frac{y}{x}\right)$$
$$r_1 = \{(x-a)^2 + y^2\}^{1/2}, \ \theta_1 = \tan^{-1}\left(\frac{y}{x-a}\right) \tag{2.185}$$
$$r_2 = \{(x+a)^2 + y^2\}^{1/2}, \ \theta_2 = \tan^{-1}\left(\frac{y}{x+a}\right)$$

The singular parts of the current densities in the neighborhood of the crack tip are

$$\begin{cases} J_x = -J_0 \left(\dfrac{a}{2r_1}\right)^2 \sin\left(\dfrac{\theta_1}{2}\right) \\ J_y = J_0 \left(\dfrac{a}{2r_1}\right)^2 \cos\left(\dfrac{\theta_1}{2}\right) \end{cases} \tag{2.186}$$

The current density J_x and the magnetic induction B_0 in the y-direction induce the electromagnetic body force $B_0 J_x$ in the z-direction (see Eq. (2.175)), and the twisting moment operates the cracked conducting Mindlin plate. The rectangular displacement components are in place of Eqs. (2.56)

$$u_x = z\Psi_x(x,y), \quad u_y = z\Psi_y(x,y), \quad u_z = \Psi_z(x,y) \tag{2.187}$$

The bending and twisting moments per unit length ($M_{xx}, M_{yy}, M_{xy} = M_{yx}$) and the vertical shear forces per unit length (Q_x, Q_y) can be expressed as Eqs. (2.60) and (2.61), respectively. Now, if we multiply Eqs. (2.175) by zdz and integrate from $-h$ to h, with the boundary condition (2.40), we obtain the results

$$M_{xx,x} + M_{yx,y} - Q_x = 0$$
$$M_{xy,x} + M_{yy,y} - Q_y = 0 \qquad (2.188)$$

$$Q_{x,x} + Q_{y,y} = 2hB_0 J_x \qquad (2.189)$$

Substitution of Eqs. (2.60) and (2.61) into Eqs. (2.188) and (2.189) yields

$$\frac{S}{2}\{(1-v)(\Psi_{x,xx} + \Psi_{x,yy}) + (1+v)\Phi_{,x}\} - \Psi_x - \Psi_{z,x} = 0$$
$$\frac{S}{2}\{(1-v)(\Psi_{y,xx} + \Psi_{y,yy}) + (1+v)\Phi_{,y}\} - \Psi_y - \Psi_{z,y} = 0 \qquad (2.190)$$

$$\Psi_{z,xx} + \Psi_{z,yy} + \Phi = -\frac{12B_0 J_x}{\pi^2 \mu} \qquad (2.191)$$

For a traction-free crack, the quantities M_{yy}, M_{yx}, Q_y must each vanish for $|x| < a, y = 0$. Consequently, Eqs. (2.190) and (2.191) are to be solved subjected to the following mixed boundary conditions:

$$M_{yy}(x,0) = 0 \qquad (0 \leq |x| < \infty) \qquad (2.192)$$

$$\begin{cases} M_{yx}(x,0) = 0 & (0 \leq |x| < a) \\ \Psi_x(x,0) = 0 & (a \leq |x| < \infty) \end{cases} \qquad (2.193)$$

$$\begin{cases} Q_y(x,0) = 0 & (0 \leq |x| < a) \\ \Psi_z(x,0) = 0 & (a \leq |x| < \infty) \end{cases} \qquad (2.194)$$

$$\Psi_x(x,l) = 0 \qquad (0 \leq |x| < \infty) \qquad (2.195)$$

$$\Psi_y(x,l) = 0 \qquad (0 \leq |x| < \infty) \qquad (2.196)$$

$$\Psi_z(x,l) = 0 \qquad (0 \leq |x| < \infty) \qquad (2.197)$$

A solution of the crack problem is obtained by the method of dual integral equations, and the result is expressed in terms of a system of simultaneous Fredholm integral equations. The twisting moment intensity factor K_{II} and the shear force intensity factor K_{III} are defined by

$$K_{II} = \lim_{x \to a^+} \{2\pi(x-a)\}^{1/2} M_{yx}(x,0) \qquad (2.198)$$

$$K_{III} = \lim_{x \to a^+} \{2\pi(x-a)\}^{1/2} Q_y(x,0) \qquad (2.199)$$

To examine the twisting moment intensity factor and shear force intensity factor, we show some numerical examples for $v = 0.3$. Figure 2.22 exhibits the variation of the normalized twisting moment intensity factor $K_{II}/B_0 J_0 a^{7/2}$ against the plate thickness to crack length ratio h/a for the plate width to crack length ratios $l/a = 1, 5, 10$. The twisting moment intensity factor increases with increasing h/a. Also, a larger value of l/a tends to increase the twisting

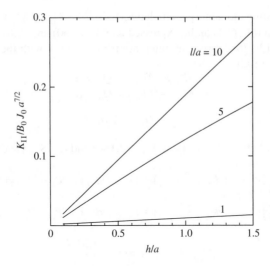

Figure 2.22 Twisting moment intensity factor versus plate thickness to crack length ratio

moment intensity factor. Figure 2.23 shows a plot of the normalized shear force intensity factor $K_{III}/B_0 J_0 a^{5/2}$ versus h/a for $l/a = 5$. Increasing h/a tends to increase the shear force intensity factor value.

As seen in this section, the twist moment and shear force intensity factors increase with the plate width to crack length ratio l/a, depending on the perturbation of the electric current field, plate thickness to crack length ratio h/a, and so on. The larger values of the parameters l/a and h/a will give the worst scenario for design purposes.

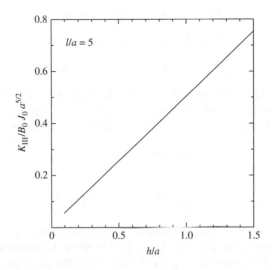

Figure 2.23 Shear force intensity factor versus plate thickness to crack length ratio

2.5 Summary

Magnetoelastic vibrations and waves of the conducting materials have been examined. The influence of magnetic field on the dynamic singular stresses of the cracked conducting materials has also been displayed graphically and discussed in detail. It has been shown that the use of conducting materials in the superconducting structures should involve the study of the effect of the magnetic field on the fracture and deformation. In addition, the results under the large electric current and strong magnetic field have been presented, and the effect of the electromagnetic force on the mechanical behavior of cracked materials has been found to be important.

Although we do not consider here the dynamic problems of magneto-thermoelasticity, we can find some Refs [27, 28]. Also, the propagation of Rayleigh waves in materials permeated by the magnetic field has been investigated on the basis of a linear theory of magneto-thermoelasticity with thermal relaxation [29].

References

[1] A. Sommerfeld, *Electrodynamics*, Academic Press, New York, 1971.

[2] J. W. Dunkin and A. C. Eringen, "On the propagation of waves in an electromagnetic elastic solid," *Int. J. Eng. Sci.* **1**(4), 461 (1963).

[3] S. A. Ambartsumian, G. E. Bagdasarian and M. V. Belubekian, "On the three-dimensional problem of magnetoelastic plate vibrations," *J. Appl. Math. Mech.* **35**(2), 184 (1971).

[4] S. A. Ambartsumian, G. E. Bagdasarian and M. V. Belubekian, "On the equations of magnetoelastic thin plate," *J. Appl. Math. Mech.* **39**(5), 922 (1975).

[5] R. D. Mindlin, "Influence of rotatory inertia and shear on flexural motions of isotropic, elastic plates," *ASME J. Appl. Mech.* **18**(1), 31 (1951).

[6] E. S. Suhubi, "Small torsional oscillations of a circular cylinder with finite electric conductivity in a constant axial magnetic field," *Int. J. Eng. Sci.* **2**(5), 441 (1965).

[7] Y. Shindo, H. Tamura and K. Horiguchi, "Transient response of cracked materials in a strong magnetic field," *Fusion Eng. Des.* **9**, 81 (1989).

[8] Y. Shindo, "Magneto-mechanical behavior of first wall materials," *Fusion Eng. Des.* **16**, 317 (1991).

[9] Y. Murakami (Ed.), *Stress Intensity Factors Handbook*, Vols. **1, 2**, Pergamon Press, Oxford, 1987.

[10] Y. Murakami (Ed.), *Stress Intensity Factors Handbook*, Vol. **3**, Pergamon Press, Oxford, 1993.

[11] Y. Murakami (Ed.), *Stress Intensity Factors Handbook*, Vols. **4, 5**, The Society of Materials Science, Elsevier Science Ltd., Tokyo, Oxford, 2001.

[12] Y. Shindo I. Ohnishi and S. Tohyama, "Flexural wave scattering at a through crack in a conducting plate under a uniform magnetic field," *ASME J. Appl. Mech.* **64**(4), 828 (1997).

[13] Y. Shindo and S. Toyama, "The scattering of oblique flexural waves of a cracked conducting plate in a uniform magnetic field," *Acta Mech.* **128**(1/2), 15 (1998).

[14] Y. Shindo, I. Ohnishi and S. Suzuki, "Dynamic singular moments in a perfectly conducting plate with a through crack under a magnetic field," *Mechanics of Electromagnetic Materials and Structures*, J. S. Lee, G. A. Maugin and Y. Shindo (eds.), ASME, New York, p. 87 (1993).

[15] P. F. Joseph and F. Erdogan, "Surface crack problems in plates," *Int. J. Fract.* **41**(2), 105 (1989).

[16] I. N. Sneddon, *Fourier Transforms*, McGraw-Hill, New York, 1951.

[17] Y. Shindo and S. Tohyama, "Scattering of oblique flexural waves by a through crack in a conducting mindlin plate in a uniform magnetic field," *Int. J. Solids Struct.* **35**(17), 2183 (1998).

[18] Y. Shindo, I. Ohnishi and S. Toyama, "Dynamic singular moments in a perfectly conducting mindlin plate with a through crack under a magnetic field," *ASME J. Appl. Mech.* **67**(4), 503 (2000).

[19] Y. Shindo, "Diffraction of antiplane shear waves by a finite crack in the presence of the magnetic field," *ZAMM* **56**(1), 33 (1976).

[20] Y. Shindo, "Diffraction of normal compression waves by a penny-shaped crack in the presence of an axial magnetic field," *Int. J. Eng. Sci.* **17**(5), 651 (1979).

[21] Y. Shindo and A. Takeuchi, "Electromagnetic twisting of a penny-shaped crack in an elastic conducting cylinder," *Theor. Appl. Fract. Mech.* **8**(3), 213 (1987).

[22] Y. Shindo and A. Takeuchi, "Singular stresses of a finite crack in an elastic conducting strip under electromagnetic force," *Fusion Eng. Des.* **6**(4), 199 (1988).

[23] Y. Shindo and H. Tamura, "Singular twisting moment in a cracked thin plate under an electric current flow and a magnetic field," *Eng. Fract. Mech.* **31**(4), 617 (1988).

[24] Y. Shindo, H. Tamura and K. Horiguchi, "Transient response of an elastic conductor with a penny-shaped crack under electromagnetic force," *Theor. Appl. Fract. Mech.* **10**(3), 191 (1988).

[25] J. A. Stratton, *Electromagnetic Theory*, McGraw-Hill Book Company Inc., New York, 1941.

[26] Y. Shindo and Y. Atobe, "Electromagnetic twisting of a mindlin plate with a through crack," *Int. J. Solids Struct.* **29**(23), 2855 (1992).

[27] W. Nowacki, *Dynamic Problems of Thermoelasticity*, Noordhoff, Leiden, 1975.

[28] W. Nowacki, "Coupled fields in mechanics of solids," *Theoretical and Applied Mechanics*, W. T. Koiter (ed.), North-Holland Publishing Company, Amsterdam, p. 171 (1977).

[29] S. Tomita and Y. Shindo, "Rayleigh waves in magneto-thermoelastic solids with thermal relaxation," *Int. J. Eng. Sci.* **17**(2), 227 (1979).

3

Dielectric/Ferroelectric Material Systems and Structures

The topics of mechanics in dielectric/ferroelectric material systems and structures are selected and introduced in the two parts of this chapter. Mechanical failure of insulators surrounding conductors carrying a high voltage current is a well-known phenomenon among insulation engineers. In Part 3.1, first, static electroelastic crack mechanics of dielectric materials is introduced. Second, electroelastic vibrations and waves of dielectric materials are discussed. Finally, the dynamic interactions between electric and elastic fields in cracked dielectric materials are reported.

In most of the applications as sensors and actuators in the field of smart structures and devices, piezoelectric ceramics and composites are subjected to both high mechanical stresses and intense electric fields; hence, it is important with regard to reliability and durability to investigate the fracture and deformation behavior of piezoelectric material systems. In Part 3.2, first, the electromechanical response of piezoelectric laminates is examined, and the effects of electric fields and polarization switching/domain wall motion on the bending behavior are discussed. The electromechanical field concentrations near electrodes in piezoelectric devices are also examined. Furthermore, cryogenic and high-temperature electromechanical responses of piezoelectric devices are described. Next, the results of the electric field dependence of cracking of piezoelectric material systems are reported.

Part 3.1   Dielectrics

Dielectric materials undergo deformation and polarization when subjected to mechanical loading and electric field. Figure 3.1 shows the electroelastic interactions of dielectric materials. Toupin [1] considered the isotropic elastic dielectrics and obtained the form of the constitutive relations for the local stress and electric field. The following references are also extended research articles on the subject: [2, 3]. In this part, mechanics of dielectrics is introduced.

Electromagneto-Mechanics of Material Systems and Structures, First Edition. Yasuhide Shindo.
© 2015 John Wiley & Sons Singapore Pte Ltd. Published 2015 by John Wiley & Sons Singapore Pte Ltd.

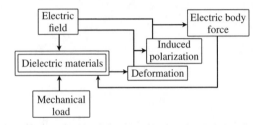

Figure 3.1   Electroelastic interactions of dielectric materials

3.1   Basic Equations of Electroelasticity

Let us now take the rectangular Cartesian coordinates $x_i(O\text{-}x_1, x_2, x_3)$. We consider small perturbations characterized by the displacement vector $\mathbf{u}$ produced in the material and decompose the electric field intensity vector $\mathbf{E}$, the polarization vector $\mathbf{P}$, and the electric displacement vector $\mathbf{D}$ into those representing the rigid body state, indicated by subscript 0, and those for the deformed state, denoted by lowercase letters:

$$\mathbf{E} = \mathbf{E}_0 + \mathbf{e}, \quad \mathbf{P} = \mathbf{P}_0 + \mathbf{p}, \quad \mathbf{D} = \mathbf{D}_0 + \mathbf{d} \tag{3.1}$$

It is assumed that the deformation will be small even with large electric fields and the second terms will have only a minor influence on the total fields. The quasilinear formulations will then be linearized with respect to these unknown deformed state quantities.

The linearized field equations are given by

$$\sigma^L_{ji,j} + E_{0i,j}p_j + P_{0j}e_{i,j} = \rho u_{i,tt} \tag{3.2}$$

$$\varepsilon_{ijk}E_{0k,j} = 0, \quad D_{0i,i} = 0 \tag{3.3}$$

$$\varepsilon_{ijk}e_{k,j} = 0, \quad d_{i,i} = 0 \tag{3.4}$$

where u_i is the displacement vector component, σ^L_{ij} is the local stress tensor component, $E_{0i}, P_{0i}, D_{0i}, e_i, p_i, d_i$ are the components of $\mathbf{E}_0, \mathbf{P}_0, \mathbf{D}_0, \mathbf{e}, \mathbf{p}, \mathbf{d}$, respectively, ρ is the mass density, ε_{ijk} is the permutation symbol, a comma followed by an index denotes partial differentiation with respect to the space coordinate x_i or the time t, and the summation convention over repeated indices is used.

The linearized constitutive equations can be written as

$$\sigma^L_{ij} = \lambda u_{k,k}\delta_{ij} + \mu(u_{i,j} + u_{j,i}) + A_1(E_{0k}E_{0k} + 2E_{0k}e_k)\delta_{ij}$$
$$+ A_2(E_{0i}E_{0j} + E_{0i}e_j + E_{0j}e_i) \tag{3.5}$$

$$\sigma^M_{ij} = \epsilon_0\epsilon_r(E_{0i}E_{0j} + E_{0i}e_j + E_{0j}e_i) - \frac{1}{2}\epsilon_0(E_{0k}E_{0k} + 2E_{0k}e_k)\delta_{ij} \tag{3.6}$$

$$D_{0i} = \epsilon_0 E_{0i} + P_{0i} = \epsilon_0\epsilon_r E_{0i}, \quad d_i = \epsilon_0 e_i + p_i = \epsilon_0\epsilon_r e_i \tag{3.7}$$

$$E_{0i} = \frac{1}{\epsilon_0\eta}P_{0i}, \quad e_i = \frac{1}{\epsilon_0\eta}p_i \tag{3.8}$$

where σ_{ij}^M is the Maxwell stress tensor component, λ and μ are the Lamé constants, A_1 and A_2 are the electrostrictive coefficients, $\epsilon_0 = 8.85 \times 10^{-12}$ C/Vm is the permittivity of free space, $\epsilon_r = 1 + \eta$ is the specific permittivity, η is the electric susceptibility, and δ_{ij} is the Kronecker delta.

The linearized boundary conditions are obtained as

$$[\![\sigma_{ji}^L]\!]n_j + \frac{1}{2\epsilon_0}\{(P_{0k}n_k)^2 + 2P_{0k}P_l n_k n_l\}n_i = 0 \tag{3.9}$$

$$\varepsilon_{ijk}n_j[\![E_{0k}]\!] = 0$$
$$[\![D_{0i}]\!]n_i = 0 \tag{3.10}$$

$$\varepsilon_{ijk}(n_j[\![e_k]\!] - n_l u_{l,j}[\![E_{0k}]\!]) = 0$$
$$[\![d_i]\!]n_i - [\![D_{0i}]\!]u_{i,j}n_j = 0 \tag{3.11}$$

where n_i is the component of outer unit vector $\mathbf{n}$ normal to an undeformed material as shown in Fig. 2.2 and $[\![f_i]\!]$ means the jump in any field quantity f_i across the boundary, that is, $[\![f_i]\!] = f_i^e - f_i$. The superscript e denotes the quantity outside the material.

3.2   Static Electroelastic Crack Mechanics

Elastic dielectrics such as insulating materials have been reported to have poor mechanical properties. Mechanical failure of insulators is also a well-known phenomenon. Therefore, understanding the crack behavior of the elastic dielectrics will provide useful information to insulation designers. Kurlandzka [4] investigated the crack problem of an elastic dielectric material subjected to an electrostatic field. Pak and Herrmann [5, 6] also derived a material force in the form of a path-independent integral for the elastic dielectric material, which is related to the energy release rate. In this section, the planar problem for a dielectric material with a crack under a uniform electric field is dealt with, and the stress intensity factor and energy release rate are given.

3.2.1   Infinite Dielectric Materials

Let us now consider a Griffith crack, which is located in the interior of an infinite elastic dielectric material. We consider a rectangular Cartesian coordinate system (x, y, z) such that the crack is placed on the x-axis from $-a$ to a as shown in Fig. 3.2 and assume plane strain normal to the z-axis. A uniform tensile stress σ_∞ is applied at infinity. By changing the angle θ_0, different mixed-mode conditions at the crack tip can be obtained. Also, a uniform electric field E_0 is applied normal to the crack surface.

By substituting the first of Eqs. (3.7) into the second of Eq. (3.3), the electric field equations for the rigid body state can be expressed as

$$E_{0x,x}^e + E_{0y,y}^e = 0, \quad E_{0x,x} + E_{0y,y} = 0 \tag{3.12}$$

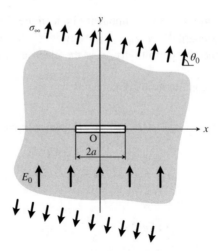

Figure 3.2   An infinite elastic dielectric material with a crack

where the superscript e stands for the electric field quantity in the void inside the crack. The solutions for the rigid body state satisfying the field equation (3.3) and boundary condition (3.10) can, with the aid of the first of Eqs. (3.7) and the first of Eqs. (3.8), be written as

$$E^e_{0y} = \epsilon_r E_0, \quad D^e_{0y} = \epsilon_0 \epsilon_r E_0, \quad P^e_{0y} = 0$$
$$E_{0y} = E_0, \quad D_{0y} = \epsilon_0 \epsilon_r E_0, \quad P_{0y} = \epsilon_0 \eta E_0 \tag{3.13}$$

The local and Maxwell stresses become

$$\sigma^L_{xx} = \lambda(u_{x,x} + u_{y,y}) + 2\mu u_{x,x} + A_1(E_0^2 + 2E_0 e_y)$$
$$\sigma^L_{yy} = \lambda(u_{x,x} + u_{y,y}) + 2\mu u_{y,y} + (A_1 + A_2)(E_0^2 + 2E_0 e_y) \tag{3.14}$$
$$\sigma^L_{xy} = \mu(u_{x,y} + u_{y,x}) + A_2 E_0 e_x$$

$$\sigma^M_{xx} = -\frac{1}{2}\epsilon_0(E_0^2 + 2E_0 e_y)$$
$$\sigma^M_{yy} = \left(\epsilon_0\epsilon_r - \frac{1}{2}\epsilon_0\right)(E_0^2 + 2E_0 e_y) \tag{3.15}$$
$$\sigma^M_{xy} = \epsilon_0\epsilon_r E_0 e_x$$

By substituting Eqs. (3.14) into Eq. (3.2) and considering the fourth and sixth of Eqs. (3.13), the governing equations in the x and y directions, without inertia, are given by

$$u_{x,xx} + u_{x,yy} + \frac{1}{1-2v}(u_{x,x} + u_{y,y})_{,x} + \frac{2A_1}{\mu}E_0 e_{y,x} + \frac{A_3}{\mu}E_0 e_{x,y} = 0$$

$$u_{y,xx} + u_{y,yy} + \frac{1}{1-2v}(u_{x,x} + u_{y,y})_{,y} + \frac{2A_1 + A_2 + A_3}{\mu}E_0 e_{y,y} + \frac{A_2}{\mu}E_0 e_{x,x} = 0 \tag{3.16}$$

where v is the Poisson's ratio and $A_3 = A_2 + \epsilon_0 \eta$. By using the second of Eqs. (3.7), the electric field equations (3.4) for the perturbed state are found to be

$$
\begin{aligned}
e^e_{x,y} - e^e_{y,x} &= 0, \quad e^e_{x,x} + e^e_{y,y} = 0 \\
e_{x,y} - e_{y,x} &= 0, \quad e_{x,x} + e_{y,y} = 0
\end{aligned}
\tag{3.17}
$$

The electric field equations (3.17) are satisfied by introducing an electric potential ϕ such that

$$
\begin{aligned}
e^e_x &= -\phi^e_{,x}, \quad e^e_y = -\phi^e_{,y}, \\
\phi^e_{,xx} + \phi^e_{,yy} &= 0, \\
e_x &= -\phi_{,x}, \quad e_y = -\phi_{,y}, \\
\phi_{,xx} + \phi_{,yy} &= 0
\end{aligned}
\tag{3.18}
$$

The governing equations become

$$
u_{x,xx} + u_{x,yy} + \frac{1}{1 - 2v}(u_{x,x} + u_{y,y})_{,x} - \frac{2A_1 + A_3}{\mu} E_0 \phi_{,xy} = 0
$$

$$
u_{y,xx} + u_{y,yy} + \frac{1}{1 - 2v}(u_{x,x} + u_{y,y})_{,y} - \frac{2A_1 + A_2 + A_3}{\mu} E_0 \phi_{,yy} - \frac{A_2}{\mu} E_0 \phi_{,xx} = 0
\tag{3.19}
$$

When the uniform mechanical loading at infinity forms an angle of θ_0 with the crack surface, the loading will cause mixed-mode deformation so that both mode I and mode II crack problems are considered. The problem will be split into two parts. From Eqs. (3.9) and (3.11), the mixed boundary conditions on the x-axis can, with the aid of the second of Eqs. (3.8), be obtained as

Mode I:

$$
\sigma^L_{yx}(x, 0) = 0 \qquad\qquad (0 \leq |x| < \infty) \tag{3.20}
$$

$$
\begin{cases}
\sigma^L_{yy}(x, 0) = \epsilon_0 \eta^2 \left\{ \dfrac{E_0^2}{2} - E_0 \phi_{,y}(x, 0) \right\} - \sigma_\infty \sin^2 \theta_0 & (0 \leq |x| < a) \\
u_y(x, 0) = 0 & (a \leq |x| < \infty)
\end{cases}
\tag{3.21}
$$

$$
\begin{cases}
\phi_{,x}(x, 0) = -\eta E_0 u_{y,x}(x, 0) + \phi^e_{,x}(x, 0) & (0 \leq |x| < a) \\
\phi(x, 0) = 0 & (a \leq |x| < \infty)
\end{cases}
\tag{3.22}
$$

$$
\epsilon_r \phi_{,y}(x, 0) = \phi^e_{,y}(x, 0) \qquad\qquad (0 \leq |x| < a) \tag{3.23}
$$

Mode II:

$$
\sigma^L_{yy}(x, 0) = 0 \qquad\qquad (0 \leq |x| < \infty) \tag{3.24}
$$

$$
\begin{cases}
\sigma^L_{xy}(x, 0) = -\sigma_\infty \sin \theta_0 \cos \theta_0 & (0 \leq |x| < a) \\
u_x(x, 0) = 0 & (a \leq |x| < \infty)
\end{cases}
\tag{3.25}
$$

$$\begin{cases} \phi_{,x}(x,0) = -\eta E_0 u_{y,x}(x,0) + \phi^e_{,x}(x,0) & (0 \le |x| < a) \\ \phi_{,y}(x,0) = 0 & (a \le |x| < \infty) \end{cases} \tag{3.26}$$

$$\epsilon_r \phi_{,y}(x,0) = \phi^e_{,y}(x,0) \qquad\qquad\qquad (0 \le |x| < a) \tag{3.27}$$

Fourier transforms [7] are used to reduce the mixed boundary value problem to two simultaneous dual integral equations. The integral equations are then solved exactly [8]. The displacements near the crack tip under mode I loading can be written as

$$u_{xs} = \frac{K_I}{2z_s\mu}\left(\frac{r_1}{2\pi}\right)^{1/2} \Bigg\{ 2(1-2v) - \left\{ (1-2v)\left(\frac{A_1}{\epsilon_0} + \eta\right) - 2(1-v)\frac{A_2}{\epsilon_0} \right\} E^2_\mu \eta$$

$$+ \left[2 + \left\{ (1-2v)\left(\frac{A_1}{\epsilon_0} + \eta\right) + 2(1-v)\frac{A_2}{\epsilon_0} \right\} E^2_\mu \eta \right] \sin^2\left(\frac{\theta_1}{2}\right) \Bigg\} \cos\left(\frac{\theta_1}{2}\right)$$

$$u_{ys} = \frac{K_I}{2z_s\mu}\left(\frac{r_1}{2\pi}\right)^{1/2} \Bigg\{ 4(1-v) + \left[2 + \left\{ (1-2v)\left(\frac{A_1}{\epsilon_0} + \eta\right) \right. \right.$$

$$\left. \left. + 2(1-v)\frac{A_2}{\epsilon_0} \right\} E^2_\mu \eta \right] \cos^2\left(\frac{\theta_1}{2}\right) \Bigg\} \sin\left(\frac{\theta_1}{2}\right) \tag{3.28}$$

where the subscript s stands for the symmetric part, $x = r_1 \cos\theta_1 + a, y = r_1 \sin\theta_1$ as defined in Fig. 3.3, and

$$E^2_\mu = \frac{\epsilon_0}{\mu} E^2_0 \tag{3.29}$$

$$z_s = 1 + \frac{1}{2}\left\{ (1-2v)\left(2\frac{A_1}{\epsilon_0} + \eta\right) + 2(1-v)\left(\frac{A_2}{\epsilon_0} + \eta + 1\right) \right\} \eta E^2_\mu \tag{3.30}$$

The mode I electric stress intensity factor K_I is defined by

$$K_I = \lim_{x\to a^+} \{2\pi(x-a)\}^{1/2}\{\sigma^L_{yys}(x,0) + \sigma^M_{yys}(x,0)\} \tag{3.31}$$

The result is

$$K_I = \sigma_\infty(\pi a)^{1/2}\frac{z_s}{y_0}\left\{ \frac{(2A_1/\epsilon_0 + 2A_2/\epsilon_0 - \eta^2)}{2\sigma_\infty/\mu}E^2_\mu + \sin^2\theta_0 \right\} \tag{3.32}$$

where

$$y_0 = 1 + \frac{1}{2}\left\{ (1-2v)(2A_1 + \epsilon_0\eta) + 2(1-v)(A_2 - \epsilon_0\eta^2 - \epsilon_0\eta) \right\}\frac{\eta}{\mu}E^2_0 \tag{3.33}$$

The singular parts of local stresses, Maxwell stresses, and electric fields are

$$\sigma^L_{xxs} = \frac{K_I}{2z_s(2\pi r_1)^{1/2}} \Bigg\{ \left[2 + \left\{ \frac{2(1-2v)A_1}{\epsilon_0} + \frac{2(1-v)A_2}{\epsilon_0} - \eta \right\} \eta E^2_\mu \right]$$

$$- \left[2 + \left\{ (1-2v)\left(2\frac{A_1}{\epsilon_0} + \eta\right) + \frac{2(1-v)A_2}{\epsilon_0} \right\} \eta E^2_\mu \right] \sin\left(\frac{\theta_1}{2}\right) \sin\left(\frac{3\theta_1}{2}\right) \Bigg\} \cos\left(\frac{\theta_1}{2}\right)$$

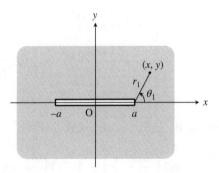

Figure 3.3 Coordinate system used to express crack tip solutions

$$\sigma_{xys}^L = \frac{K_I}{2z_s(2\pi r_1)^{1/2}} \left[2 + \left\{(1-2v)\left(2\frac{A_1}{\epsilon_0} + \eta\right) + \frac{2(1-v)A_2}{\epsilon_0}\right\}\eta E_\mu^2\right]$$

$$\times \sin\left(\frac{\theta_1}{2}\right)\cos\left(\frac{\theta_1}{2}\right)\cos\left(\frac{3\theta_1}{2}\right)$$

$$\sigma_{yys}^L = \frac{K_I}{2z_s(2\pi r_1)^{1/2}} \left\{\left[2 + \left\{\frac{2(1-2v)A_1}{\epsilon_0} + \frac{2(1-v)A_2}{\epsilon_0} - \eta\right\}\eta E_\mu^2\right]\right.$$

$$\left. + \left[2 + \left\{(1-2v)\left(2\frac{A_1}{\epsilon_0} + \eta\right) + \frac{2(1-v)A_2}{\epsilon_0}\right\}\eta E_\mu^2\right]\sin\left(\frac{\theta_1}{2}\right)\sin\left(\frac{3\theta_1}{2}\right)\right\}\cos\left(\frac{\theta_1}{2}\right)$$

(3.34)

$$\sigma_{xxs}^M = -\frac{K_I}{z_s(2\pi r_1)^{1/2}}(1-v)\eta E_\mu^2\cos\left(\frac{\theta_1}{2}\right)$$

$$\sigma_{xys}^M = -\frac{K_I}{z_s(2\pi r_1)^{1/2}}(1-v)\eta \epsilon_r E_\mu^2\sin\left(\frac{\theta_1}{2}\right) \qquad (3.35)$$

$$\sigma_{yys}^M = \frac{K_I}{z_s(2\pi r_1)^{1/2}}(1-v)(1+2\eta)\eta E_\mu^2\cos\left(\frac{\theta_1}{2}\right)$$

$$E_{xs} = -\frac{K_I}{z_s\mu(2\pi r_1)^{1/2}}(1-v)\eta E_0\sin\left(\frac{\theta_1}{2}\right)$$

(3.36)

$$E_{ys} = \frac{K_I}{z_s\mu(2\pi r_1)^{1/2}}(1-v)\eta E_0\cos\left(\frac{\theta_1}{2}\right)$$

The displacements near the crack tip under mode II loading are also given by

$$u_{xa} = \frac{K_{II}}{z_a\mu}\left(\frac{r_1}{2\pi}\right)^{1/2}\left\{2(1-v) + \left\{\frac{(1-2v)A_1}{\epsilon_0} + \frac{(2-3v)A_2}{\epsilon_0} - (1-v)\eta\right\}\eta E_\mu^2\right.$$

$$\left. + \left[1 + \left\{\frac{(1-2v)A_1}{\epsilon_0} + \frac{(1-v)A_2}{\epsilon_0} - v\eta\right\}\eta E_\mu^2\right]\cos^2\left(\frac{\theta_1}{2}\right)\right\}\sin\left(\frac{\theta_1}{2}\right)$$

$$u_{ya} = \frac{K_{\mathrm{II}}}{z_a\mu}\left(\frac{r_1}{2\pi}\right)^{1/2}\left\{-(1-2v)+\left[1+\left\{\frac{(1-2v)A_1}{\epsilon_0}\right.\right.\right.$$
$$\left.\left.\left.+\frac{(1-v)A_2}{\epsilon_0}-v\eta\right\}\eta E_\mu^2\right]\sin^2\left(\frac{\theta_1}{2}\right)\right\}\cos\left(\frac{\theta_1}{2}\right)$$

$$(3.37)$$

where the subscript a stands for the antisymmetric part, and

$$z_a = 1+\left\{(1-2v)\left(\frac{A_1}{\epsilon_0}-\frac{1}{2}\right)+(1-v)\left(\frac{A_2}{\epsilon_0}-\eta\right)\right\}\eta E_\mu^2 \qquad (3.38)$$

The singular parts of local stresses, Maxwell stresses, and electric fields are

$$\sigma_{xxa}^L = -\frac{K_{\mathrm{II}}}{z_a(2\pi r_1)^{1/2}}\left\{2\left[1+\left\{\frac{(1-2v)A_1}{\epsilon_0}+\frac{(1-v)A_2}{\epsilon_0}-\frac{1}{2}\eta\right\}\eta E_\mu^2\right]\right.$$
$$\left.+\left[1+\left\{\frac{(1-2v)A_1}{\epsilon_0}+\frac{(1-v)A_2}{\epsilon_0}-v\eta\right\}\eta E_\mu^2\right]\cos\left(\frac{\theta_1}{2}\right)\cos\left(\frac{3\theta_1}{2}\right)\right\}\sin\left(\frac{\theta_1}{2}\right)$$

$$\sigma_{xya}^L = \frac{K_{\mathrm{II}}}{z_a(2\pi r_1)^{1/2}}\left\{\left[1+\left\{\frac{(1-2v)A_1}{\epsilon_0}+\frac{(1-v)A_2}{\epsilon_0}-\frac{1}{2}\eta\right\}\eta E_\mu^2\right]\right.$$
$$\left.-\left[1+\left\{\frac{(1-2v)A_1}{\epsilon_0}+\frac{(1-v)A_2}{\epsilon_0}-v\eta\right\}\eta E_\mu^2\right]\sin\left(\frac{\theta_1}{2}\right)\sin\left(\frac{3\theta_1}{2}\right)\right\}\cos\left(\frac{\theta_1}{2}\right)$$

$$\sigma_{yya}^L = \frac{K_{\mathrm{II}}}{z_a(2\pi r_1)^{1/2}}\left[1+\left\{\frac{(1-2v)A_1}{\epsilon_0}+\frac{(1-v)A_2}{\epsilon_0}-v\eta\right\}\eta E_\mu^2\right]$$
$$\times\sin\left(\frac{\theta_1}{2}\right)\cos\left(\frac{\theta_1}{2}\right)\cos\left(\frac{3\theta_1}{2}\right)$$

$$(3.39)$$

$$\sigma_{xxa}^M = \frac{K_{\mathrm{II}}}{2z_a(2\pi r_1)^{1/2}}(1-2v)E_\mu^2\sin\left(\frac{\theta_1}{2}\right)$$

$$\sigma_{xya}^M = -\frac{K_{\mathrm{II}}}{2z_a(2\pi r_1)^{1/2}}(1-2v)(1+\eta)E_\mu^2\cos\left(\frac{\theta_1}{2}\right) \qquad (3.40)$$

$$\sigma_{yya}^M = -\frac{K_{\mathrm{II}}}{2z_a(2\pi r_1)^{1/2}}(1-2v)(1+2\eta)E_\mu^2\sin\left(\frac{\theta_1}{2}\right)$$

$$E_{xa} = \frac{K_{\mathrm{II}}}{z_a\mu(2\pi r_1)^{1/2}}(1-2v)\eta E_0\cos\left(\frac{\theta_1}{2}\right)$$

$$(3.41)$$

$$E_{ya} = -\frac{K_{\mathrm{II}}}{z_a\mu(2\pi r_1)^{1/2}}(1-2v)\eta E_0\sin\left(\frac{\theta_1}{2}\right)$$

The mode II electric stress intensity factor K_{II} is defined by

$$K_{\mathrm{II}} = \lim_{x \to a^+} \{2\pi(x-a)\}^{1/2} \{\sigma_{xya}^L(x,0) + \sigma_{xya}^M(x,0)\} \tag{3.42}$$

$$= (\pi a)^{1/2} \frac{2(1-v)z_a}{G_1 + \eta G_2 E_0} \sigma_\infty \cos\theta_0 \sin\theta_0 \tag{3.43}$$

where

$$\begin{aligned}
G_1 &= 2(1-v) + \frac{\{(1-2v)A_1 + (2-3v)A_2 - (1-v)\epsilon_0\eta\}\eta}{\mu}E_0^2 \\
G_2 &= \frac{(1-2v)\{(1-2v)A_1 - vA_2\}}{\mu}E_0
\end{aligned} \tag{3.44}$$

A path-independent integral for elastic dielectric materials is given by Pak and Herrmann [5]

$$J = \int_{\Gamma_0} [(\rho\Sigma + \Phi)\delta_{jx} - (\sigma_{ij}^L + \sigma_{ij}^M)u_{i,x} + D_j E_x]n_j \, d\Gamma \tag{3.45}$$

where Γ_0 is a contour closing a crack tip as shown in Fig. 3.4, n_i is the component of the outer unit vector $\mathbf{n}$ normal to Γ_0, Σ is the stored energy function of deformation and polarization, and

$$\Phi = -\frac{1}{2}\epsilon_r\phi_{,i}\phi_{,i} + \phi_{,i}P_i \tag{3.46}$$

$$\rho\Sigma(E_{ij}, P_i/\rho) = \frac{\rho^2}{2\epsilon_0\eta}\delta_{ij}\Pi_i\Pi_j + \left\{\frac{\lambda}{2}\delta_{ij}\delta_{kl} + \frac{\mu}{2}(\delta_{ik}\delta_{jl} + \delta_{il}\delta_{jk})\right\}E_{ij}E_{kl} \tag{3.47}$$

$$\begin{aligned}
E_{ij} &= \frac{1}{2}(\delta_{Ai}\delta_{Bj}u_{A,B} + \delta_{Bi}\delta_{Aj}u_{B,A}) \\
\Pi_i &= (\delta_{Ai} + \delta_{Bi}u_{A,B})\frac{P_A}{\rho}
\end{aligned} \tag{3.48}$$

If all the electric field quantities are made to vanish, then Eq. (3.45) reduces to the total force on the singularities and inhomogeneities within Γ_0 [9] or the J-integral [10] for elastic materials.

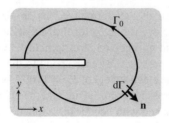

Figure 3.4 Typical contour Γ_0 for evaluation of path-independent line integral

Path independence is convenient if no dissipate forces are present, and the concept applies only to mode I crack extension [11]. This path-independent integral gives the energy release rate. Evaluating the path-independent integral J with the solutions given in Eqs. (3.28), (3.34) – (3.36) along a small circular contour enclosing the crack tip, we obtain the following J value under mode I loading:

$$J = \frac{1}{(1 - 2v)} \frac{K_I^2}{128 \mu z_s^2} \{64(1 - v)(1 - 2v) + C_{2s} E_\mu^2 + C_{1s} E_\mu^4\} \tag{3.49}$$

where

$$
\begin{aligned}
C_{1s} &= 2k_{2s}^2 + k_{3s}^2 + 4(1 - 2v)k_{1s}k_{3s} + 2(1 - 2v)k_{1s}k_{2s} + (1 + 4v)k_{3s}k_{2s} \\
&\quad + 4(1 - v)(1 - 2v)\eta(k_{2s} - 2\eta k_{3s}) \\
C_{2s} &= 4(1 - 2v)[3k_{1s} - 4vk_{2s} - 3k_{3s} \\
&\quad + 2(1 - v)\{12 - 16v + (7 - 8v)\eta - 8(1 - v)\eta^2\}\eta]
\end{aligned}
\tag{3.50}
$$

$$
\begin{aligned}
k_{1s} &= \left\{ \frac{2(1 - 2v)A_1}{\epsilon_0} + \frac{2(1 - v)A_2}{\epsilon_0} - \eta \right\} \eta \\
k_{2s} &= \left\{ (1 - 2v)\left(2\frac{A_1}{\epsilon_0} + \eta\right) + \frac{2(1 - v)A_2}{\epsilon_0} \right\} \eta \\
k_{3s} &= \left\{ (1 - 2v)\left(2\frac{A_1}{\epsilon_0} + \eta\right) - \frac{2(1 - v)A_1}{\epsilon_0} \right\} \eta
\end{aligned}
\tag{3.51}
$$

To examine the electroelastic interactions on the electric stress intensity factor and J-integral, we consider a polymethylmethacrylate (PMMA) material with $a = 1$ mm. The material properties of PMMA are as follows:

$$
\begin{aligned}
\mu &= 1.1 \times 10^9 \ \text{N/m}^2 \\
v &= 0.4 \\
A_1 &= 0, \quad A_2 = 3.61\epsilon_0 \\
\eta &= 2
\end{aligned}
\tag{3.52}
$$

Figure 3.5 shows the variation of the normalized mode I electric stress intensity factor $K_I/\sigma_\infty(\pi a)^{1/2}$ against the angle of loading θ_0 for PMMA under the normalized applied stress $\sigma_\mu = \sigma_\infty/\mu = 0.01, 0.02, \infty$ and normalized electric field $E_\mu = 0.1$. Also shown is the result under $E_\mu = 0$. For PMMA, $\sigma_\mu = 0.01, 0.02$ correspond to the stress of $\sigma_\infty = 11, 22$ MN/m^2, and $E_\mu = 0.1$ corresponds to the electric field of $E_0 = 1100$ MV/m. Applying the electric field increases the mode I electric stress intensity factor, and the effect of electric field on the stress intensity factor is more pronounced with decreasing applied stress. The mode I electric stress intensity factor also increases with increasing the angle θ_0 and attains its maximum value at $\theta_0 = \pi/2$. Figure 3.6 gives the variation of the normalized J-integral $J/\{\pi a(1 - v)\sigma_\infty^2/2v$ under the same conditions shown in Fig. 3.5. Figure 3.7 shows the variation of the normalized mode II electric stress intensity factor $K_{II}/\sigma_\infty(\pi a)^{1/2}$ against the angle of loading θ_0 for PMMA under $E_\mu = 0, 0.1$. The mode II electric stress intensity factor peaks at $\theta_0 = \pi/4$.

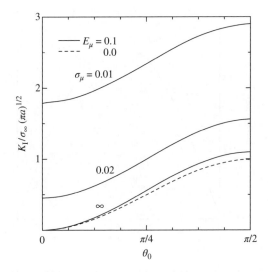

Figure 3.5 Mode I electric stress intensity factor versus angle of loading

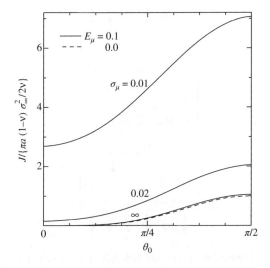

Figure 3.6 J-integral versus angle of loading

3.2.2 Dielectric Strip

We consider an elastic dielectric strip of width $2h$, which contains a central crack of length $2a$ aligned with its plane normal to the free edges as shown in Fig. 3.8. A rectangular Cartesian coordinate system (x, y, z), with origin at the center of the crack, is used. The x-axis is directed along the line of the crack and y-axis along the direction of the perpendicular bisector of the crack. The edges of the dielectric strip are therefore the lines with $x = \pm h$, while the crack

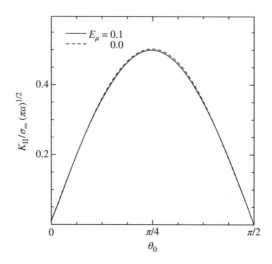

Figure 3.7   Mode II electric stress intensity factor versus angle of loading

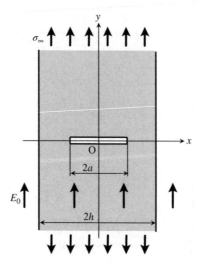

Figure 3.8   An elastic dielectric strip with a crack

occupies the segment $-a < x < a, y = 0$. We assume plane strain normal to the z-axis, and the strip is subjected to a uniform tensile stress σ_∞ and a uniform electric field E_0.

The solutions for the rigid body state satisfying the field equation (3.3) and boundary condition (3.10) can, with the aid of the first of Eqs. (3.7) and the first of Eqs. (3.8), be written as

$$
\left.
\begin{aligned}
E^e_{0y} &= E_0, \quad D^e_{0y} = \epsilon_0 E_0, \quad P^e_{0y} = 0 && (|x| > h) \\
E_{0y} &= E_0, \quad D_{0y} = \epsilon_0 \epsilon_r E_0, \quad P_{0y} = \epsilon_0 \eta E_0 && (|x| \le h) \\
E^e_{0y} &= \epsilon_r E_0, \quad D^e_{0y} = \epsilon_0 \epsilon_r E_0, \quad P^e_{0y} = 0 && (0 \le |x| < a, y = 0)
\end{aligned}
\right\}
\qquad (3.53)
$$

From Eqs. (3.9) and (3.11), the mixed boundary conditions on the x-axis can, with the aid of the second of Eqs. (3.8), be obtained as

$$\sigma_{yx}^L(x, 0) = 0 \qquad (0 \leq |x| \leq h) \tag{3.54}$$

$$\begin{cases} \sigma_{yy}^L(x, 0) = \epsilon_0 \eta^2 \left\{ \dfrac{E_0^2}{2} - E_0 \phi_{,y}(x, 0) \right\} & (0 \leq |x| < a) \\ u_y(x, 0) = 0 & (a \leq |x| \leq h) \end{cases} \tag{3.55}$$

$$\begin{cases} \phi_{,x}(x, 0) = -\eta E_0 u_{y,x}(x, 0) + \phi_{,x}^e(x, 0) & (0 \leq |x| < a) \\ \phi(x, 0) = 0 & (a \leq |x| \leq h) \end{cases} \tag{3.56}$$

$$\epsilon_r \phi_{,y}(x, 0) = \phi_{,y}^e(x, 0) \qquad (0 \leq |x| < a) \tag{3.57}$$

$$\begin{aligned} &\sigma_{xx}^L(\pm h, y) = 0 \\ &\sigma_{xy}^L(\pm h, y) = 0 \\ &\epsilon_r \phi_{,x}(\pm h, y) = -\eta E_0 u_{x,y}(\pm h, y) + \phi_{,x}^e(\pm h, y) \\ &\phi_{,y}(\pm h, y) = \phi_{,y}^e(\pm h, y) \end{aligned} \tag{3.58}$$

Fourier transforms are used to reduce the problem to the solution of a pair of dual integral equations. The solution of the dual integral equations is then expressed in terms of a Fredholm integral equation of the second kind. The mode I stress intensity factor can be found from Eq. (3.31).

Figure 3.9 shows the normalized mode I electric stress intensity factor $K_I/\sigma_\infty(\pi a)^{1/2}$ versus crack length to strip width ratio a/h for PMMA under the normalized applied stress

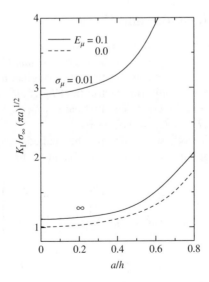

Figure 3.9 Mode I electric stress intensity factor versus crack length to strip width ratio

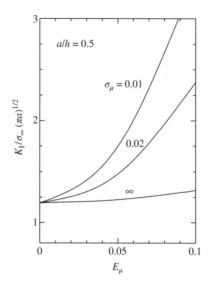

Figure 3.10   Mode I electric stress intensity factor versus electric field

$\sigma_\mu = \sigma_\infty/\mu = 0.01, \infty$ and normalized electric field $E_\mu = 0.1$. Also shown is the result under $E_\mu = 0$. The mode I electric stress intensity factor increases as the a/h ratio increases. Figure 3.10 presents the variation of the normalized mode I electric stress intensity factor $K_I/\sigma_\infty(\pi a)^{1/2}$ with normalized electric field E_μ for PMMA under $\sigma_\mu = 0.01, 0.02, \infty$ for $a/h = 0.5$. The electric field increases the mode I stress intensity factor, and the effect of electric field on the stress intensity factor is more pronounced with decreasing applied stress.

3.3   Electroelastic Vibrations and Waves

In this section, the electroelastic vibrations and waves of a dielectric material are discussed. Consider an elastic dielectric strip with width $2h$ in a rectangular Cartesian coordinate system (x, y, z) as shown in Fig. 3.11. Let electroelastic waves originating at $y = \infty$ be directed at an angle $\pi/2$ with the x-axis. A uniform electric field E_0 is also applied. We consider two possible cases of electric field direction. The first case is $E_{0x} = E_0, E_{0y} = E_{0z} = 0$ (Case I), and the second is $E_{0y} = E_0, E_{0x} = E_{0z} = 0$ (Case II).

The solutions for the rigid body state satisfying the field equation (3.3) and boundary condition (3.10) can, with the aid of the first of Eqs. (3.7) and the first of Eqs. (3.8), be written as

$$\left.\begin{array}{l} E^e_{0x} = E_0, \quad D^e_{0x} = \epsilon_0 E_0, \quad P^e_{0x} = 0 \qquad (|z| > h) \\[2mm] E_{0x} = \dfrac{E_0}{\epsilon_r}, \quad D_{0x} = \epsilon_0 E_0, \quad P_{0x} = \dfrac{\epsilon_0 \eta}{\epsilon_r} E_0 \ (|z| \le h) \end{array}\right\} \text{(Case I)} \qquad (3.59)$$

$$\left.\begin{array}{l} E^e_{0y} = E_0, \quad D^e_{0y} = \epsilon_0 E_0, \quad P^e_{0y} = 0 \qquad (|z| > h) \\[2mm] E_{0y} = E_0, \quad D_{0y} = \epsilon_0 \epsilon_r E_0, \quad P_{0y} = \epsilon_0 \eta E_0 \ (|z| \le h) \end{array}\right\} \text{(Case II)} \qquad (3.60)$$

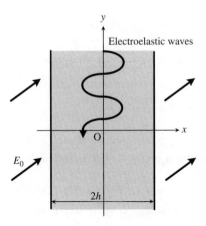

Figure 3.11 An elastic dielectric strip and electroelastic waves

We assume plane strain perpendicular to the z-axis. Substituting Eqs. (3.14) into Eq. (3.2), considering the fourth and sixth of Eqs. (3.59) or (3.60), and using the fourth and fifth of Eqs. (3.18), we obtain the displacement equations of motion as

$$u_{x,xx} + u_{x,yy} + \frac{1}{1-2v}(u_{x,x} + u_{y,y})_{,x} - \frac{2A_1 + A_2 + A_3}{\mu\epsilon_r}E_0\phi_{,xx} - \frac{A_2}{\mu\epsilon_r}E_0\phi_{,yy}$$
$$= \frac{1}{c_2^2}u_{x,tt} \tag{3.61}$$

$$u_{y,xx} + u_{y,yy} + \frac{1}{1-2v}(u_{x,x} + u_{y,y})_{,y} - \frac{2A_1 + A_3}{\mu\epsilon_r}E_0\phi_{,xy} = \frac{1}{c_2^2}u_{y,tt}$$

(Case I)

$$u_{x,xx} + u_{x,yy} + \frac{1}{1-2v}(u_{x,x} + u_{y,y})_{,x} - \frac{2A_1 + A_3}{\mu}E_0\phi_{,xy} = \frac{1}{c_2^2}u_{x,tt}$$

$$u_{y,xx} + u_{y,yy} + \frac{1}{1-2v}(u_{x,x} + u_{y,y})_{,y} - \frac{2A_1 + A_2 + A_3}{\mu}E_0\phi_{,yy} - \frac{A_2}{\mu}E_0\phi_{,xx} \tag{3.62}$$
$$= \frac{1}{c_2^2}u_{y,tt} \quad \text{(Case II)}$$

where $c_2 = (\mu/\rho)^{1/2}$ is the shear wave velocity. From Eqs. (3.9) and (3.11), with the second of Eqs. (3.8), it is clear that

$$\mu\{u_{x,y}(h, y, t) + u_{y,x}(h, y, t)\} - \frac{A_2}{\epsilon_r\rho}E_0\phi_{,y}(h, y, t) = 0$$

$$(\lambda + 2\mu)u_{x,x}(h, y, t) + \lambda u_{y,y}(h, y, t) - \frac{2A_1 + 2A_2 - \epsilon_0\eta^2}{\epsilon_r\rho}E_0\phi_{,x}(h, y, t) = 0$$

$$\phi^e_{,y}(h, y, t) - \phi_{,y}(h, y, t) = \frac{\eta}{\epsilon_r}E_0u_{x,y}(h, y, t) \tag{3.63}$$

$$\phi^e_{,x}(h, y, t) - \epsilon_r\phi_{,x}(h, y, t) = 0$$

(Case I)

$$\mu\{u_{x,y}(h,y,t) + u_{y,x}(h,y,t)\} - \frac{A_2}{\epsilon_r \rho}E_0\phi_{,x}(h,y,t) = 0$$

$$(\lambda + 2\mu)u_{x,x}(h,y,t) + \lambda u_{y,y}(h,y,t) - \frac{A_1}{\epsilon_r \rho}E_0\phi_{,y}(h,y,t) = 0$$

$$\phi^e_{,y}(h,y,t) - \phi_{,y}(h,y,t) = 0$$

$$\phi^e_{,x}(h,y,t) - \phi_{,x}(h,y,t) = -\eta E_0 u_{y,x}(h,y,t)$$

(3.64)

(Case II)

We can write the solutions for u_x, u_y, ϕ, and ϕ^e in the form

$$u_x = u_{x0}\exp\{px - i(ky + \omega t)\}$$
$$u_y = u_{y0}\exp\{px - i(ky + \omega t)\}$$
$$\phi = \phi_0\exp\{px - i(ky + \omega t)\}$$
$$\phi^e = \phi^e_0\exp\{px - i(ky + \omega t)\}$$

(3.65)

where u_{x0}, u_{y0}, ϕ_0, and ϕ^e_0 are the amplitudes of the time harmonic waves, p is the modification factor of the wave amplitude with respect to the width, k is the wave number, and ω is the angular frequency. Substitution of Eqs. (3.65) into Eqs. (3.61), (3.62), and the third and sixth of Eqs. (3.18) yields the following algebraic equations:

$$\begin{bmatrix} a_{11} & a_{12} & a_{13} & 0 \\ a_{21} & a_{22} & a_{23} & 0 \\ 0 & 0 & a_{33} & 0 \\ 0 & 0 & 0 & a_{44} \end{bmatrix}\begin{bmatrix} u_x \\ u_y \\ \phi \\ \phi^e \end{bmatrix} = 0$$

(3.66)

where

$$a_{11} = (\lambda + 2\mu)p^2 - \mu k^2 + \rho\omega^2$$
$$a_{12} = a_{21} = i(\lambda + \mu)kp$$
$$a_{13} = -i\frac{(2A_1 + A_3)kp}{\epsilon_r}E_0$$
$$a_{22} = \mu p^2 - (\lambda + 2\mu)k^2 + \rho\omega^2$$
$$a_{23} = -A_2 p^2 + (2A_1 + A_2 + A_3)k^2$$
$$a_{33} = a_{44} = p^2 - k^2$$

(3.67)

A nontrivial solution of Eq. (3.66) will exist when p is related to k and ω such that the determinant of the coefficient matrix is zero, that is, when

$$\left\{\left(\frac{p}{k}\right)^2 - 1\right\}\left[\frac{c_1^2}{c_2^2}\left(\frac{p}{k}\right)^4 + \left\{\left(1 + \frac{c_1^2}{c_2^2}\right)\left(\frac{\omega}{kc_2}\right)^2 - 2\frac{c_1^2}{c_2^2}\right\}\left(\frac{p}{k}\right)^2\right.$$

$$\left. + \left(\frac{\omega}{kc_2}\right)^4 - \left(1 + \frac{c_1^2}{c_2^2}\right)\left(\frac{\omega}{kc_2}\right)^2 + \frac{c_1^2}{c_2^2}\right] = 0 \qquad (3.68)$$

where $c_1 = \{(\lambda + 2\mu)/\rho\}^{1/2}$ is the longitudinal wave velocity. The roots of this equation are

$$\frac{p}{k} = \pm 1, \ \pm\lambda_1, \ \pm\lambda_2 \tag{3.69}$$

where

$$\lambda_1^2 = 1 - \left(\frac{\omega}{kc_1}\right)^2, \ \lambda_2^2 = 1 - \left(\frac{\omega}{kc_2}\right)^2 \tag{3.70}$$

The solutions take the form

$$
\begin{bmatrix} u_x \\ u_y \\ \phi \\ \phi^e \end{bmatrix} =
\begin{bmatrix}
b_{11} & b_{12} & -i & -b_{11} & -b_{12} & i & 0 \\
1 & 1 & 1 & 1 & 1 & 1 & 0 \\
0 & 0 & b_{33} & 0 & 0 & -b_{33} & 0 \\
0 & 0 & 0 & 0 & 0 & 0 & 1
\end{bmatrix}
$$
$$
\times \begin{bmatrix}
A_{11}\exp(\lambda_1 kx)\exp\{-i(ky+\omega t)\} \\
A_{12}\exp(\lambda_2 kx)\exp\{-i(ky+\omega t)\} \\
A_{13}\exp(kx)\exp\{-i(ky+\omega t)\} \\
B_{11}\exp(-\lambda_1 kx)\exp\{-i(ky+\omega t)\} \\
B_{12}\exp(-\lambda_2 kx)\exp\{-i(ky+\omega t)\} \\
B_{13}\exp(-kx)\exp\{-i(ky+\omega t)\} \\
B_{14}\exp(-kx)\exp\{-i(ky+\omega t)\}
\end{bmatrix} \text{(Case I)} \tag{3.71}
$$

$$
\begin{bmatrix} u_x \\ u_y \\ \phi \\ \phi^e \end{bmatrix} =
\begin{bmatrix}
b_{11} & b_{12} & -i & -b_{11} & -b_{12} & i & 0 \\
1 & 1 & 1 & 1 & 1 & 1 & 0 \\
0 & 0 & b_{33} & 0 & 0 & -b_{33} & 0 \\
0 & 0 & 0 & 0 & 0 & 0 & 1
\end{bmatrix}
$$
$$
\times \begin{bmatrix}
A_{11}\exp(\lambda_1 kx)\exp\{-i(ky+\omega t)\} \\
A_{12}\exp(\lambda_2 kx)\exp\{-i(ky+\omega t)\} \\
A_{13}\exp(kx)\exp\{-i(ky+\omega t)\} \\
B_{11}\exp(-\lambda_1 kx)\exp\{-i(ky+\omega t)\} \\
B_{12}\exp(-\lambda_2 kx)\exp\{-i(ky+\omega t)\} \\
B_{13}\exp(-kx)\exp\{-i(ky+\omega t)\} \\
B_{14}\exp(-kx)\exp\{-i(ky+\omega t)\}
\end{bmatrix} \text{(Case II)} \tag{3.72}
$$

where $A_{11}, A_{12}, A_{13}, B_{11}, B_{12}, B_{13}$, and B_{14} are arbitrary, and

$$b_{1j} = i\frac{\lambda_j^2 - c_1^2/c_2^2 + (\omega/kc_2)^2}{\lambda_j(c_1^2/c_2^2 - 1)} \quad (j = 1, 2) \quad \text{(Case I)} \tag{3.73}$$

$$b_{33} = \frac{-i}{2A_1 + A_3}\frac{\mu\epsilon_r}{E_0}\left(\frac{\omega}{kc_2}\right)^2$$

$$b_{1j} = i\frac{\lambda_j^2 - c_1^2/c_2^2 + (\omega/kc_2)^2}{\lambda_j(c_1^2/c_2^2 - 1)} \quad (j = 1, 2)$$

$$\text{(Case II)} \tag{3.74}$$

$$b_{33} = \frac{-i\mu}{2A_1 + A_3}\frac{1}{E_0}\left(\frac{\omega}{kc_2}\right)^2$$

For convenience in Eqs. (3.71) and (3.72), we set

$$A_{1j} = \frac{1}{2}(A_j + B_j), \quad B_{1j} = \frac{1}{2}(A_j - B_j) \quad (j = 1, 2, 3) \tag{3.75}$$

Hence

$$
\begin{aligned}
u_x &= \{A_1 b_{11} \sinh(\lambda_1 kx) + A_2 b_{12} \sinh(\lambda_2 kx) - iA_3 \sinh(kx) + B_1 b_{11} \cosh(\lambda_1 kx) \\
&\quad + B_2 b_{12} \cosh(\lambda_2 kx) - iB_3 \cosh(kx)\} \exp\{-i(ky + \omega t)\} \\
u_y &= \{A_1 \cosh(\lambda_1 kx) + A_2 \cosh(\lambda_2 kx) + A_3 \cosh(kx) + B_1 \sinh(\lambda_1 kx) \\
&\quad + B_2 \sinh(\lambda_2 kx) + B_3 \sinh(kx)\} \exp\{-i(ky + \omega t)\} \\
\phi &= \{A_3 b_{33} \sinh(kx) + B_3 b_{33} \cosh(kx)\} \exp\{-i(ky + \omega t)\} \\
\phi^e &= B_{14} \exp(-kx) \exp\{-i(ky + \omega t)\}
\end{aligned}
\tag{3.76}
$$

$$\text{(Case I)}$$

$$
\begin{aligned}
u_x &= \{A_1 b_{11} \sinh(\lambda_1 kx) + A_2 b_{12} \sinh(\lambda_2 kx) - iA_3 \sinh(kx) + B_1 b_{11} \cosh(\lambda_1 kx) \\
&\quad + B_2 b_{12} \cosh(\lambda_2 kx) - iB_3 \cosh(kx)\} \exp\{-i(ky + \omega t)\} \\
u_y &= \{A_1 \cosh(\lambda_1 kx) + A_2 \cosh(\lambda_2 kx) + A_3 \cosh(kx) + B_1 \sinh(\lambda_1 kx) \\
&\quad + B_2 \sinh(\lambda_2 kx) + B_3 \sinh(kx)\} \exp\{-i(ky + \omega t)\} \\
\phi &= \{A_3 b_{33} \cosh(kx) + B_3 b_{33} \sinh(kx)\} \exp\{-i(ky + \omega t)\} \\
\phi^e &= B_{14} \exp(-kx) \exp\{-i(ky + \omega t)\}
\end{aligned}
\tag{3.77}
$$

$$\text{(Case II)}$$

Upon substituting Eqs. (3.76) into Eqs. (3.63) and assuming the symmetric motion of the strip relative to the middle surface, we obtain

$$
\begin{bmatrix}
\alpha_{11} & \alpha_{12} & \alpha_{13} & 0 \\
\alpha_{21} & \alpha_{22} & \alpha_{23} & 0 \\
\alpha_{31} & \alpha_{32} & \alpha_{33} & \alpha_{34} \\
0 & 0 & \alpha_{43} & \alpha_{44}
\end{bmatrix}
\begin{bmatrix}
A_1 \sinh(\lambda_1 kh) \\
A_2 \sinh(\lambda_2 kh) \\
A_3 \sinh(kh) \\
B_{14} \exp(-kh)
\end{bmatrix}
= 0
\qquad \text{(Case I)} \tag{3.78}
$$

where

$$\alpha_{1j} = \lambda_j + ib_{1j} \quad (j = 1, 2)$$

$$\alpha_{13} = -\frac{A_2/\epsilon_0}{2A_1/\epsilon_0 + A_2/\epsilon_0 + \eta}\left(\frac{\omega}{kc_2}\right)^2 + 2$$

$$\alpha_{2j} = \frac{i\left(c_1^2/c_2^2 - 2\right) + (c_1^2/c_2^2)b_{1j}\lambda_j}{\tanh(\lambda_j kh)} \quad (j = 1, 2)$$

$$\alpha_{23} = i \left\{ \frac{2A_1/\epsilon_0 + A_2/\epsilon_0 - \eta^2}{(2A_1/\epsilon_0 + A_2/\epsilon_0 + \eta)\tanh(kh)} \left(\frac{\omega}{kc_2}\right)^2 - \frac{2}{\tanh(kh)} \right\}$$

$$\text{(Case I)} \qquad (3.79)$$

$$\alpha_{3j} = -\frac{\eta}{\epsilon_r^2} E_\mu^2 b_{1j} \quad (j = 1, 2)$$

$$\alpha_{33} = i \left\{ \frac{1}{2A_1/\epsilon_0 + A_2/\epsilon_0 + \eta} \left(\frac{\omega}{kc_2}\right)^2 + \frac{\eta}{\epsilon_r^2} E_\mu^2 \right\}$$

$$\alpha_{34} = \alpha_{44} = \frac{1}{\mu\epsilon_r} E_0$$

$$\alpha_{43} = \frac{-i\epsilon_r}{(2A_1/\epsilon_0 + A_2/\epsilon_0 + \eta)\tanh(kh)} \left(\frac{\omega}{kc_2}\right)^2$$

Substituting Eqs. (3.77) into Eq. (3.64) and assuming the symmetric motion of the strip relative to the middle surface, we also get

$$\begin{bmatrix} \alpha_{11} & \alpha_{12} & \alpha_{13} & 0 \\ \alpha_{21} & \alpha_{22} & \alpha_{23} & 0 \\ 0 & 0 & \alpha_{33} & \alpha_{34} \\ \alpha_{41} & \alpha_{42} & \alpha_{43} & \alpha_{44} \end{bmatrix} \begin{bmatrix} A_1 \sinh(\lambda_1 kh) \\ A_2 \sinh(\lambda_2 kh) \\ A_3 \sinh(kh) \\ B_{14} \exp(-kh) \end{bmatrix} = 0 \qquad \text{(Case II)} \qquad (3.80)$$

where

$$\alpha_{1j} = \lambda_j + ib_{1j} \quad (j = 1, 2)$$

$$\alpha_{13} = \frac{A_2/\epsilon_0}{2A_1/\epsilon_0 + A_2/\epsilon_0 + \eta} \left(\frac{\omega}{kc_2}\right)^2 + 2$$

$$\alpha_{2j} = \frac{i\left(c_1^2/c_2^2 - 2\right) + (c_1^2/c_2^2)b_{1j}\lambda_j}{\tanh(\lambda_j kh)} \quad (j = 1, 2)$$

$$\alpha_{23} = i \left\{ \frac{A_1/\epsilon_0}{(2A_1/\epsilon_0 + A_2/\epsilon_0 + \eta)\tanh(kh)} \left(\frac{\omega}{kc_2}\right)^2 - \frac{2}{\tanh(kh)} \right\}$$

$$\text{(Case II)} \qquad (3.81)$$

$$\alpha_{33} = \frac{1}{(2A_1/\epsilon_0 + A_2/\epsilon_0 + \eta)\tanh(kh)} \left(\frac{\omega}{kc_2}\right)^2$$

$$\alpha_{34} = \alpha_{44} = \frac{1}{\mu} E_0$$

$$\alpha_{4j} = \eta E_\mu^2 \lambda_j \quad (j = 1, 2)$$

$$\alpha_{43} = -\frac{\epsilon_r}{2A_1/\epsilon_0 + A_2/\epsilon_0 + \eta} \left(\frac{\omega}{kc_2}\right)^2 + \eta E_\mu^2$$

The respective frequency equations are obtained by setting the determinants of the coefficients of Eqs. (3.78) and (3.80) equal to zero. Thus, symmetric waves have frequencies that are solutions of

$$(\alpha_{11}\alpha_{22} - \alpha_{12}\alpha_{21})(\alpha_{33} - \alpha_{43}) + (\alpha_{12}\alpha_{23} - \alpha_{13}\alpha_{22})\alpha_{31}$$

$$+ (\alpha_{13}\alpha_{21} - \alpha_{11}\alpha_{23})\alpha_{32} = 0 \quad \text{(Case I)} \quad (3.82)$$

and

$$(\alpha_{11}\alpha_{22} - \alpha_{12}\alpha_{21})(\alpha_{43} - \alpha_{33}) + (\alpha_{12}\alpha_{23} - \alpha_{13}\alpha_{22})\alpha_{41}$$

$$+ (\alpha_{13}\alpha_{21} - \alpha_{11}\alpha_{23})\alpha_{42} = 0 \quad \text{(Case II)} \quad (3.83)$$

For the antisymmetric motion, we have

$$\begin{bmatrix} \alpha_{11} & \alpha_{12} & \alpha_{13} & 0 \\ \alpha_{21} & \alpha_{22} & \alpha_{23} & 0 \\ \alpha_{31} & \alpha_{32} & \alpha_{33} & \alpha_{34} \\ 0 & 0 & \alpha_{43} & \alpha_{44} \end{bmatrix} \begin{bmatrix} B_1 \cosh(\lambda_1 kh) \\ B_2 \cosh(\lambda_2 kh) \\ B_3 \cosh(kh) \\ B_{14} \exp(-kh) \end{bmatrix} = 0 \quad \text{(Case I)} \quad (3.84)$$

with

$$\alpha_{1j} = \lambda_j + ib_{1j} \quad (j = 1, 2)$$

$$\alpha_{13} = -\frac{A_2/\epsilon_0}{2A_1/\epsilon_0 + A_2/\epsilon_0 + \eta} \left(\frac{\omega}{kc_2}\right)^2 + 2$$

$$\alpha_{2j} = \left\{ i\left(\frac{c_1^2}{c_2^2} - 2\right) + \frac{c_1^2}{c_2^2}b_{1j}\lambda_j \right\} \tanh(\lambda_j kh) \quad (j = 1, 2)$$

$$\alpha_{23} = i\left\{ \frac{2A_1/\epsilon_0 + A_2/\epsilon_0 - \eta^2}{2A_1/\epsilon_0 + A_2/\epsilon_0 + \eta} \left(\frac{\omega}{kc_2}\right)^2 - 2 \right\} \tanh(kh)$$

$$\alpha_{3j} = -\frac{\eta}{\epsilon_r^2}E_\mu^2 b_{1j} \quad (j = 1, 2) \hspace{3cm} \text{(Case I)} \quad (3.85)$$

$$\alpha_{33} = i\left\{ \frac{1}{2A_1/\epsilon_0 + A_2/\epsilon_0 + \eta} \left(\frac{\omega}{kc_2}\right)^2 + \frac{\eta}{\epsilon_r^2}E_\mu^2 \right\}$$

$$\alpha_{34} = \alpha_{44} = \frac{1}{\mu\epsilon_r}E_0$$

$$\alpha_{43} = \frac{-i\epsilon_r}{2A_1/\epsilon_0 + A_2/\epsilon_0 + \eta} \left(\frac{\omega}{kc_2}\right)^2 \tanh(kh)$$

and

$$\begin{bmatrix} \alpha_{11} & \alpha_{12} & \alpha_{13} & 0 \\ \alpha_{21} & \alpha_{22} & \alpha_{23} & 0 \\ 0 & 0 & \alpha_{33} & \alpha_{34} \\ \alpha_{41} & \alpha_{42} & \alpha_{43} & \alpha_{44} \end{bmatrix} \begin{bmatrix} B_1 \cosh(\lambda_1 kh) \\ B_2 \cosh(\lambda_2 kh) \\ B_3 \cosh(kh) \\ B_{14} \exp(-kh) \end{bmatrix} = 0 \quad \text{(Case II)} \quad (3.86)$$

with

$$\alpha_{1j} = \lambda_j + ib_{1j} \quad (j = 1, 2)$$

$$\alpha_{13} = \frac{A_2/\epsilon_0}{2A_1/\epsilon_0 + A_2/\epsilon_0 + \eta} \left(\frac{\omega}{kc_2}\right)^2 + 2$$

$$\alpha_{2j} = \left\{ i\left(\frac{c_1^2}{c_2^2} - 2\right) + \frac{c_1^2}{c_2^2} b_{1j}\lambda_j \right\} \tanh(\lambda_j kh) \quad (j = 1, 2)$$

$$\alpha_{23} = i\left\{ \frac{A_1/\epsilon_0}{2A_1/\epsilon_0 + A_2/\epsilon_0 + \eta} \left(\frac{\omega}{kc_2}\right)^2 - 2 \right\} \tanh(kh) \qquad \text{(Case II)} \qquad (3.87)$$

$$\alpha_{33} = \frac{1}{2A_1/\epsilon_0 + A_2/\epsilon_0 + \eta} \left(\frac{\omega}{kc_2}\right)^2 \tanh(kh)$$

$$\alpha_{34} = \alpha_{44} = \frac{1}{\mu} E_0$$

$$\alpha_{4j} = \eta E_\mu^2 \lambda_j \quad (j = 1, 2)$$

$$\alpha_{43} = -\frac{\epsilon_r}{2A_1/\epsilon_0 + A_2/\epsilon_0 + \eta} \left(\frac{\omega}{kc_2}\right)^2 + \eta E_\mu^2$$

Frequencies are obtained by solving the equations resulting from by setting the determinants of the coefficients of Eqs. (3.84) and (3.86) equal to zero. These equations, respectively, reduce to

$$(\alpha_{11}\alpha_{22} - \alpha_{12}\alpha_{21})(\alpha_{33} - \alpha_{43}) + (\alpha_{12}\alpha_{23} - \alpha_{13}\alpha_{22})\alpha_{31}$$

$$+ (\alpha_{13}\alpha_{21} - \alpha_{11}\alpha_{23})\alpha_{32} = 0 \qquad \text{(Case I)} \qquad (3.88)$$

and

$$(\alpha_{11}\alpha_{22} - \alpha_{12}\alpha_{21})(\alpha_{43} - \alpha_{33}) + (\alpha_{12}\alpha_{23} - \alpha_{13}\alpha_{22})\alpha_{41}$$

$$+ (\alpha_{13}\alpha_{21} - \alpha_{11}\alpha_{23})\alpha_{42} = 0 \qquad \text{(Case II)} \qquad (3.89)$$

In Fig. 3.12, a plot of the dimensionless phase velocity ω/kc_2 versus the dimensionless wave number kh is given for the first symmetric mode in PMMA strip under normalized electric field $E_\mu = 0, 0.1$ (Case I). As the wavelength decreases (kh increases), the phase velocity decreases. Little difference is found between the results under $E_\mu = 0, 0.1$. Figure 3.13 shows a plot of the dimensionless phase velocity ω/kc_2 versus dimensionless wave number kh for the first antisymmetric mode in PMMA strip under normalized electric field $E_\mu = 0, 0.1$ (Case I). As the wavelength decreases (kh increases), the phase velocity increases. Figures 3.14 and 3.15 show, respectively, the dimensionless phase velocity ω/kc_2 versus the dimensionless wave number kh for the first symmetric and antisymmetric modes in PMMA strip under normalized electric field $E_\mu = 0, 0.1$ (Case II). In contrast to Case I, the phase velocity for Case II depends on the electric field.

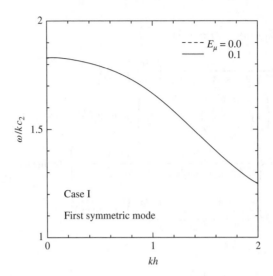

Figure 3.12   Phase velocity of symmetric wave (Case I)

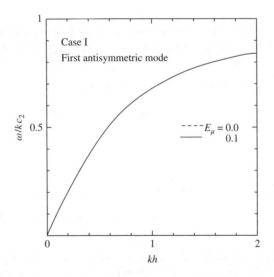

Figure 3.13   Phase velocity of antisymmetric wave (Case I)

3.4   Dynamic Electroelastic Crack Mechanics

In this section, the dynamic electroelastic crack problem for a dielectric material is considered, and the scattering of in-plane compressional (P) and shear (SV) waves by a crack in an infinite dielectric material is examined. Let a Griffith crack be located in the interior of an infinite elastic dielectric material. We consider a rectangular Cartesian coordinate system (x, y, z) such

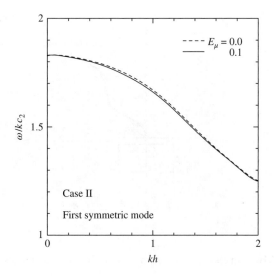

Figure 3.14　Phase velocity of symmetric wave (Case II)

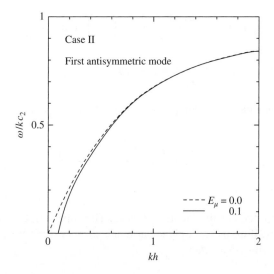

Figure 3.15　Phase velocity of antisymmetric waves (Case II)

that the crack is placed on the x-axis from $-a$ to a as shown in Fig. 3.16 and assume plane strain normal to the z-axis. A uniform electric field E_0 is applied normal to the crack surface.

The solutions for the rigid body state can be obtained as Eqs. (3.13). The equations of motion are given by Eq. (3.62). The displacement components can be written in terms of two scalar potentials $\varphi_e(x, y, t)$ and $\psi_e(x, y, t)$ as

$$u_x = \varphi_{e,x} + \psi_{e,y}, \quad u_y = \varphi_{e,y} - \psi_{e,x} \tag{3.90}$$

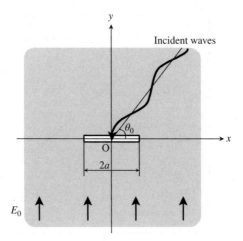

Figure 3.16   An elastic dielectric material with a crack and incident waves

The equations of motion (3.62) become

$$\varphi_{e,xx} + \varphi_{e,yy} - \frac{2A_1 + A_2 + A_3}{\mu}\left(\frac{c_2}{c_1}\right)^2 E_0\phi_{,y} = \frac{1}{c_1^2}\varphi_{e,tt}$$

$$\psi_{e,xx} + \psi_{e,yy} + \frac{A_2}{\mu}E_0\phi_{,x} = \frac{1}{c_2^2}\psi_{e,tt} \tag{3.91}$$

Let an incident plane harmonic compression wave (P-wave) be directed at an angle θ_0 with the x-axis so that

$$\varphi_e^i = \varphi_{e0}\exp\left\{-i\omega\left(t + \frac{x\cos\theta_0 + y\sin\theta_0}{c_1}\right)\right\} \qquad \text{(P-wave)} \qquad (3.92)$$

$$\psi_e^i = 0$$

where φ_{e0} is the amplitude of the incident P-wave. The superscript i stands for the incident component. Similarly, if an incident plane harmonic shear wave (SV-wave) impinges on the crack at an angle θ_0 with x-axis, then

$$\varphi_e^i = 0$$

$$\psi_e^i = \psi_{e0}\exp\left\{-i\omega\left(t + \frac{x\cos\theta_0 + y\sin\theta_0}{c_2}\right)\right\} \qquad \text{(SV-wave)} \qquad (3.93)$$

where ψ_{e0} is the amplitude of the incident SV-wave. In view of the harmonic time variation of the incident waves given by Eqs. (3.92) and (3.93), the field quantities will all contain the time factor $\exp(-i\omega t)$, which will henceforth be dropped.

The problem may be split into two parts: one symmetric (opening mode, mode I) and the other skew-symmetric (sliding mode, mode II). Hence, the boundary conditions for the scattered fields are

Mode I:

$$\sigma_{yx}^{L}(x,0) = 0 \qquad (0 \leq |x| < \infty) \qquad (3.94)$$

$$\begin{cases} \sigma_{yy}^{L}(x,0) = -\epsilon_0 \eta^2 E_0 \phi_{,y} - p_j \exp(-ik_j x \cos\theta_0) & (j=1,2) \quad (0 \leq |x| < a) \\ u_y(x,0) = 0 & (a \leq |x| < \infty) \end{cases} \qquad (3.95)$$

$$\begin{cases} \phi_{,x}(x,0) = -\eta E_0 u_{y,x}(x,0) + \phi_{,x}^{e}(x,0) & (0 \leq |x| < a) \\ \phi(x,0) = 0 & (a \leq |x| < \infty) \end{cases} \qquad (3.96)$$

Mode II:

$$\sigma_{yy}^{L}(x,0) = 0 \qquad (0 \leq |x| < \infty) \qquad (3.97)$$

$$\begin{cases} \sigma_{yx}^{L}(x,0) = -q_j \exp(-ik_j x \cos\theta_0) & (j=1,2) \quad (0 \leq |x| < a) \\ u_x(x,0) = 0 & (a \leq |x| < \infty) \end{cases} \qquad (3.98)$$

$$\begin{cases} \phi_{,x}(x,0) = -\eta E_0 u_{y,x}(x,0) + \phi_{,x}^{e}(x,0) & (0 \leq |x| < a) \\ \phi_{,y}(x,0) = 0 & (a \leq |x| < \infty) \end{cases} \qquad (3.99)$$

where the subscripts $j=1$ and 2 correspond to the incident P- and SV-waves, $p_1 = \mu k_2^2 \varphi_{e0} \{1 - 2(c_2/c_1)^2 \cos^2\theta_0\}$, $p_2 = \mu k_2^2 \psi_{e0} \sin 2\theta_0$, $q_1 = \mu k_2^2 \varphi_{e0} \sigma^2 \sin 2\theta_0$, $q_2 = \mu k_2^2 \psi_{e0} \cos 2\theta_0$, and $k_1 = \omega/c_1$ and $k_2 = \omega/c_2$ are the longitudinal and shear wave numbers, respectively.

By the use of Fourier transforms, we reduce the problem of solving two simultaneous dual integral equations. The solution of the dual integral equations is then expressed in terms of a pair of coupled Fredholm integral equations of the second kind having a kernel that is a finite integral [12]. The mode I dynamic electric stress intensity factor K_{I} is found to be

$$K_{\mathrm{I}} = |K_{\mathrm{ID}}| + K_{\mathrm{IS}} \qquad (3.100)$$

where the dynamic electric stress intensity factor K_{ID} is defined by

$$K_{\mathrm{ID}} = \lim_{x \to a^+} \{2\pi(x-a)\}^{1/2} \{\sigma_{yy}^{L}(x,0,t) + \sigma_{yy}^{M}(x,0,t)\} \qquad (3.101)$$

Form Eq. (3.32), the electric stress intensity factor K_{IS} in Eq. (3.100) is obtained as

$$K_{\mathrm{IS}} = \mu(\pi a)^{1/2} \frac{z_s}{y_0} \frac{2A_1/\epsilon_0 + 2A_2/\epsilon_0 - \eta^2}{2} E_\mu^2 \qquad (3.102)$$

Figure 3.17 shows the variation of the normalized mode I dynamic electric stress intensity factor $|K_{\mathrm{I}}/p_{0\mathrm{P}}(\pi a)^{1/2}|$ (where $p_{0\mathrm{P}} = \mu k_2^2 \varphi_{e0}$) against the normalized frequency $a\omega/c_2$ subjected to P-waves for the normalized electric field $E_\mu = 0.0, 0.1$ and the angle of incidence $\theta_0 = \pi/4, \pi/2$. In the calculations, $\sigma_{\mu\mathrm{P}} = p_{0\mathrm{P}}/\mu = 0.02$ is chosen. Generally speaking, the mode I dynamic electric stress intensity factor tends to increase with the frequency reaching a peak and then to decrease in magnitude. The peak values of $|K_{\mathrm{I}}/p_{0\mathrm{P}}(\pi a)^{1/2}|$ under $E_\mu = 0.0$, 0.1 are 1.36, 2.41 for $\theta_0 = \pi/2$ [13]. The electric fields have large effect on the mode I dynamic electric stress intensity factor. On the other hand, the effect of electric field on the mode II dynamic electric stress intensity factor is small (not shown).

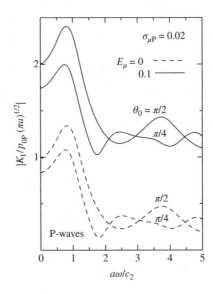

Figure 3.17   Mode I dynamic electric stress intensity factor versus electric field

3.5   Summary

An effort has been made to introduce the static electroelastic crack mechanics of dielectric materials. Electroelastic vibrations and waves of dielectric materials have also been examined, and then some of the dynamic interactions between electric and elastic fields in cracked dielectric materials have been discussed. It can be seen that the electric field effects are important in the discussion of the fracture mechanics parameters such as the electric stress intensity factor of the dielectric materials.

Part 3.2   Piezoelectricity

An investigation of the behavior of materials in interaction with electric fields has become increasingly popular in recent years because of its important and growing device applications in the industry. Electrostriction [3] and piezoelectricity [14] are manifestations of these interactions. One particular branch is piezoelectricity, which is a reversible, inherently anisotropic, and electromechanical phenomenon. In a piezoelectric material, a mechanical stress produces an electric polarization and hence an electric field; this is referred to as the direct piezoelectric effect. Conversely, when an electric field is applied to a piezoelectric material, stress and strain occur; this is called the inverse piezoelectric effect. The relations between the fields are illustrated by the diagram in Fig. 3.18 [15]. Both the piezoelectric effects have been discovered by the Curie brothers in 1880. In this part, mechanics of piezoelectric materials and structures is introduced.

3.6 Piezomechanics and Basic Equations

3.6.1 Linear Theory

Let us now take the rectangular Cartesian coordinates $x_i(O\text{-}x_1, x_2, x_3)$. The field equations are given by Tiersten [16]

$$\sigma_{ji,j} = \rho u_{i,tt} \tag{3.103}$$

$$\varepsilon_{ijk}E_{k,j} = 0, \quad D_{i,i} = 0 \tag{3.104}$$

where σ_{ij}, u_i, E_i, and D_i are the components of stress tensor, displacement vector, electric field intensity vector, and electric displacement vector, respectively, ρ is the mass density, ε_{ijk} is the permutation symbol, a comma followed by an index denotes partial differentiation with respect to the space coordinate x_i or the time t, and the summation convention over repeated indices is used.

The most important class of ferroelectric materials is the perovskite oxides ABO_3 (e.g., $PbTiO_3$), as shown in Fig. 3.19. A central B atom displaces off-center with respect to surrounding O atoms, so that the unit cell possesses a spontaneous polarization P^s and a spontaneous strain γ^s aligned with the dipole moment of the charge distribution. Piezoelectric behavior is induced by the poling process, that is, by applying a high voltage to the material. The electric field aligns dipoles along the field lines. This alignment results in the remanent polarization vector component P_i^r, and the remanently polarized state has a remanent strain tensor component ε_{ij}^r. Constitutive relations can be written as [16, 17]

$$\varepsilon_{ij} = s_{ijkl}^E \sigma_{kl} + d_{kij}E_k + \varepsilon_{ij}^r \tag{3.105}$$

$$D_i = d_{ikl}\sigma_{kl} + \epsilon_{ik}^T E_k + P_i^r \tag{3.106}$$

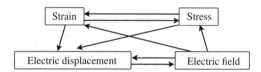

Figure 3.18   Relations between the electrical and mechanical fields

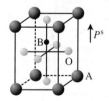

Figure 3.19   Crystal structure of the perovskite ferroelectric ABO_3

where ε_{ij} is the strain tensor component, and s_{ijkl}^{E}, d_{kij}, and ϵ_{ik}^{T} are the elastic compliance at constant electric field, direct (or inverse) piezoelectric coefficient, and permittivity at constant stress, respectively. The quantities ε_{ij}^{r} and P_i^{r} are taken to be due entirely to polarization switching. Valid symmetry conditions for the material constants are

$$s_{ijkl} = s_{jikl} = s_{ijlk} = s_{klij}, \quad d_{kij} = d_{kji}, \quad \epsilon_{ij}^{T} = \epsilon_{ji}^{T} \tag{3.107}$$

Here and in the following, we have dropped the superscript E on the elastic compliance. σ_{ij} and D_i are related to ε_{ij} and E_i by

$$\sigma_{ij} = c_{ijkl}^{E}(\varepsilon_{kl} - \varepsilon_{kl}^{r}) - e_{kij}E_k \tag{3.108}$$

$$D_i = e_{ikl}(\varepsilon_{kl} - \varepsilon_{kl}^{r}) + \epsilon_{ik}^{S}E_k + P_i^{r} \tag{3.109}$$

where c_{ijkl}^{E} is the elastic stiffness at constant electric field, e_{ikl} is the piezoelectric constant, and ϵ_{ik}^{S} is the permittivity at constant strain, which satisfy the following symmetry relations:

$$c_{ijkl} = c_{jikl} = c_{ijlk} = c_{klij}, \quad e_{kij} = e_{kji}, \quad \epsilon_{ik} = \epsilon_{ki} \tag{3.110}$$

Here and in the following, we have dropped the superscript E on the elastic stiffness and the superscript S on the permittivity.

The mechanical boundary condition is obtained as

$$[\![\sigma_{ji}]\!]n_j = 0 \tag{3.111}$$

where n_i is the component of the outer unit vector $\mathbf{n}$ normal to the surface and $[\![\ ]\!]$ means the jump of the quantity across the surface. The electrical boundary condition is

$$\begin{aligned} \varepsilon_{ijk}n_j[\![E_k]\!] &= 0 \\ [\![D_i]\!]n_i &= 0 \end{aligned} \tag{3.112}$$

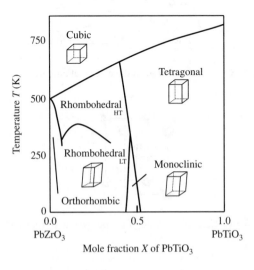

Figure 3.20   PZT Phase diagram

The most extensively studied and technologically important solid solution is lead zirconate titanate $Pb(Zr_{1-X}, Ti_X)O_3$ (known as PZT). The standard phase diagram of PZT is shown in Fig. 3.20 [18, 19]. As the temperature T is decreased, the PZT ceramic undergoes a paraelectric-to-ferroelectric phase transition, and the cubic unit cell is distorted depending on the mole fraction X of $PbTiO_3$. In the Zr-rich region, the paraelectric phase changes to the orthorhombic phase. An intermediate monoclinic phase exists between the Zr-rich rhombohedral perovskite phase and the Ti-rich tetragonal perovskite phase. Compositions between Zr/Ti ratio 90/10 and Zr/Ti ratio 65/35 reveal a ferroelectric-to-ferroelectric transition between rhombohedral space groups. This transition involves the oxygen octahedral tilt. The morphotropic phase boundary (MPB) between the tetragonal and rhombohedral/monoclinic phases is associated with the favorable piezoelectric properties of PZT ceramics [19]. For PZT ceramics poled in the x_3-direction at room temperature (RT), the constitutive relations (3.105) and (3.106) are written in the following form:

$$
\begin{Bmatrix} \varepsilon_{11} \\ \varepsilon_{22} \\ \varepsilon_{33} \\ 2\varepsilon_{23} \\ 2\varepsilon_{31} \\ 2\varepsilon_{12} \end{Bmatrix} =
\begin{bmatrix} s_{11} & s_{12} & s_{13} & 0 & 0 & 0 \\ s_{12} & s_{11} & s_{13} & 0 & 0 & 0 \\ s_{13} & s_{13} & s_{33} & 0 & 0 & 0 \\ 0 & 0 & 0 & s_{44} & 0 & 0 \\ 0 & 0 & 0 & 0 & s_{44} & 0 \\ 0 & 0 & 0 & 0 & 0 & s_{66} \end{bmatrix}
\begin{Bmatrix} \sigma_{11} \\ \sigma_{22} \\ \sigma_{33} \\ \sigma_{23} \\ \sigma_{31} \\ \sigma_{12} \end{Bmatrix}
$$

$$
+ \begin{bmatrix} 0 & 0 & d_{31} \\ 0 & 0 & d_{31} \\ 0 & 0 & d_{33} \\ 0 & d_{15} & 0 \\ d_{15} & 0 & 0 \\ 0 & 0 & 0 \end{bmatrix}
\begin{Bmatrix} E_1 \\ E_2 \\ E_3 \end{Bmatrix} +
\begin{Bmatrix} \varepsilon_{11}^r \\ \varepsilon_{22}^r \\ \varepsilon_{33}^r \\ 2\varepsilon_{23}^r \\ 2\varepsilon_{31}^r \\ 2\varepsilon_{12}^r \end{Bmatrix} \tag{3.113}
$$

$$
\begin{Bmatrix} D_1 \\ D_2 \\ D_3 \end{Bmatrix} =
\begin{bmatrix} 0 & 0 & 0 & 0 & d_{15} & 0 \\ 0 & 0 & 0 & d_{15} & 0 & 0 \\ d_{31} & d_{31} & d_{33} & 0 & 0 & 0 \end{bmatrix}
\begin{Bmatrix} \sigma_{11} \\ \sigma_{22} \\ \sigma_{33} \\ \sigma_{23} \\ \sigma_{31} \\ \sigma_{12} \end{Bmatrix}
$$

$$
+ \begin{bmatrix} \epsilon_{11}^T & 0 & 0 \\ 0 & \epsilon_{11}^T & 0 \\ 0 & 0 & \epsilon_{33}^T \end{bmatrix}
\begin{Bmatrix} E_1 \\ E_2 \\ E_3 \end{Bmatrix} +
\begin{Bmatrix} P_1^r \\ P_2^r \\ P_3^r \end{Bmatrix} \tag{3.114}
$$

where

$$
\varepsilon_{23} = \varepsilon_{32}, \ \varepsilon_{31} = \varepsilon_{13}, \ \varepsilon_{12} = \varepsilon_{21} \tag{3.115}
$$

$$
\sigma_{23} = \sigma_{32}, \ \sigma_{31} = \sigma_{13}, \ \sigma_{12} = \sigma_{21} \tag{3.116}
$$

$$\varepsilon_{23}^r = \varepsilon_{32}^r, \ \varepsilon_{31}^r = \varepsilon_{13}^r, \ \varepsilon_{12}^r = \varepsilon_{21}^r \tag{3.117}$$

$$s_{11} = s_{1111} = s_{2222}, \ s_{12} = s_{1122}, \ s_{13} = s_{1133} = s_{2233}, \ s_{33} = s_{3333}$$
$$s_{44} = 4s_{2323} = 4s_{3131}, \ s_{66} = 4s_{1212} = 2(s_{11} - s_{12}) \tag{3.118}$$

$$d_{15} = 2d_{131} = 2d_{223}, \ d_{31} = d_{311} = d_{322}, \ d_{33} = d_{333} \tag{3.119}$$

The constitutive relations (3.108) and (3.109) are

$$
\begin{Bmatrix} \sigma_{11} \\ \sigma_{22} \\ \sigma_{33} \\ \sigma_{23} \\ \sigma_{31} \\ \sigma_{12} \end{Bmatrix} =
\begin{bmatrix}
c_{11} & c_{12} & c_{13} & 0 & 0 & 0 \\
c_{12} & c_{11} & c_{13} & 0 & 0 & 0 \\
c_{13} & c_{13} & c_{33} & 0 & 0 & 0 \\
0 & 0 & 0 & c_{44} & 0 & 0 \\
0 & 0 & 0 & 0 & c_{44} & 0 \\
0 & 0 & 0 & 0 & 0 & c_{66}
\end{bmatrix}
\begin{Bmatrix} \varepsilon_{11} - \varepsilon_{11}^r \\ \varepsilon_{22} - \varepsilon_{22}^r \\ \varepsilon_{33} - \varepsilon_{33}^r \\ 2(\varepsilon_{23} - \varepsilon_{23}^r) \\ 2(\varepsilon_{31} - \varepsilon_{31}^r) \\ 2(\varepsilon_{12} - \varepsilon_{12}^r) \end{Bmatrix}
$$

$$
-
\begin{bmatrix}
0 & 0 & e_{31} \\
0 & 0 & e_{31} \\
0 & 0 & e_{33} \\
0 & e_{15} & 0 \\
e_{15} & 0 & 0 \\
0 & 0 & 0
\end{bmatrix}
\begin{Bmatrix} E_1 \\ E_2 \\ E_3 \end{Bmatrix} \tag{3.120}
$$

$$
\begin{Bmatrix} D_1 \\ D_2 \\ D_3 \end{Bmatrix} =
\begin{bmatrix}
0 & 0 & 0 & 0 & e_{15} & 0 \\
0 & 0 & 0 & e_{15} & 0 & 0 \\
e_{31} & e_{31} & e_{33} & 0 & 0 & 0
\end{bmatrix}
\begin{Bmatrix} \varepsilon_{11} - \varepsilon_{11}^r \\ \varepsilon_{22} - \varepsilon_{22}^r \\ \varepsilon_{33} - \varepsilon_{33}^r \\ 2(\varepsilon_{23} - \varepsilon_{23}^r) \\ 2(\varepsilon_{31} - \varepsilon_{31}^r) \\ 2(\varepsilon_{12} - \varepsilon_{12}^r) \end{Bmatrix}
$$

$$
+
\begin{bmatrix}
\epsilon_{11} & 0 & 0 \\
0 & \epsilon_{11} & 0 \\
0 & 0 & \epsilon_{33}
\end{bmatrix}
\begin{Bmatrix} E_1 \\ E_2 \\ E_3 \end{Bmatrix} +
\begin{Bmatrix} P_1^r \\ P_2^r \\ P_3^r \end{Bmatrix} \tag{3.121}
$$

where

$$c_{11} = c_{1111} = c_{2222}, \ c_{12} = c_{1122}, \ c_{13} = c_{1133} = c_{2233}, \ c_{33} = c_{3333}$$
$$c_{44} = c_{2323} = c_{3131}, \ c_{66} = c_{1212} = \frac{1}{2}(c_{11} - c_{12}) \tag{3.122}$$

$$e_{15} = e_{131} = e_{223}, \ e_{31} = e_{311} = e_{322}, \ e_{33} = e_{333} \tag{3.123}$$

The relations among the material constants are given by

$$c_{11} = \frac{s_{11}s_{33} - s_{13}^2}{(s_{11} - s_{12})\{s_{33}(s_{11} + s_{12}) - 2s_{13}^2\}}$$

$$c_{12} = \frac{-(s_{12}s_{33} - s_{13}^2)}{(s_{11} - s_{12})\{s_{33}(s_{11} + s_{12}) - 2s_{13}^2\}}$$

$$c_{13} = \frac{-s_{33}}{s_{33}(s_{11} + s_{12}) - 2s_{13}^2} \tag{3.124}$$

$$c_{33} = \frac{s_{11} + s_{12}}{s_{33}(s_{11} + s_{12}) - 2s_{13}^2}$$

$$c_{44} = \frac{1}{s_{44}}$$

$$c_{66} = \frac{1}{s_{66}}$$

$$\begin{aligned} e_{15} &= d_{15}c_{44} \\ e_{31} &= d_{31}(c_{11} + c_{12}) + d_{33}c_{13} \\ e_{33} &= 2d_{31}c_{13} + d_{33}c_{33} \end{aligned} \tag{3.125}$$

$$\begin{aligned} \epsilon_{11} &= \epsilon_{11}^T - d_{15}^2 c_{44} \\ \epsilon_{33} &= \epsilon_{33}^T - (2d_{31}e_{31} + d_{33}e_{33}) \end{aligned} \tag{3.126}$$

3.6.2 Model of Polarization Switching

Polarization switching occurs when an applied electric field exceeds the coercive electric field E_c and leads to changes in the remanent strain ε_{ij}^r and remanent polarization P_i^r. Some nonlinear constitutive models have been developed for ferroelectric materials. Chen and Tucker [20] presented a dynamic macroscopic phenomenological theory for the existence of butterfly and hysteresis loops in ferroelectricity. This is, perhaps, first quantitative indication of this phenomenon. Hwang et al. [21] used a work energy criterion to determine the critical loading level at which polarization switching occurs. Figure 3.21 illustrates several possibilities. A direct current (DC) electric field may rotate the poling direction by either 180° or 90°, but a stress may only rotate it by 90° [22]. The criterion states that a polarization switches when the combined electrical and mechanical work exceeds a critical value, that is,

$$\sigma_{ij}\Delta\varepsilon_{ij}^s + E_i\Delta P_i^s \geq 2P^sE_c \tag{3.127}$$

where $\Delta\varepsilon_{ij}^s$ and ΔP_i^s are the changes in the strain and polarization due to switching. Considering a ferroelectric domain with its spontaneous polarization forming an angle 0 with the x_3-axis,

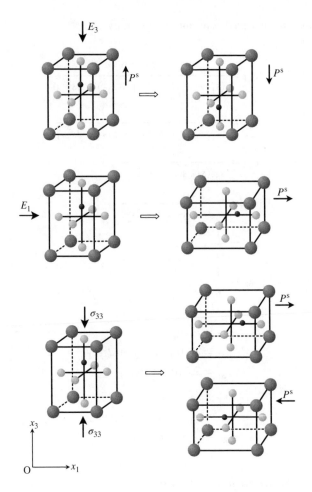

Figure 3.21   Polarization switching induced by electromechanical loads

the values of $\Delta\varepsilon_{ij}^{s} = \varepsilon_{ij}^{r}$ and $\Delta P_i^s = P_i^r$ for 180° switching can be expressed as

$$\Delta\varepsilon_{11}^{s} = 0, \quad \Delta\varepsilon_{22}^{s} = 0, \quad \Delta\varepsilon_{33}^{s} = 0$$
$$\Delta\varepsilon_{12}^{s} = 0, \quad \Delta\varepsilon_{23}^{s} = 0, \quad \Delta\varepsilon_{31}^{s} = 0 \tag{3.128}$$

$$\Delta P_1^s = 0, \quad \Delta P_2^s = 0, \quad \Delta P_3^s = -2P^s \tag{3.129}$$

For 90° switching in the $x_3 x_1$ plane, the changes are

$$\Delta\varepsilon_{11}^{s} = \gamma^s, \quad \Delta\varepsilon_{22}^{s} = 0, \quad \Delta\varepsilon_{33}^{s} = -\gamma^s$$
$$\Delta\varepsilon_{12}^{s} = 0, \quad \Delta\varepsilon_{23}^{s} = 0, \quad \Delta\varepsilon_{31}^{s} = 0 \tag{3.130}$$

$$\Delta P_1^s = \pm P^s, \quad \Delta P_2^s = 0, \quad \Delta P_3^s = -P^s \tag{3.131}$$

For 90° switching in the x_2x_3 plane, we have

$$\Delta\varepsilon^s_{11} = 0, \quad \Delta\varepsilon^s_{22} = \gamma^s, \quad \Delta\varepsilon^s_{33} = -\gamma^s$$
$$\Delta\varepsilon^s_{12} = 0, \quad \Delta\varepsilon^s_{23} = 0, \quad \Delta\varepsilon^s_{31} = 0 \tag{3.132}$$

$$\Delta P^s_1 = 0, \quad \Delta P^s_2 = \pm P^s, \quad \Delta P^s_3 = -P^s \tag{3.133}$$

The constitutive equations (3.105) and (3.106) after polarization switching are

$$\varepsilon_{ij} = s_{ijkl}\sigma_{kl} + d^s_{kij}E_k + \varepsilon^r_{ij} \tag{3.134}$$

$$D_i = d^s_{ikl}\sigma_{kl} + \epsilon^T_{ik}E_k + P^r_i \tag{3.135}$$

The new direct piezoelectric coefficient d^s_{ikl} is

$$d^s_{ikl} = \{d_{33}n_{Pi}n_{Pk}n_{Pl} + d_{31}(n_{Pi}\delta_{kl} - n_{Pi}n_{Pk}n_{Pl})$$
$$+ \frac{1}{2}d_{15}(\delta_{ik}n_{Pl} - 2n_{Pi}n_{Pk}n_{Pl} + \delta_{il}n_{Pk})\} \tag{3.136}$$

where n_{Pi} is the component of unit vector $\mathbf{n}_P$ in the poling direction, δ_{ij} is the Kronecker delta, and d_{15}, d_{31}, d_{33} are the shear, transverse, and longitudinal direct piezoelectric coefficients when the ceramics are completely poled. The new piezoelectric constants e^s_{ikl} are related to the elastic stiffnesses and direct piezoelectric coefficients by

$$
\begin{aligned}
e^s_{111} &= d^s_{111}c_{11} + d^s_{122}c_{12} + d^s_{133}c_{13}\\
e^s_{122} &= d^s_{111}c_{12} + d^s_{122}c_{11} + d^s_{133}c_{13}\\
e^s_{133} &= d^s_{111}c_{13} + d^s_{122}c_{13} + d^s_{133}c_{33}\\
e^s_{123} &= 2d^s_{123}c_{44}\\
e^s_{131} &= 2d^s_{131}c_{44}\\
e^s_{112} &= 2d^s_{112}c_{66}\\
e^s_{211} &= d^s_{211}c_{11} + d^s_{222}c_{12} + d^s_{233}c_{13}\\
e^s_{222} &= d^s_{211}c_{12} + d^s_{222}c_{11} + d^s_{233}c_{13}\\
e^s_{233} &= d^s_{211}c_{13} + d^s_{222}c_{13} + d^s_{233}c_{33}\\
e^s_{223} &= 2d^s_{223}c_{44}\\
e^s_{231} &= 2d^s_{231}c_{44}\\
e^s_{212} &= 2d^s_{212}c_{66}\\
e^s_{311} &= d^s_{311}c_{11} + d^s_{322}c_{12} + d^s_{333}c_{13}\\
e^s_{322} &= d^s_{311}c_{12} + d^s_{322}c_{11} + d^s_{333}c_{13}\\
e^s_{333} &= d^s_{311}c_{13} + d^s_{322}c_{13} + d^s_{333}c_{33}\\
e^s_{323} &= 2d^s_{323}c_{44}\\
e^s_{331} &= 2d^s_{331}c_{44}\\
e^s_{312} &= 2d^s_{312}c_{66}
\end{aligned}
\tag{3.137}
$$

It is assumed that elastic compliances and permittivities of the piezoelectric materials remain unchanged after 180° or 90° polarization switching occurs and only direct piezoelectric coefficients vary with switching.

Later, Huber and Fleck [23] discussed the following three models: (i) a self-consistent polycrystal model; (ii) a crystal plasticity model; (iii) a rate-independent phenomenological model, and they compared the predictions of each model for the material response in multiaxial electrical loading with the measured responses. Sun and Achuthan [24] reviewed several existing polarization switching criteria and proposed a new criterion in terms of internal energy density for combined electromechanical loading.

3.6.3 Model of Domain Wall Motion

The piezoelectric effect consists of two parts, the intrinsic and the extrinsic effects. The intrinsic contributions are from the relative ion shift that preserves the ferroelectric crystal structure, caused by the electric field, and the remaining extrinsic contributions are the result of the elastic deformation caused by the motions of domain walls [25, 26]. Consider the domain structure in piezoelectric materials as shown in Fig. 3.22. For simplicity here, we assume that the direction of the applied alternating current (AC) electric field $E_0 \exp(i\omega t)$ is parallel to the direction of spontaneous polarization P^s in one of the domains; E_0 is the AC electric field amplitude and ω is the angular frequency. A domain wall displacement Δl gives rise to the following changes of the strain and polarization of this basic unit:

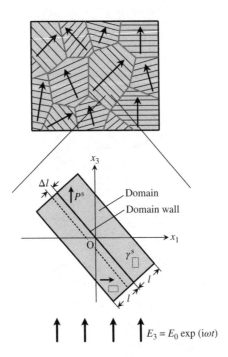

Figure 3.22 Illustrations of grains in piezoelectric ceramics and basic unit of a crystallite with a displaceable domain wall

$$\Delta\varepsilon_{11}^m = -\frac{\Delta l}{l}\gamma^s, \quad \Delta\varepsilon_{22}^m = 0, \quad \Delta\varepsilon_{33}^m = \frac{\Delta l}{l}\gamma^s$$
$$\Delta\varepsilon_{12}^m = 0, \quad \Delta\varepsilon_{23}^m = 0, \Delta\varepsilon_{31}^m = 0 \tag{3.138}$$

$$\Delta P_1^m = -\frac{\Delta l}{l}P^s, \quad \Delta P_2^m = 0, \quad \Delta P_3^m = \frac{\Delta l}{l}P^s \tag{3.139}$$

where $\Delta\varepsilon_{ij}^m$ and ΔP_i^m are the changes in the strain and polarization due to domain wall motion, and l is the domain width. The equation of motion of the domain wall may be written as [27, 28]

$$m_D\Delta l_{,tt} + \beta\Delta l_{,t} + f_D\Delta l = -\frac{\partial W}{\partial\Delta l}l \tag{3.140}$$

where m_D is the effective mass per unit area of the wall, β is the damping constant of the wall!maximum strain motion, f_D represents the force constant for the domain wall motion process, both per unit area, and $W = -(\sigma_{ij}\Delta\varepsilon_{ij}^m + E_i\Delta P_i^m)/2$ is the induced energy density. Damping may be occasioned by coupling with lattice vibrations and other causes, but at present we set $\beta = 0$ purely for convenience. The frequency ω of the applied AC electric field is usually much smaller than the resonance frequency of the domain wall, $\omega_0 = (f_D/m_D)^{1/2}$, which is in the GHz range. So, we can neglect the first term on the left-hand side of Eq. (3.140). Setting $\Delta l = \Delta l_0 \exp(i\omega t)$, from Eqs. (3.138)–(3.140), we have

$$\Delta l_0 = \frac{1}{2f_D}\{\gamma^s(\sigma_{33} - \sigma_{11}) + P^s(E_3 - E_1)\}\exp(-i\omega t) \tag{3.141}$$

The changes of the strains and polarization due to the domain wall displacement Δl of the ceramics according to Fig. 3.22 are given by

$$\Delta\varepsilon_{11}^m = \Delta s_{11}^m\sigma_{11} + \Delta s_{13}^m\sigma_{33} + \Delta d_{31}^m E_3$$
$$\Delta\varepsilon_{33}^m = \Delta s_{13}^m\sigma_{11} + \Delta s_{33}^m\sigma_{33} + \Delta d_{33}^m E_3 \tag{3.142}$$
$$\Delta P_3^m = \Delta d_{31}^m\sigma_{11} + \Delta d_{33}^m\sigma_{33} + \Delta\epsilon_{33}^{Tm} E_3$$

where

$$\Delta s_{11}^m = \frac{(\gamma^s)^2}{2lf_D}$$

$$\Delta s_{13}^m = -\frac{(\gamma^s)^2}{2lf_D}$$

$$\Delta s_{33}^m = \frac{(\gamma^s)^2}{2lf_D}$$

$$\Delta d_{31}^m = -\frac{\gamma^s P^s}{2lf_D} \tag{3.143}$$

$$\Delta d_{33}^m = \frac{\gamma^s P^s}{2lf_D}$$

$$\Delta\epsilon_{33}^{Tm} = \frac{(P^s)^2}{2lf_D}$$

The strains $\varepsilon_{11}, \varepsilon_{33}$ and electric displacement D_3 become

$$\varepsilon_{11} = s_{11}^m \sigma_{11} + s_{12} \sigma_{22} + s_{13}^m \sigma_{33} + d_{31}^m E_3$$
$$\varepsilon_{33} = s_{13}^m \sigma_{11} + s_{13}^m \sigma_{22} + s_{33}^m \sigma_{33} + d_{33}^m E_3 \qquad (3.144)$$
$$D_3 = d_{31}^m \sigma_{11} + d_{31}^m \sigma_{22} + d_{33}^m \sigma_{33} + \epsilon_{33}^{Tm} E_3$$

where

$$s_{11}^m = s_{11} + \Delta s_{11}^m$$
$$s_{13}^m = s_{13} + \Delta s_{13}^m$$
$$s_{33}^m = s_{33} + \Delta s_{33}^m$$
$$d_{31}^m = d_{31} + \Delta d_{31}^m \qquad (3.145)$$
$$d_{33}^m = d_{33} + \Delta d_{33}^m$$
$$\epsilon_{33}^{Tm} = \epsilon_{33}^T + \Delta \epsilon_{33}^{Tm}$$

Experimental studies on PZTs have shown that 45–70% of direct piezoelectric coefficients and permittivities may originate from the extrinsic contributions [26, 29]. The extrinsic permittivity $\Delta \epsilon_{33}^T$ was approximately estimated by Li et al. [30] as two-thirds of the measured value. Narita et al. [31] proposed the following equation to describe $\Delta \epsilon_{33}^T$ in terms of AC electric field amplitude E_0 and coercive electric field E_c:

$$\Delta \epsilon_{33}^T = \epsilon_{33}^T \frac{2E_0}{3E_c} \qquad (3.146)$$

By substituting Eq. (3.146) into the sixth of Eq. (3.143), $lf_D = 3(P^s)^2 E_c / (4\epsilon_{33}^T E_0)$ is obtained. By eliminating lf_D, the changes of the elastic compliances $\Delta s_{11}^m, \Delta s_{13}^m, \Delta s_{33}^m$ and direct piezoelectric coefficients $\Delta d_{31}^m, \Delta d_{33}^m$ in Eq. (3.143) can be rewritten in terms of AC electric field amplitude and so on.

3.6.4   Classical Lamination Theory

Fiber-reinforced polymer (FRP) composite materials with thermopiezoelectric layers have varied potential in many applications especially in aerospace structures. By integrating thermopiezoelectric layers and advanced composite materials, the potential exists for forming high-strength, high-stiffness, lightweight structures capable of self-monitoring and self-controlling. For the thermopiezoelectric materials, the constitutive relations, Eqs. (3.105) and (3.106), of the piezoelectric materials need to be modified and can be rewritten as [32]

$$\varepsilon_{ij} = s_{ijkl}^{E\Theta} \sigma_{kl} + d_{kij}^{\Theta} E_k + \alpha_{ij}^E \Theta + \varepsilon_{ij}^r \qquad (3.147)$$

$$D_i = d_{ikl}^{\Theta} \sigma_{kl} + \epsilon_{ik}^{T\Theta} E_k + p_i^T \Theta + P_i^r \qquad (3.148)$$

where $\Theta = T - T_R$ is the temperature rise from the stress-free reference temperature T_R; $s_{ijkl}^{E\Theta}, d_{kij}^{\Theta}$, and $\epsilon_{ik}^{T\Theta}$ are the elastic compliance at constant electric field and temperature, direct piezoelectric coefficient at constant temperature, permittivity at constant stress and temperature, respectively; α_{ij}^E is the coefficient of thermal expansion at constant electric field; and p_i^T is the pyroelectric constant at constant stress. Similarly, the relations (3.108) and (3.109) can be rewritten as

$$\sigma_{ij} = c_{ijkl}^{E\Theta}(\varepsilon_{kl} - \varepsilon_{kl}^r) - e_{kij}^{\Theta}E_k - \lambda_{ij}^E\Theta \tag{3.149}$$

$$D_i = e_{ikl}^{\Theta}(\varepsilon_{kl} - \varepsilon_{kl}^r) + \epsilon_{ik}^{S\Theta}E_k + p_i^S\Theta + P_i^r \tag{3.150}$$

where $c_{ijkl}^{E\Theta}, e_{kij}^{\Theta}$, and $\epsilon_{ik}^{S\Theta}$ are the elastic stiffness at constant electric field and temperature, piezoelectric constant at constant temperature, permittivity at constant strain and temperature, respectively; λ_{ij}^E is the stress–temperature coefficient at constant electric field; and p_i^S is the pyroelectric constant at constant strain.

Consider a laminated plate constructed of N layers of the thermopiezoelectric materials and FRPs, as shown in Fig. 3.23. The principal material coordinates are x_1, x_2, x_3. Let the coordinate axes x and y be such that they are in the middle plane of the hybrid laminate and the $z = x_3$ axis is normal to this plane. θ_1^p is the angle between the lamina x-axis and lamina principal x_1-axis. a and b are the side lengths in the x and y directions. The total thickness is $2h$ and the kth layer has thickness $h_k = z_k - z_{z-1}(k = 1, \ldots, N)$, where $z_0 = -h$ and $z_N = h$. Here, a large uniform electric field is applied to one or more layers. It is therefore assumed that the electric field resulting from variations in stress and temperature (the so-called direct piezoelectric effect) is insignificant compared with the applied electric field [33].

The stress equations of motion, considering the damping effect of the structure, are as follows:

$$\sigma_{xx,x} + \sigma_{yx,y} + \sigma_{zx,z} = \rho u_{x,tt} + Cu_{x,t}$$
$$\sigma_{xy,x} + \sigma_{yy,y} + \sigma_{zy,z} = \rho u_{y,tt} + Cu_{y,t} \tag{3.151}$$
$$\sigma_{xz,x} + \sigma_{yz,y} + \sigma_{zz,z} = \rho u_{z,tt} + Cu_{z,t}$$

where C is the viscoelastic damping coefficient. The thermopiezoelectric polymer, such as polyvinylidene fluoride (PVDF), is supplied in the form of a film. It is flexible and shows large

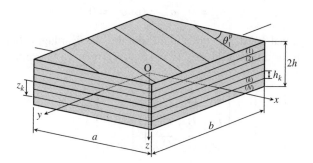

Figure 3.23 A laminated plate

compliance, and this feature makes it suitable for applications in soft sensors and actuators. For PVDF polymer, the constitutive relation (3.149) reduces to

$$
\begin{Bmatrix} \sigma_{11} \\ \sigma_{22} \\ \sigma_{33} \\ \sigma_{23} \\ \sigma_{31} \\ \sigma_{12} \end{Bmatrix} = \begin{bmatrix} c_{11} & c_{12} & c_{13} & 0 & 0 & 0 \\ c_{12} & c_{22} & c_{23} & 0 & 0 & 0 \\ c_{13} & c_{23} & c_{33} & 0 & 0 & 0 \\ 0 & 0 & 0 & c_{44} & 0 & 0 \\ 0 & 0 & 0 & 0 & c_{55} & 0 \\ 0 & 0 & 0 & 0 & 0 & c_{66} \end{bmatrix} \begin{Bmatrix} \varepsilon_{11} - \varepsilon_{11}^{\mathrm{r}} \\ \varepsilon_{22} - \varepsilon_{22}^{\mathrm{r}} \\ \varepsilon_{33} - \varepsilon_{33}^{\mathrm{r}} \\ 2(\varepsilon_{23} - \varepsilon_{23}^{\mathrm{r}}) \\ 2(\varepsilon_{31} - \varepsilon_{31}^{\mathrm{r}}) \\ 2(\varepsilon_{12} - \varepsilon_{12}^{\mathrm{r}}) \end{Bmatrix}
$$

$$
- \begin{bmatrix} 0 & 0 & e_{31} \\ 0 & 0 & e_{32} \\ 0 & 0 & e_{33} \\ 0 & e_{24} & 0 \\ e_{15} & 0 & 0 \\ 0 & 0 & 0 \end{bmatrix} \begin{Bmatrix} E_1 \\ E_2 \\ E_3 \end{Bmatrix} - \begin{Bmatrix} \lambda_1 \\ \lambda_2 \\ \lambda_3 \\ 0 \\ 0 \\ 0 \end{Bmatrix} \Theta \tag{3.152}
$$

where

$$
\begin{aligned}
& c_{11} = c_{1111}, \; c_{12} = c_{1122} = c_{2211}, \; c_{13} = c_{1133} = c_{3311}, c_{22} = c_{2222} \\
& c_{23} = c_{2233} = c_{3322}, \; c_{33} = c_{3333}, c_{44} = c_{2323}, \; c_{55} = c_{3131}, \; c_{66} = c_{1212}
\end{aligned} \tag{3.153}
$$

$$
e_{15} = e_{131}, \; e_{24} = e_{223}, \; e_{31} = e_{311}, \; e_{32} = e_{322}, \; e_{33} = e_{333} \tag{3.154}
$$

$$
\lambda_1 = \lambda_{11}, \; \lambda_2 = \lambda_{22}, \; \lambda_3 = \lambda_{33} \tag{3.155}
$$

Here and in the following, we have dropped the superscripts E and Θ. When $c_{22} = c_{11}, c_{23} = c_{13}, c_{55} = c_{44}, e_{32} = e_{31}, e_{24} = e_{15}, \lambda_1 = \lambda_2 = \lambda_3 = 0$, the constitutive relation (3.152) becomes Eq. (3.120) for PZT materials.

For a lamina in the xy plane, a plane stress state is defined by setting

$$
\sigma_{zz} = 0, \; \sigma_{zx} = 0, \; \sigma_{zy} = 0 \tag{3.156}
$$

in the three-dimensional constitutive relation. From the classical lamination theory [34], the constitutive relation for a typical layer k ($k = 1, \dots, N$), referred to arbitrary plate axes x, y, and z, becomes

$$
\begin{Bmatrix} \sigma_{xx} \\ \sigma_{yy} \\ \sigma_{xy} \end{Bmatrix}_k = \begin{bmatrix} \overline{Q}_{11} & \overline{Q}_{12} & \overline{Q}_{16} \\ \overline{Q}_{12} & \overline{Q}_{22} & \overline{Q}_{26} \\ \overline{Q}_{16} & \overline{Q}_{26} & \overline{Q}_{66} \end{bmatrix}_k \begin{Bmatrix} \varepsilon_{xx} \\ \varepsilon_{yy} \\ 2\varepsilon_{xy} \end{Bmatrix}_k
$$

$$
- \begin{bmatrix} 0 & 0 & \overline{e}_{31} \\ 0 & 0 & \overline{e}_{32} \\ 0 & 0 & \overline{e}_{36} \end{bmatrix}_k \begin{Bmatrix} E_x \\ E_y \\ E_z \end{Bmatrix}_k - \begin{Bmatrix} \overline{\lambda}_1 \\ \overline{\lambda}_2 \\ \overline{\lambda}_6 \end{Bmatrix}_k (\Theta)_k \tag{3.157}
$$

Here and in the following, we have dropped the remanent strains. For $\overline{Q}_{ij}, \overline{e}_{ij}$ and $\overline{\lambda}_i$, we have

$$\overline{Q}_{11} = Q_{11} \cos^4 \theta_1^{\mathrm{p}} + 2(Q_{12} + 2Q_{66}) \sin^2 \theta_1^{\mathrm{p}} \cos^2 \theta_1^{\mathrm{p}} + Q_{22} \sin^4 \theta_1^{\mathrm{p}}$$
$$\overline{Q}_{12} = (Q_{11} + Q_{22} - 4Q_{66}) \sin^2 \theta_1^{\mathrm{p}} \cos^2 \theta_1^{\mathrm{p}} + Q_{12}(\sin^4 \theta_1^{\mathrm{p}} + \cos^4 \theta_1^{\mathrm{p}})$$
$$\overline{Q}_{16} = (Q_{11} - Q_{12} - 2Q_{66}) \sin \theta_1^{\mathrm{p}} \cos^3 \theta_1^{\mathrm{p}} + (Q_{12} - Q_{22} + 2Q_{66}) \sin^3 \theta_1^{\mathrm{p}} \cos \theta_1^{\mathrm{p}}$$
$$\overline{Q}_{22} = Q_{11} \sin^4 \theta_1^{\mathrm{p}} + 2(Q_{12} + 2Q_{66}) \sin^2 \theta_1^{\mathrm{p}} \cos^2 \theta_1^{\mathrm{p}} + Q_{22} \cos^4 \theta_1^{\mathrm{p}}$$
$$\overline{Q}_{26} = (Q_{11} - Q_{12} - 2Q_{66}) \sin^3 \theta_1^{\mathrm{p}} \cos \theta_1^{\mathrm{p}} + (Q_{12} - Q_{22} + 2Q_{66}) \sin \theta_1^{\mathrm{p}} \cos^3 \theta_1^{\mathrm{p}}$$
$$\overline{Q}_{66} = (Q_{11} + Q_{22} - 2Q_{12} - 2Q_{66}) \sin^2 \theta_1^{\mathrm{p}} \cos^2 \theta_1^{\mathrm{p}} + Q_{66}(\sin^4 \theta_1^{\mathrm{p}} + \cos^4 \theta_1^{\mathrm{p}})$$
$$\text{(3.158)}$$

$$\overline{e}_{31} = e_{31} \cos^2 \theta_1^{\mathrm{p}} + e_{32} \sin^2 \theta_1^{\mathrm{p}}$$
$$\overline{e}_{32} = e_{31} \sin^2 \theta_1^{\mathrm{p}} + e_{32} \cos^2 \theta_1^{\mathrm{p}}$$
$$\overline{e}_{36} = 2(e_{31} - e_{32}) \sin \theta_1^{\mathrm{p}} \cos \theta_1^{\mathrm{p}}$$
$$\text{(3.159)}$$

$$\overline{\lambda}_1 = \lambda_1 \cos^2 \theta_1^{\mathrm{p}} + \lambda_2 \sin^2 \theta_1^{\mathrm{p}}$$
$$\overline{\lambda}_2 = \lambda_1 \sin^2 \theta_1^{\mathrm{p}} + \lambda_2 \cos^2 \theta_1^{\mathrm{p}}$$
$$\overline{\lambda}_6 = 2(\lambda_1 - \lambda_2) \sin \theta_1^{\mathrm{p}} \cos \theta_1^{\mathrm{p}}$$
$$\text{(3.160)}$$

The Q_{ij}, the so-called reduced stiffnesses, are

$$Q_{11} = \left(c_{11} - \frac{c_{13}^2}{c_{33}} \right) = \frac{E_{11}}{1 - \nu_{12}\nu_{21}}$$

$$Q_{12} = \left(c_{12} - \frac{c_{13}c_{23}}{c_{33}} \right) = \frac{\nu_{12}E_{22}}{1 - \nu_{12}\nu_{21}} = \frac{\nu_{21}E_{11}}{1 - \nu_{12}\nu_{21}}$$

$$Q_{22} = \left(c_{22} - \frac{c_{23}^2}{c_{33}} \right) = \frac{E_{22}}{1 - \nu_{12}\nu_{21}}$$

$$Q_{66} = c_{66} = G_{12}$$

$$\text{(3.161)}$$

in which E_{11}, E_{22} are the x_1, x_2-axis Young's moduli, respectively, ν_{12}, ν_{21} are the Poisson's ratios, and G_{12} is the shear modulus. The Poisson's ratio ν_{12} reflects shrinkage (expansion) in the x_2-direction, due to tensile (compressive) stress in the x_1-direction. Additionally, the elastic moduli and Poisson's ratios are related by

$$\frac{\nu_{12}}{E_{11}} = \frac{\nu_{21}}{E_{22}} \tag{3.162}$$

The piezoelectric constants and stress–temperature coefficients are, respectively, related to the direct piezoelectric coefficients and coefficients of thermal expansion using the reduced stiffnesses by

$$e_{31} = d_{31}Q_{11} + d_{32}Q_{12}$$
$$e_{32} = d_{31}Q_{12} + d_{32}Q_{22}$$
$$\text{(3.163)}$$

$$\lambda_1 = \alpha_{11}Q_{11} + \alpha_{22}Q_{12}$$
$$\lambda_2 = \alpha_{11}Q_{12} + \alpha_{22}Q_{22} \tag{3.164}$$

Here and in the following, we have dropped the superscript E on the coefficient of thermal expansion.

The rectangular displacement components u_x, u_y, u_z may be expressed as follows:

$$u_x = u_x^0 - zw_{,x}, \quad u_y = u_y^0 - zw_{,y}, \quad u_z = w(x, y, t) \tag{3.165}$$

where u_x^0, u_y^0 are the x-, y-components of the midplane displacement vector $\mathbf{u}^0$ and $w(x, y, t)$ represents the deflection of the middle plane of the composite plate. The strains are defined in terms of displacements as

$$\varepsilon_{xx} = u_{x,x}$$
$$\varepsilon_{yy} = u_{y,y} \tag{3.166}$$
$$\varepsilon_{xy} = \frac{1}{2}(u_{x,y} + u_{y,x})$$

Thus, the strain variations within the laminate are related to the midplane deflection $w(x, y, t)$ by the expressions

$$\varepsilon_{xx} = u_{x,x}^0 - zw_{,xx} = \varepsilon_{xx}^0 - zw_{,xx}$$
$$\varepsilon_{yy} = u_{y,y}^0 - zw_{,yy} = \varepsilon_{yy}^0 - zw_{,yy} \tag{3.167}$$
$$\varepsilon_{xy} = \frac{1}{2}(u_{x,y}^0 + u_{y,x}^0) - zw_{,xy} = \varepsilon_{xy}^0 - zw_{,xy}$$

where $\varepsilon_{xx}^0, \varepsilon_{yy}^0$, and ε_{xy}^0 are the midplane strains. The stress boundary conditions on the plate surfaces are

$$\sigma_{zx} = \sigma_{zy} = \sigma_{zz} = 0 \ (z = \pm h) \tag{3.168}$$

The resultant forces and moments per unit length of the cross section of the laminate are defined as

$$\begin{Bmatrix} N_{xx} \\ N_{yy} \\ N_{xy} \end{Bmatrix} = \int_{-h}^{h} \begin{Bmatrix} \sigma_{xx} \\ \sigma_{yy} \\ \sigma_{xy} \end{Bmatrix} dz = \sum_{k=1}^{N} \int_{z_{k-1}}^{z_k} \begin{Bmatrix} \sigma_{xx} \\ \sigma_{yy} \\ \sigma_{xy} \end{Bmatrix}_k dz \tag{3.169}$$

$$\begin{Bmatrix} M_{xx} \\ M_{yy} \\ M_{xy} \end{Bmatrix} = \int_{-h}^{h} \begin{Bmatrix} \sigma_{xx} \\ \sigma_{yy} \\ \sigma_{xy} \end{Bmatrix} z\,dz = \sum_{k=1}^{N} \int_{z_{k-1}}^{z_k} \begin{Bmatrix} \sigma_{xx} \\ \sigma_{yy} \\ \sigma_{xy} \end{Bmatrix}_k z\,dz \tag{3.170}$$

The vertical shear forces per unit length are

$$\begin{Bmatrix} Q_x \\ Q_y \end{Bmatrix} = \int_{-h}^{h} \begin{Bmatrix} \sigma_{zx} \\ \sigma_{zy} \end{Bmatrix} dz = \sum_{k=1}^{N} \int_{z_{k-1}}^{z_k} \begin{Bmatrix} \sigma_{zx} \\ \sigma_{zy} \end{Bmatrix}_k dz \tag{3.171}$$

Now if we multiply the first and second of Eqs. (3.151) by $z\,dz$ and integrate from $-h$ to h, taking into account the boundary conditions (3.168) and neglecting the h^3 terms, we shall obtain the results

$$M_{xx,x} + M_{yx,y} - Q_x = 0$$
$$M_{xy,x} + M_{yy,y} - Q_y = 0 \tag{3.172}$$

If Eqs. (3.151) is multiplied by dz and integrated from $-h$ to h, taking into account the boundary condition (3.168) and neglecting the h^3 terms, we obtain

$$N_{xx,x} + N_{yx,y} = \rho_L h u^0_{x,tt} + C_{eq} u^0_{x,t}$$
$$N_{yy,y} + N_{xy,x} = \rho_L h u^0_{y,tt} + C_{eq} u^0_{y,t} \tag{3.173}$$
$$Q_{x,x} + Q_{y,y} = \rho_L h w_{,tt} + C_{eq} w_{,t}$$

where

$$C_{eq} = \sum_{k=1}^{N} C_k h_k \tag{3.174}$$

$$\rho_L = \sum_{k=1}^{N} \frac{\rho_k h_k}{h} \tag{3.175}$$

Eliminating Q_x, Q_y from Eqs. (3.172) and the third of Eqs. (3.173), we have

$$M_{xx,xx} + 2M_{xy,xy} + M_{yy,yy} = \rho_L h w_{,tt} + C_q w_{,t} \tag{3.176}$$

From Eq. (3.157), with Eq. (3.167), the constitutive relation for a typical layer k ($k = 1, \ldots, N$) become

$$
\begin{Bmatrix} \sigma_{xx} \\ \sigma_{yy} \\ \sigma_{xy} \end{Bmatrix}_k =
\begin{bmatrix} \bar{Q}_{11} & \bar{Q}_{12} & \bar{Q}_{16} \\ \bar{Q}_{12} & \bar{Q}_{22} & \bar{Q}_{26} \\ \bar{Q}_{16} & \bar{Q}_{26} & \bar{Q}_{66} \end{bmatrix}_k
\begin{Bmatrix} u^0_{x,x} - zw_{,xx} \\ u^0_{y,y} - zw_{,yy} \\ u^0_{x,y} + u^0_{y,x} - 2zw_{,xy} \end{Bmatrix}
$$
$$
- \begin{bmatrix} 0 & 0 & \bar{e}_{31} \\ 0 & 0 & \bar{e}_{32} \\ 0 & 0 & \bar{e}_{36} \end{bmatrix}_k
\begin{Bmatrix} E_x \\ E_y \\ E_z \end{Bmatrix}_k
- \begin{Bmatrix} \bar{\lambda}_1 \\ \bar{\lambda}_2 \\ \bar{\lambda}_6 \end{Bmatrix}_k (\Theta)_k \tag{3.177}
$$

Substituting Eq. (3.177) into Eqs. (3.169) and (3.170) gives

$$
\begin{Bmatrix} N_{xx} \\ N_{yy} \\ N_{xy} \end{Bmatrix} =
\begin{bmatrix} A_{11} & A_{12} & A_{16} \\ A_{12} & A_{22} & A_{26} \\ A_{16} & A_{26} & A_{66} \end{bmatrix}
\begin{Bmatrix} u^0_{x,x} \\ u^0_{y,y} \\ 2u^0_{x,y} \end{Bmatrix}
+ \begin{bmatrix} B_{11} & B_{12} & B_{16} \\ B_{12} & B_{22} & B_{26} \\ B_{16} & B_{26} & B_{66} \end{bmatrix}
\begin{Bmatrix} w_{,xx} \\ w_{,yy} \\ 2w_{,xy} \end{Bmatrix}
$$
$$
- \begin{Bmatrix} N^E_{xx} \\ N^E_{yy} \\ N^E_{xy} \end{Bmatrix}
- \begin{Bmatrix} N^\Theta_{xx} \\ N^\Theta_{yy} \\ N^\Theta_{xy} \end{Bmatrix} \tag{3.178}
$$

$$
\left\{
\begin{array}{c}
M_{xx} \\
M_{yy} \\
M_{xy}
\end{array}
\right\}
=
\begin{bmatrix}
B_{11} & B_{12} & B_{16} \\
B_{12} & B_{22} & B_{26} \\
B_{16} & B_{26} & B_{66}
\end{bmatrix}
\left\{
\begin{array}{c}
u^0_{x,x} \\
u^0_{y,y} \\
2u^0_{x,y}
\end{array}
\right\}
+
\begin{bmatrix}
D_{11} & D_{12} & D_{16} \\
D_{12} & D_{22} & D_{26} \\
D_{16} & D_{26} & D_{66}
\end{bmatrix}
\left\{
\begin{array}{c}
w_{,xx} \\
w_{,yy} \\
2w_{,xy}
\end{array}
\right\}
$$

$$
-
\left\{
\begin{array}{c}
M^{\mathrm{E}}_{xx} \\
M^{\mathrm{E}}_{yy} \\
M^{\mathrm{E}}_{xy}
\end{array}
\right\}
-
\left\{
\begin{array}{c}
M^{\Theta}_{xx} \\
M^{\Theta}_{yy} \\
M^{\Theta}_{xy}
\end{array}
\right\}
\qquad (3.179)
$$

where the electric force and moment resultants are defined as

$$
\left\{
\begin{array}{c}
N^{\mathrm{E}}_{xx} \\
N^{\mathrm{E}}_{yy} \\
N^{\mathrm{E}}_{xy}
\end{array}
\right\}
=
\sum_{k=1}^{N} \int_{z_{k-1}}^{z_k}
\left\{
\begin{array}{c}
\bar{e}_{31} \\
\bar{e}_{32} \\
\bar{e}_{36}
\end{array}
\right\}_k
(E_z)_k \, dz
\qquad (3.180)
$$

$$
\left\{
\begin{array}{c}
M^{\mathrm{E}}_{xx} \\
M^{\mathrm{E}}_{yy} \\
M^{\mathrm{E}}_{xy}
\end{array}
\right\}
=
\sum_{k=1}^{N} \int_{z_{k-1}}^{z_k}
\left\{
\begin{array}{c}
\bar{e}_{31} \\
\bar{e}_{32} \\
\bar{e}_{36}
\end{array}
\right\}_k
(E_z)_k z \, dz
\qquad (3.181)
$$

and the thermal force and moment resultants are

$$
\left\{
\begin{array}{c}
N^{\Theta}_{xx} \\
N^{\Theta}_{yy} \\
N^{\Theta}_{xy}
\end{array}
\right\}
=
\sum_{k=1}^{N} \int_{z_{k-1}}^{z_k}
\left\{
\begin{array}{c}
\bar{\lambda}_1 \\
\bar{\lambda}_2 \\
\bar{\lambda}_6
\end{array}
\right\}_k
(\Theta)_k \, dz
\qquad (3.182)
$$

$$
\left\{
\begin{array}{c}
M^{\Theta}_{xx} \\
M^{\Theta}_{yy} \\
M^{\Theta}_{xy}
\end{array}
\right\}
=
\sum_{k=1}^{N} \int_{z_{k-1}}^{z_k}
\left\{
\begin{array}{c}
\bar{\lambda}_1 \\
\bar{\lambda}_2 \\
\bar{\lambda}_6
\end{array}
\right\}_k
(\Theta)_k z \, dz
\qquad (3.183)
$$

In addition, A_{ij}, B_{ij}, D_{ij} are the standard extensional, coupling, and bending stiffnesses, given by

$$
(A_{ij}, B_{ij}, D_{ij}) = \sum_{k=1}^{N} \int_{z_{k-1}}^{z_k} (\bar{Q}_{ij})_k (1, z, z^2) dz \quad (i,j = 1, 2, 6)
\qquad (3.184)
$$

From Eqs. (3.172), with Eq. (3.179), we have

$$
\begin{aligned}
Q_x =\ & (B_{11} + B_{16})u^0_{x,xx} + 2(B_{16} + B_{66})u^0_{x,xy} + (B_{12} + B_{26})u^0_{y,xy} \\
& - \{D_{11}w_{,xxx} + 3D_{16}w_{,xxy} + (D_{12} + 2D_{66})w_{,xyy} + D_{26}w_{,yyy}\} \\
& - (M^{\mathrm{E}}_{xx,x} + M^{\mathrm{E}}_{xy,y} + M^{\Theta}_{xx,x} + M^{\Theta}_{xy,y}) \\
Q_y =\ & (B_{16} + B_{12})u^0_{x,xy} + 2(B_{26} + B_{66})u^0_{x,yy} + (B_{22} + B_{26})u^0_{y,yy} \\
& - \{D_{16}w_{,xxx} + (D_{12} + 2D_{66})w_{,xxy} + 3D_{26}w_{,xyy} + D_{22}w_{,yyy}\} \\
& - (M^{\mathrm{E}}_{xy,x} + M^{\mathrm{E}}_{yy,y} + M^{\Theta}_{xy,x} + M^{\Theta}_{yy,y})
\end{aligned}
\qquad (3.185)
$$

From the first and second of Eqs. (3.173) and Eq. (3.176), with Eqs. (3.178) and (3.179), we obtain

$$A_{11}u^0_{x,xx} + 2A_{16}u^0_{x,xy} + A_{66}u^0_{x,yy} + A_{16}u^0_{y,xx} + (A_{12}+A_{66})u^0_{y,xy} + A_{26}u^0_{y,yy}$$
$$- B_{11}w_{,xxx} - 3B_{16}w_{,xxy} - (B_{12}+2B_{66})w_{,xyy} - B_{26}w_{,yyy}$$
$$- (N^E_{xx,x} + N^E_{xy,y} + N^\Theta_{xx,x} + N^\Theta_{xy,y})$$
$$- \rho_L hu^0_{x,tt} - C_{eq}u^0_{x,t} = 0$$

$$A_{16}u^0_{x,xx} + (A_{12}+A_{66})u^0_{x,xy} + A_{26}u^0_{x,yy} + A_{66}u^0_{y,xx} + 2A_{26}u^0_{y,xy} + A_{22}u^0_{y,yy}$$
$$- B_{16}w_{,xxx} - (B_{12}+2B_{66})w_{,xxy} - 3B_{26}w_{,xyy} - B_{22}w_{,yyy}$$
$$- (N^E_{xy,x} + N^E_{yy,y} + N^\Theta_{xy,x} + N^\Theta_{yy,y})$$
$$- \rho_L hu^0_{y,tt} - C_{eq}u^0_{y,t} = 0$$

$$\tag{3.186}$$

$$B_{11}u^0_{x,xxx} + 3B_{16}u^0_{x,xxy} + (B_{12}+2B_{66})u^0_{x,xyy} + B_{26}u^0_{x,yyy}$$
$$+ B_{16}u^0_{y,xxx} + (B_{12}+2B_{66})u^0_{y,xxy} + 3B_{26}u^0_{y,xyy} + B_{22}u^0_{y,yyy}$$
$$- D_{11}w_{,xxxx} - 4D_{16}w_{,xxxy} - 2(D_{12}+2D_{66})w_{,xxyy}$$
$$- 4D_{26}w_{,xyyy} - D_{22}w_{,yyyy}$$
$$- (M^E_{xx,xx} + 2M^E_{xy,xy} + M^E_{yy,yy}) - (M^\Theta_{xx,xx} + 2M^\Theta_{xy,xy} + M^\Theta_{yy,yy})$$
$$- \rho_L hw_{,tt} - C_{eq}w_{,t} = 0 \tag{3.187}$$

Equations (3.186) and (3.187) are the basic equations of linear bending theory for thermopiezo-electric laminates.

If we assume a symmetric laminated plate, we obtain

$$B_{ij} = 0 \quad (i,j = 1,2,6) \tag{3.188}$$

Assuming a symmetric cross-ply panel having angles of 0° or 90°, we have

$$A_{16} = A_{26} = D_{16} = D_{26} = 0 \tag{3.189}$$

Equations (3.186) and (3.187) can then be rewritten in the form

$$A_{11}u^0_{x,xx} + A_{66}u^0_{x,yy} + (A_{12}+A_{66})u^0_{y,xy}$$
$$- N^E_{xx,x} - N^\Theta_{xx,x} - \rho_L hu^0_{x,tt} - C_{eq}u^0_{x,t} = 0$$
$$(A_{12}+A_{66})u^0_{x,xy} + A_{66}u^0_{y,xx} + A_{22}u^0_{y,yy}$$
$$- N^E_{yy,y} - N^\Theta_{yy,y} - \rho_L hu^0_{y,tt} - C_{eq}u^0_{y,t} = 0$$

$$\tag{3.190}$$

$$D_{11}w_{,xxxx} + 2(D_{12}+2D_{66})w_{,xxyy} + D_{22}w_{,yyyy}$$
$$+ M^E_{xx,xx} + M^E_{yy,yy} + M^\Theta_{xx,xx} + M^\Theta_{yy,yy}$$
$$+ \rho_L hw_{,tt} + C_{eq}w_{,t} = 0 \tag{3.191}$$

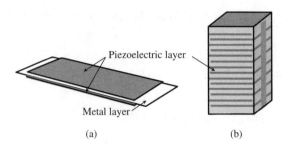

(a)   (b)

Figure 3.24 Schematic drawing of (a) bimorph-type bending device and (b) multilayer stacked device

Equation (3.191) is the basic equation of linear bending theory for symmetric thermopiezo-electric laminates.

3.7   Bending of Piezoelectric Laminates

Piezoelectric devices can be classified into some categories: unimorph or bimorph-type bending device, multilayer stacked device, and so on. Figure 3.24(a) and (b) shows bimorph-type bending and multilayer stacked devices, respectively. Bending mode devices can produce a large displacement but the generative force is small. With multilayer devices, a large generative force can be produced but a displacement level is usually much lower than bending mode devices. In this section, bending behavior of the piezoelectric laminates is discussed.

3.7.1   Bimorphs

A considerable amount of research has been devoted to the development of piezoelectric devices that exhibit amplified electromechanical displacement. Many of these devices are bimorph-type structures. Piezoelectric bimorphs come in different versions due to differences in manufacturing methods. Some existing forms [35] are shown in Fig. 3.25. The bimorphs (a) and (b) have no electrode between the upper and lower piezoelectric layers, and the electric field that drives the bimorph is generated by the top and bottom electrodes. This electric field is equal to the voltage V_0 divided by the total distance $2h$ between the electrodes. On the other hand, the bimorph (c) has an intermediate electrode. The electric field across the piezoelectric layers is equal to the voltage V_0 divided by the electrode distance h and is therefore twice the field of the bimorph (a) or (b). The resulting bending moment of the type (c) bimorph is twice that of type (a) or (b) when all other parameters are held constant. Type (a) is often called a series or antiparallel bimorph, as is type (b), while type (c) is called a parallel bimorph. Type (d) consists of a piezoelectric layer with electrodes and a nonpiezoelectric layer and is called a unimorph or monomorph. Here, we consider the piezoelectric bimorphs.

A laminated beam of length a and width b, with integrated piezoelectric layers, is shown in Fig. 3.26. Let the coordinate axes $x = x_1$ and $y = x_2$ be chosen such that they coincide with the middle plane of the laminated beam and the $z = x_3$ axis is normal to this plane. PZT

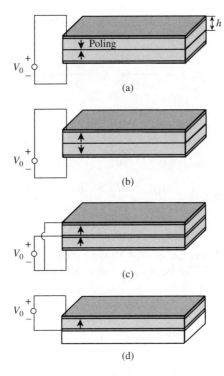

Figure 3.25 Schematic drawing of (a) inward series bimorph, (b) outward series bimorph, (c) parallel bimorph, and (d) unimorph or monomorph

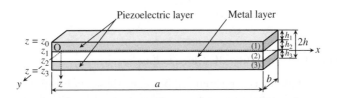

Figure 3.26 A piezoelectric/metal/piezoelectric laminate

layers poled in z-direction are added to the upper and lower surfaces of a metal layer. The total thickness is $2h$ and the kth layer has thickness $h_k = z_k - z_{k-1} (k = 1, 2, 3)$, where $z_0 = -h$ and $z_3 = h$. Here, it is assumed that the electric field resulting from variations in stress (the so-called direct piezoelectric effect) is insignificant compared with the applied electric field.

In the piezoelectric laminated beam, the layers are free to expand vertically, which implies that the stress $(\sigma_{zz})_k$ is equal to zero. Because the beam is considered to be long and slender, we also assume that the stress $(\sigma_{yy})_k$ is zero. As the electric field $(E_z)_k$ is parallel to the poling, no shear stresses will develop, and we may conclude that no shear strains are present, which will reduce the nontrivial stresses to $(\sigma_{xx})_k$. It is assumed that 90° switching would not occur.

The lamina constitutive equation for the kth layer with respect to the reference axes of the laminate (x,z) can be written as

$$(\sigma_{xx})_k = \frac{1}{(s_{11})_k}(\varepsilon_{xx})_k - \frac{(d^s_{31})_k}{(s_{11})_k}(E_z)_k \tag{3.192}$$

where

$$\begin{aligned} (s_{11})_k &= s_{11} \quad (k = 1, 3) \\ (s_{11})_2 &= s^{elas}_{11} \end{aligned} \tag{3.193}$$

$$\begin{aligned} (d^s_{31})_k &= \begin{cases} d_{31} & (E_z > -E_c) \\ -d_{31} & (E_z \le -E_c) \end{cases} \quad (k = 1, 3) \\ (d^s_{31})_2 &= 0 \end{aligned} \tag{3.194}$$

and s^{elas}_{11} is an elastic compliance (inverse of Young's modulus) of the metal. The superscript "elas" means elastic material. The bending stiffness per unit length D_{11} of the piezoelectric laminated beam is expressed as

$$D_{11} = \sum_{k=1}^{3} \int_{z_{k-1}}^{z_k} \frac{1}{(s_{11})_k} z^2 \, dz \tag{3.195}$$

The bending moment per unit length M^E_{xx} is given by

$$M^E_{xx} = \sum_{k=1}^{3} \int_{z_{k-1}}^{z_k} \frac{(d^s_{31})_k}{(s_{11})_k}(E_z)_k z \, dz \tag{3.196}$$

3.7.1.1   Cantilever Beam under DC Electric Field

Consider a cantilever piezoelectric laminated beam, as shown in Fig. 3.27(a), that is fixed at one end $(x = 0)$ and subjected to an external monomorph drive DC electric field [36]. The differential equation for the deflection $w(x)$ can be expressed as

$$D_{11}w_{,xx} = -M^E_{xx} \quad \text{or} \quad D_{11}w_{,xxx} = 0 \quad \text{or} \quad D_{11}w_{,xxxx} = 0 \tag{3.197}$$

The deflection can be found by solving any one of the three preceding differential equations. The choice usually depends upon which equation provides the most efficient solution. Here, the first of Eqs. (3.197) can be integrated to find the deflection w. The boundary conditions are given by

$$w(0) = 0, \quad w_{,x}(0) = 0 \tag{3.198}$$

From Eq. (3.198), the solution for Eq. (3.197) is obtained as

$$w = -\frac{M^E_{xx}}{2D_{11}}x^2 \tag{3.199}$$

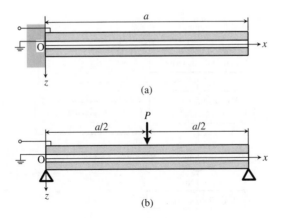

Figure 3.27 Schematic drawing of (a) cantilever beam and (b) simply supported beam

The tip $(x = a)$ deflection w_{tip} becomes

$$w_{tip} = -\frac{M^E_{xx}}{2D_{11}}a^2 \tag{3.200}$$

We can also obtain the deflection of the cantilever piezoelectric laminated beam by a finite element analysis (FEA). The polarization switching is defined for each element. Each element consists of many grains, and each grain is modeled as a uniformly polarized cell that contains a single domain, that is, each grain is equivalent to a uniformly polarized single domain. The model neglects the interaction among different grains (domains). In reality, this is not true, but the assumption does not affect the macroscopic piezoelectric behavior. The polarization of each grain (domain) initially aligns with the z-direction; so, initially each element with many grains (domains) is assumed to be poled in the z-direction. The boundary conditions are applied, and the electromechanical fields of each element are computed from the FEA. The switching criterion of Eq. (3.127) is checked for every element for the model to see if switching will occur. After all possible polarization switches have occurred, the new direct piezoelectric coefficient, Eq. (3.136), of each element is used, and the electromechanical fields are recalculated. This procedure is repeated until convergence is obtained. The detailed flowchart of the finite element procedure is shown in Fig. 3.28. We use the commercial finite element code ANSYS [37]. Each element is defined by eight-node three-dimensional (3D) coupled field solid for the PZT layers and eight-node 3D structural solid for the metal layer.

The spontaneous polarization P^s and strain γ^s are assigned the following representative values:

$$P^s = 0.3 \ \text{C/m}^2$$
$$\gamma^s = 0.004 \tag{3.201}$$

Simulations are also run with the spontaneous polarization and strain values varying, and the results are not shown here since the change of these values does not impact the results.

In order to validate the predictions, we measure the tip deflection w_{tip} of the commercial bimorph actuator. Consider a monomorph drive by DC electric field in the top piezoelectric layer as shown in Fig. 3.27(a). The present actuator is made of PZT P-7B and Fe-48Ni (Murata

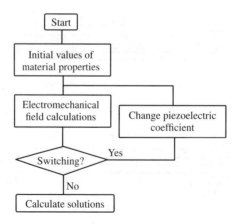

Figure 3.28 Flowchart of the FEA

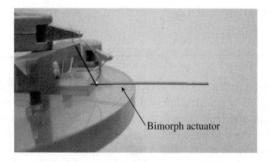

Figure 3.29 A cantilever bimorph actuator and experimental setup

Manufacturing Co. Ltd. Japan). Figure 3.29 shows a specimen and experimental setup. The PZT P-7B of thickness $h_1 = h_3 = 0.2$ mm has Ag paste electrodes on both sides. The thickness of the Fe-48Ni is $h_2 = 0.2$ mm. The specimen has a length of $a = 43$ mm, a width of $b = 2$ mm, and a total thickness of $2h = h_1 + h_2 + h_3 = 0.6$ mm. The material properties of P-7B are

$$s_{11} = 16.7 \times 10^{-12} \text{ m}^2/\text{N}$$
$$s_{12} = -5.9 \times 10^{-12} \text{ m}^2/\text{N}$$
$$s_{13} = -7.5 \times 10^{-12} \text{ m}^2/\text{N}$$
$$s_{33} = 18.8 \times 10^{-12} \text{ m}^2/\text{N}$$
$$s_{44} = 38.8 \times 10^{-12} \text{ m}^2/\text{N}$$
$$d_{15} = 592 \times 10^{-12} \text{ m/V} \qquad (3.202)$$
$$d_{31} = -303 \times 10^{-12} \text{ m/V}$$
$$d_{33} = 603 \times 10^{-12} \text{ m/V}$$
$$\epsilon_{11}^{T} = 418 \times 10^{-10} \text{ C/Vm}$$
$$\epsilon_{33}^{T} = 283 \times 10^{-10} \text{ C/Vm}$$
$$E_{c} = 0.5 \text{ MV/m}$$

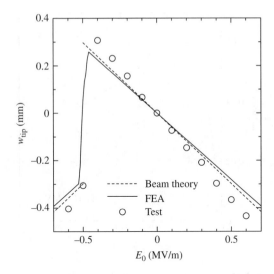

Figure 3.30 Tip deflection versus DC electric field

The elastic compliance s_{11}^{elas} and Poisson's ratio v^{elas} of Fe-48Ni are

$$s_{11}^{elas} = 4.76 \times 10^{-12} \text{ m}^2/\text{N}$$
$$v^{elas} = 0.3$$

(3.203)

A voltage V_0 is applied using a power supply for voltages up to 1.25 kV/DC. The tip deflection of the cantilever bimorph actuator is measured with a digital microscope, and the deflection versus electric field curve is recorded.

We now show some theoretical and experimental results. Figure 3.30 gives a plot of the tip deflection w_{tip} with the applied DC electric field $E_0 = (E_z)_1 = V_0/h_1$ (monomorph drive) for the P-7B/Fe-48Ni/P-7B actuator of $h_1 = h_2 = h_3 = 0.2$ mm. The dashed and solid lines represent the values of deflection predicted by the beam theory and FEA, and the open circle denotes the test data. The tip deflection rises at first when the electric field is reduced starting at zero. Polarization switching occurs in the top piezoelectric layer after the electric field is reduced to about -0.4 MV/m, and the tip deflection falls. As the electric field is increased from zero, the negative deflection is increased. Good agreement is observed between the model predictions and test data. Figure 3.31 shows the plot of the tip deflection w_{tip} obtained from the beam theory as a function of the piezoelectric layer thickness $h_1 = h_3$ for the P-7B/Fe-48Ni/P-7B actuator of $h_2 = 0.2$ mm under $E_0 = 0.1$ MV/m. The tip deflection exhibits a maximum at about $h_1 = h_3 = 0.13$ mm.

3.7.1.2 Simply Supported Beam under DC Electric Field

Consider a simply supported piezoelectric laminated beam with a concentrated load P at the center $(x = a/2)$ and an external monomorph drive DC electric field [38] as shown in

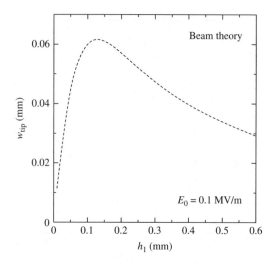

Figure 3.31 Tip deflection versus piezoelectric layer thickness

Fig. 3.27(b). The differential equation for the deflection $w(x)$ can be expressed as

$$D_{11}w_{,xx} = -\left(M_{xx}^{E} + \frac{Px}{2}\right) \qquad (0 \leq x \leq a/2) \tag{3.204}$$

The laminated beam is subject to the following boundary conditions:

$$w(0) = 0, \quad w_{,x}(a/2) = 0 \tag{3.205}$$

From Eq. (3.205), the solution for Eq. (3.204) is obtained as

$$w = -\frac{M_{xx}^{E}}{2D_{11}}x(x-a) - \frac{P}{48D_{11}}x(4x^2 - 3a^2) \qquad (0 \leq x \leq a/2) \tag{3.206}$$

The load-point ($x = a/2$) deflection w_{max} becomes

$$w_{max} = \frac{M_{xx}^{E}}{8D_{11}}a^2 + \frac{P}{48D_{11}}a^3 \tag{3.207}$$

We also obtain the deflection of the piezoelectric laminated beam by the FEA. Furthermore, we use the P-7B/Fe-48Ni/P-7B actuator and measure the load-point deflection w_{max}. A voltage V_0 is applied as shown in Fig. 3.27(b). Specimen is loaded by a concentrated load P that consists of weight. All the tests are conducted on the $a = 30\,\text{mm}$ simply supported span. The maximum deflection is measured with a digital microscope, and the deflection versus electric field curve is recorded.

Figure 3.32 shows a plot of the load-point deflection w_{max} with the applied DC electric field $E_0 = (E_z)_1 = V_0/h_1$ (monomorph drive) for the P-7B/Fe-48Ni/P-7B actuator of $h_1 = h_2 = h_3 = 0.2\,\text{mm}$ under concentrated load $P = 10\,\text{gf} = 0.098\,\text{N}$. The dashed and solid lines represent the values of deflection predicted by the beam theory and FEA, and the open circle denotes the test data. As the electric field reduced from zero, the load-point deflection decreases. As the electric field continues to be reduced, local polarization switching causes an unexpected

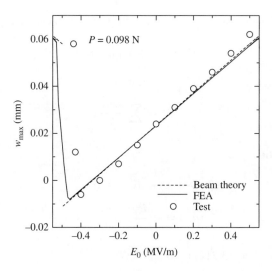

Figure 3.32 Tip deflection versus DC electric field under concentrated load

increase in the load-point deflection. As the positive electric field increases, the load-point deflection increases. The trend is similar between analyses and tests.

3.7.1.3 Cantilever Beam under AC Electric Field

Consider a cantilever piezoelectric laminated beam, as shown in Fig. 3.27(a), that is fixed at one end ($x = 0$) and subjected to an external monomorph drive AC electric field [31]. The lamina constitutive equation for the first and third layers, Eq. (3.192), can be rewritten as

$$(\sigma_{xx})_k = \frac{1}{(s_{11})_k}(\varepsilon_{xx})_k - \frac{(d_{31}^{m})_k}{(s_{11})_k}(E_z)_k \tag{3.208}$$

The bending stiffness D_{11} is given by Eq. (3.195) and the bending moment per unit length, Eq. (3.196), becomes

$$M_{xx}^{E} = \sum_{k=1}^{3} \int_{z_{k-1}}^{z_k} \frac{(d_{31}^{m})_k}{(s_{11})_k}(E_z)_k z \, dz \tag{3.209}$$

An applied voltage $V_0 \exp(i\omega t)$ induces a bending moment given by

$$M_{xx}^{E} = M_0 \exp(i\omega t) \tag{3.210}$$

where

$$M_0 = V_0 \int_{z_0}^{z_1} \frac{(d_{31}^{m})_1}{(s_{11})_1 (\varepsilon_{33}^{Tm})_1 \sum_{k=1}^{3} \{h_k/(\varepsilon_{33}^{Tm})_k\}} z \, dz \tag{3.211}$$

It is noted that $(\varepsilon_{33}^{Tm})_2 = \epsilon_0$.

The differential equation of motion for the deflection $w(x, t)$ can be expressed as

$$D_{11}w_{,xxxx} + h\rho_L w_{,tt} = 0 \tag{3.212}$$

From Eq. (3.175), we have

$$\rho_L = \sum_{k=1}^{3} \frac{\rho_k h_k}{h} \tag{3.213}$$

The boundary conditions are

$$w(0, t) = 0, \quad w_{,x}(0, t) = 0$$
$$w_{,xx}(a, t) = -\frac{M_{xx}^E}{D_{11}}, \quad w_{,xxx}(a, t) = 0 \tag{3.214}$$

The general solution of Eq. (3.212) is obtained as

$$w = \{C_1 \cos(kx) + C_2 \sin(kx) + C_3 \cosh(kx) + C_4 \sinh(kx)\} \exp(i\omega t) \tag{3.215}$$

where

$$k^2 = \left(\frac{h\rho_L}{D_{11}}\right)^{1/2} \omega \tag{3.216}$$

and $C_j(j = 1, \ldots, 4)$ are unknown coefficients. Imposing the boundary conditions (3.214) to the solution gives the following four equations:

$$C_1 + C_3 = 0$$
$$C_2 + C_4 = 0$$
$$-C_1 \cos(ka) - C_2 \sin(ka) + C_3 \cosh(ka) + C_4 \sinh(ka) = -\frac{M_0}{D_{11}k^2} \tag{3.217}$$
$$C_1 \sin(ka) - C_2 \cos(ka) + C_3 \sinh(ka) + C_4 \cosh(ka) = 0$$

Solving these four equations, one can determine the coefficients $C_j(j = 1, \ldots, 4)$. Upon substitutions, the dynamic deflection can be derived as

$$w = \frac{1}{2\{\cos(ka)\cosh(ka) + 1\}D_{11}k^2}[\{\cos(ka) + \cosh(ka)\}\{\cos(kx) - \cosh(kx)\}$$
$$+ \{\sin(ka) - \sinh(ka)\}\{\sin(kx) - \sinh(kx)\}]M_0 \exp(i\omega t) \tag{3.218}$$

The tip ($x = a$) deflection w_{tip} becomes

$$w_{tip} = -\frac{\sin(ka)\sinh(ka)}{\{\cos(ka)\cosh(ka) + 1\}D_{11}k^2}M_0 \exp(i\omega t) \tag{3.219}$$

We perform 3D finite element calculations to determine the deflection for the cantilever piezoelectric laminated beam. The commercial software ANSYS with eight-node 3D elements is used in the analysis. We assume that the piezoelectric material properties vary with the domain wall motion. When the AC electric field is parallel to the poling of the beam according to Fig. 3.27(a), we can consider only the Δs_{11}^m, Δd_{31}^m, and $\Delta \epsilon_{33}^{Tm}$. By using electric-field-dependent material properties, the model calculates the nonlinear dynamic bending response. The spontaneous polarization and strain (3.201) are used to get Eqs. (3.143).

In order to validate the predictions, we measure the amplitude of tip deflection $|w_{tip}|$ of the commercial bimorph actuator. Consider a monomorph drive by AC electric field in the top piezoelectric layer as shown in Fig. 3.27(a). The present actuator is made of PZT C-82 and Fe-42Ni (Fuji Ceramics Co. Ltd. Japan). The PZT C-82 of thickness $h_1 = h_3 = 0.27$ mm has Ni paste electrodes on both sides. The thickness of the Fe-42Ni is $h_2 = 0.1$ mm. The specimen has a length of $a = 40$ mm, a width of $b = 2$ mm, and a total thickness of $2h = h_1 + h_2 + h_3 = 0.64$ mm. The material properties of C-82 are

$$s_{11} = 16.3 \times 10^{-12} \text{ m}^2/\text{N}$$
$$s_{12} = -5.5 \times 10^{-12} \text{ m}^2/\text{N}$$
$$s_{13} = -8.1 \times 10^{-12} \text{ m}^2/\text{N}$$
$$s_{33} = 20.6 \times 10^{-12} \text{ m}^2/\text{N}$$
$$s_{44} = 45.6 \times 10^{-12} \text{ m}^2/\text{N}$$
$$d_{15} = 775 \times 10^{-12} \text{ m/V}$$
$$d_{31} = -271 \times 10^{-12} \text{ m/V} \tag{3.220}$$
$$d_{33} = 630 \times 10^{-12} \text{ m/V}$$
$$\epsilon_{11}^T = 268 \times 10^{-10} \text{ C/Vm}$$
$$\epsilon_{33}^T = 306 \times 10^{-10} \text{ C/Vm}$$
$$\rho = 7400 \text{ kg/m}^3$$
$$E_c = 0.35 \text{ MV/m}$$

The elastic compliance s_{11}^{elas}, Poisson's ratio v^{elas}, and mass density ρ^{elas} of Fe-42Ni are

$$s_{11}^{elas} = 4.76 \times 10^{-12} \text{ m}^2/\text{N}$$
$$v^{elas} = 0.302 \tag{3.221}$$
$$\rho^{elas} = 8297 \text{ kg/m}^3$$

An AC voltage $V_0 \exp(i\omega t)$ is applied by using an AC power supply. The amplitude of tip deflection of the cantilever bimorph actuator is measured by using a digital microscope, and the deflection versus electric field curve is recorded.

Figure 3.33 shows a plot of the amplitude of tip deflection $|w_{tip}|$ with the applied AC electric field amplitude $E_0 = V_0/h_1$ (monomorph drive) at frequency $f = \omega/2\pi = 60$ Hz for the C-82/Fe-42Ni/C-82 actuator. The solid and dotted lines represent the values of deflection predicted by the beam theory with and without the domain wall motion effect, and the open circle denotes the test data. A nonlinear relationship between the tip deflection and AC electric field is observed, that is, as AC electric field increases, the tip deflection increases rapidly. This

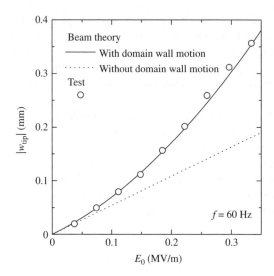

Figure 3.33   Tip deflection versus AC electric field

nonlinearity is due to the larger extrinsic contribution resulting from the increased domain wall motion under the influence of higher AC electric fields.

A linear relationship between displacement and electric field is only valid under weak field driving [39], and piezoelectric material properties provided by manufacturers are no longer applicable to describe the device performance since they are measured at a weak signal level. Our calculations show a reasonable agreement with the experimental data and support a role for domain wall motion in the nonlinear material response of the PZT ceramics under high AC electric fields. Note that this nonlinear behavior is more significant in soft PZT ceramics than in hard PZT ceramics.

3.7.2   Functionally Graded Bimorphs

Functionally graded piezoelectric materials (FGPMs) have recently been the focus of much attention. FGPMs possess continuously varying microstructure and electromechanical properties. In contrast to the sharp bimaterial interfaces that commonly arise in traditional bimorphs or unimorphs, the gradual change in properties throughout an FGPM seems to improve the resistance to interfacial delamination [40, 41]. Taya et al. [42] discussed the effect of DC electric field on the bending performance of bimorph piezoelectric actuators with functionally graded microstructure, based on the Eshelby's model and classical lamination theory. They considered two types of grading, type I where the piezoelectric properties increase toward the midplane and type II where the piezoelectric properties decrease toward the midplane, as shown in Fig. 3.34. Fang et al. [43] also studied the bending response of the type I and type II FGPM bimorphs under DC electric field using a state vector method for piezoelectricity. Here, we discuss the effect of AC electric field on the bending behavior of FGPM bimorphs.

Figure 3.35 shows the geometry of the FGPM bimorph model. Let the coordinate axes $x = x_1$ and $y = x_2$ be chosen such that they coincide with the middle plane of the bimorph

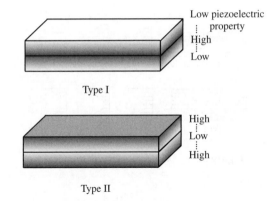

Figure 3.34 FGPM bimorphs

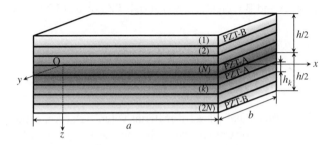

Figure 3.35 A $2N$ layer FGPM

and the $z = x_3$ axis is normal to this plane. Dimensions of the FGPM bimorph are: a the length, b the width, h the thickness. The piezoelectric materials are graded through the thickness only, from pure PZT-A at $z = 0$ to pure PZT-B at $z = \pm h/2$. PZT of each layer is poled in the z-direction. The bimorph is discretized into $2N$ layers, which are assumed to have constant material properties. The thickness of the kth layer is h_k.

In the numerical examples, the functionally graded model concerning the material inhomogeneity is employed. The material constants of the kth layer for $2N$ layer FGPM are

$$(s_{11})_k = s_{11}^A V_k + s_{11}^B (1 - V_k)$$
$$(s_{12})_k = s_{12}^A V_k + s_{12}^B (1 - V_k)$$
$$(s_{13})_k = s_{13}^A V_k + s_{13}^B (1 - V_k) \tag{3.222}$$
$$(s_{33})_k = s_{33}^A V_k + s_{33}^B (1 - V_k)$$
$$(s_{44})_k = s_{44}^A V_k + s_{44}^B (1 - V_k)$$

$$(d_{15})_k = d_{15}^A V_k + d_{15}^B (1 - V_k)$$
$$(d_{31})_k = d_{31}^A V_k + d_{31}^B (1 - V_k) \tag{3.223}$$
$$(d_{33})_k = d_{33}^A V_k + d_{33}^B (1 - V_k)$$

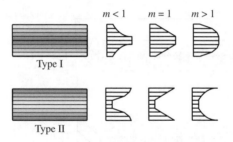

Figure 3.36   FGPM property distributions

$$(\epsilon_{11}^{T})_k = \epsilon_{11}^{TA} V_k + \epsilon_{11}^{TB}(1 - V_k)$$
$$(\epsilon_{33}^{T})_k = \epsilon_{33}^{TA} V_k + \epsilon_{33}^{TB}(1 - V_k)$$

$$(3.224)$$

where the superscripts A and B represent the materials PZT-A and PZT-B, respectively, and

$$V_k = \begin{cases} \left(\dfrac{k-1}{N-1}\right)^{1/m} & (k = 1, 2, \ldots, N) \\[2mm] \left(\dfrac{2N-k}{N-1}\right)^{1/m} & (k = N+1, N+2, \ldots, 2N) \end{cases}$$

$$(3.225)$$

In Eq. (3.225), m is the functionally graded material volume fraction exponent and governs the distribution pattern of the properties across the thickness of the FGPM. Distributions of the FGPM properties are illustrated schematically in Fig. 3.36. When $m = 1$, the property distribution is linear. If the FGPM bimorph is driven by applying the AC electric field, we must consider the extrinsic contribution from the domain wall motion. The changes of material constants due to the domain wall motion are given by Eqs. (3.143). When the AC electric field is parallel to the poling of the beams according to Fig. 3.27, we can consider only the Δs_{11}^m, Δd_{31}^m, and $\Delta \epsilon_{33}^{Tm}$. The extrinsic material constants Δs_{11}^m, Δd_{31}^m, and $\Delta \epsilon_{33}^{Tm}$ of the kth layer are

$$(\Delta s_{11}^{m})_k = \Delta s_{11}^{mA} V_k + \Delta s_{11}^{mB}(1 - V_k)$$

$$(3.226)$$

$$(\Delta d_{31}^{m})_k = \Delta d_{31}^{mA} V_k + \Delta d_{31}^{mB}(1 - V_k)$$

$$(3.227)$$

$$(\Delta \epsilon_{33}^{Tm})_k = \Delta \epsilon_{33}^{TmA} V_k + \Delta \epsilon_{33}^{TmB}(1 - V_k)$$

$$(3.228)$$

We use the commercial finite element package ANSYS to perform the 3D finite element calculations and to determine the electromechanical fields of the FGPM bimorphs. We also measure the bending performance of the FGPM bimorphs to validate the FEA.

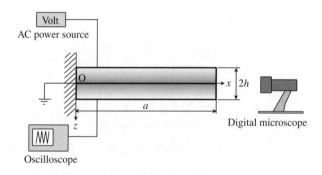

Figure 3.37   A clamped-free FGPM bimorph and experimental setup

3.7.2.1   Clamped-Free Bimorphs

Consider the clamped-free FGPM bimorphs. Figure 3.37 shows the schematic drawing of a clamped-free FGPM bimorph. The origin of the coordinate system is located at the center of the left site of the bimorph. Functionally graded piezoelectric bimorph cantilever is prepared using soft PZTs C-91 and C-6 (Fuji Ceramics Ltd. Co. Japan). Three-layer FGPMs of thickness $h_1 + h_2 + h_3 = 0.3$ mm and $h_4 + h_5 + h_6 = 0.3$ mm, created by extrusion, are respectively added to the upper and lower surfaces of an electrode film to make the six-layer inward series bimorph. The bimorph has also electrodes on both sides. The thickness of each layer is about $h_k = 0.1$ mm $(k = 1, \ldots, 6)$. The specimen has a length of 54 mm, a width of $b = 20$ mm, a thickness of $2h = 0.6$ mm, and a material volume fraction exponent of $m = 1$. The free length is $a = 40$ mm. The material properties of C-91 are

$$
\begin{aligned}
s_{11} &= 17.1 \times 10^{-12} \text{ m}^2/\text{N} \\
s_{12} &= -6.3 \times 10^{-12} \text{ m}^2/\text{N} \\
s_{13} &= -7.3 \times 10^{-12} \text{ m}^2/\text{N} \\
s_{33} &= 18.6 \times 10^{-12} \text{ m}^2/\text{N} \\
s_{44} &= 41.4 \times 10^{-12} \text{ m}^2/\text{N} \\
d_{15} &= 836 \times 10^{-12} \text{ m/V} \\
d_{31} &= -340 \times 10^{-12} \text{ m/V} \\
d_{33} &= 645 \times 10^{-12} \text{ m/V} \\
\epsilon_{11}^{\text{T}} &= 395 \times 10^{-10} \text{ C/Vm} \\
\epsilon_{33}^{\text{T}} &= 490 \times 10^{-10} \text{ C/Vm} \\
\rho &= 7800 \text{ kg/m}^3 \\
E_{\text{c}} &= 0.35 \text{ MV/m}
\end{aligned}
\tag{3.229}
$$

The material properties of C-6 are

$$s_{11} = 16.6 \times 10^{-12} \text{ m}^2/\text{N}$$
$$s_{12} = -5.1 \times 10^{-12} \text{ m}^2/\text{N}$$
$$s_{13} = -8.2 \times 10^{-12} \text{ m}^2/\text{N}$$
$$s_{33} = 20.5 \times 10^{-12} \text{ m}^2/\text{N}$$
$$s_{44} = 53.2 \times 10^{-12} \text{ m}^2/\text{N}$$

$$d_{15} = 771 \times 10^{-12} \text{ m/V}$$
$$d_{31} = -224 \times 10^{-12} \text{ m/V} \qquad (3.230)$$
$$d_{33} = 471 \times 10^{-12} \text{ m/V}$$

$$\epsilon_{11}^{\text{T}} = 203 \times 10^{-10} \text{ C/Vm}$$
$$\epsilon_{33}^{\text{T}} = 180 \times 10^{-10} \text{ C/Vm}$$

$$\rho = 7400 \text{ kg/m}^3$$

$$E_{\text{c}} = 0.45 \text{ MV/m}$$

The PZT C-91 is characterized by high piezoelectric constants and low coercive electric field. An AC voltage $V_0 \exp(i\omega t)$ is applied to the surface electrode of the upper element, while the center electrode is grounded. We calculate the amplitudes of deflection and stress for the FGPM bimorphs under AC electric field and measure the amplitude of the deflection with a digital microscope [44].

Fig 3.38 shows the amplitude of tip deflection $|w_{\text{tip}}|$ as a function of the AC voltage amplitude V_0 at frequency $f = 60$ Hz for the Type I [C-6/.../C-91]$_{\text{s}}$ and Type II [C-91/.../C-6]$_{\text{s}}$ six-layer inward series bimorphs ($m = 1$) obtained from the FEA with the domain wall motion

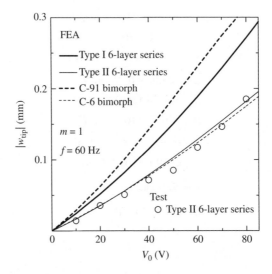

Figure 3.38   Tip deflection versus AC voltage of the clamped-free FGPM bimorphs

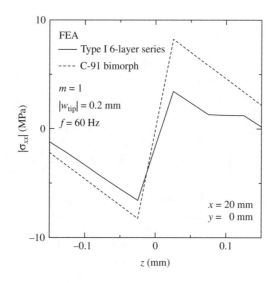

Figure 3.39 Normal stress distribution along the thickness direction for the clamped-free FGPM bimorph

effect, where $[\ ]_s$ designates symmetry about the middle surface. For comparison, the results for the traditional C-91 and C-6 bimorphs are included. Also shown are the test data for the Type II $[C-91/.../C-6]_s$ six-layer inward series bimorph ($m = 1$). The calculation results with the domain wall motion effect are in good agreement with experimental measurements. In general, the tip deflection increases as the AC voltage increases. The tip deflection of the Type I $[C-6/.../C-91]_s$ bimorph is smaller than that of the traditional C-91 bimorph. On the other hand, the tip deflection of the Type II $[C-91/.../C-6]_s$ bimorph is slightly larger than that of the traditional C-6 bimorph. The variation of normal stress amplitude $|\sigma_{xx}|$ along the thickness direction is calculated for the Type I $[C-6/.../C-91]_s$ six-layer inward series bimorph ($m = 1$) at a chosen point ($x = 20$ mm and $y = 0$ mm here) and the result is shown in Fig. 3.39. The result for the traditional C-91 bimorph is shown for comparison purposes. All calculations are done at a fixed deflection amplitude of 0.2 mm and frequency of 60 Hz. The driving voltages of the bimorph under $|w_{tip}| = 0.2$ mm are about 63.0 V for the $[C-6/.../C-91]_s$ six-layer inward series bimorph and 53.7 V for the traditional C-91 bimorph, respectively. The magnitude of the normal stress increases reaching a peak and then decreases toward the midplane as is expected. High normal stress is noted for the traditional C-91 bimorph. In other words, the functionally graded piezoelectric bimorph gives lower normal stress. Similar results for the shear stress amplitude $|\sigma_{zx}|$ are shown in Fig. 3.40. Lower shear stress is found in the functionally graded bimorph.

Further discussion on the topic is found in a recent article [45]. The self-sensing capability of the clamped-free FGPMs is evaluated using parallel bimorphs. Three-layer FGPMs of thickness $h_1 + h_2 + h_3 = 0.21$ mm and $h_4 + h_5 + h_6 = 0.21$ mm are, respectively, added to the upper and lower surfaces of an electrode film to make the six-layer parallel bimorph. The bimorph has also electrodes on both sides. The thickness of each layer is about $h_k = 0.07$ mm ($k = 1, \ldots, 6$). The specimen has a length of 55 mm, a width of $b = 20$ mm, a thickness of $2h = 0.42$ mm, and a material volume fraction exponent of $m = 1$. The free length is $a = 40$ mm. The 10-layer

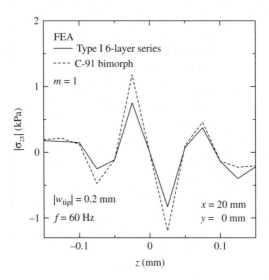

Figure 3.40   Shear stress distribution along the thickness direction for the clamped-free FGPM bimorph

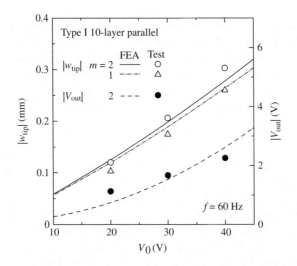

Figure 3.41   Tip deflection and output voltage versus AC voltage of the clamped-free FGPM bimorphs

parallel bimorphs ($m = 1, 2$) with same dimension are also considered. The thickness of each layer is about $h_k = 0.042$ mm ($k = 1, \ldots, 10$). An AC voltage $V_0 \exp(i\omega t)$ is applied to the surface electrode of the upper element, while the center electrode is grounded. One channel of an oscilloscope is connected directly across the lower element as shown in Fig. 3.37, and the amplitude of output voltage and the amplitude of deflection are evaluated.

Fig 3.41 shows the amplitude of tip deflection $|w_{tip}|$ as a function of AC voltage amplitude V_0 at frequency $f = 60$ Hz for the Type I $[C-6/\ldots/C-91]_s$ 10-layer parallel bimorphs ($m = 1, 2$), obtained from the FEA with the domain wall motion effect and test. Also shown is the amplitude of output voltage $|V_{out}|$ for $m = 2$. The tip deflection for $m = 2$ is larger than that for $m = 1$. Although the results are not shown here, the tip deflection increases slightly with the number of layers. As the tip deflection increases, the output voltage also increases. Agreement between FEA and test is fair. Figure 3.42 shows the amplitude of tip deflection $|w_{tip}|$ as a function of the material volume fraction exponent m for the Type I $[C-6/.../C-91]_s$ and Type II $[C-91/.../C-6]_s$ 10-layer parallel bimorphs under the AC voltage amplitude $V_0 = 30$ V at frequency $f = 60$ Hz. Larger deflection is observed for the Type I $[C-6/.../C-91]_s$ parallel bimorph, and in general, $|w_{tip}|$ increases as m increases. Figure 3.43 shows the similar results for the amplitude of output voltage $|V_{out}|$. In our results for the output voltage, larger $|V_{out}|$ can be achieved with the Type II $[C-91/.../C-6]_s$ parallel bimorph than with the Type I $[C-6/.../C-91]_s$ parallel bimorph. The output voltage of the Type II $[C-91/.../C-6]_s$ parallel bimorph decreases slightly as m increases. In Fig. 3.44, we plot the variation of normal stress amplitude $|\sigma_{xx}|$ along the thickness direction at $x = 20$ mm and $y = 0$ mm of the Type I $[C-6/.../C-91]_s$ 10-layer parallel bimorphs with different volume fraction exponents under $|w_{tip}| = 0.2$ mm at $f = 60$ Hz. The driving (output) voltages of the bimorphs with $m = 0.5, 1, 2$ under $|w_{tip}| = 0.2$ mm are about 33.4 (1.17), 31.4 (1.36), and 30.0 (1.52) V, respectively. At larger volume fraction exponent, the driving voltage becomes smaller and larger output voltage occurs, whereas the internal stress increases.

3.7.2.2 Clamped–Clamped Bimorphs

Consider the clamped–clamped FGPM bimorphs. Figure 3.45 shows the schematic drawing of a clamped–clamped FGPM bimorph. Three-layer FGPMs of thickness $h_1 + h_2 + h_3 = 0.21$ mm and $h_4 + h_5 + h_6 = 0.21$ mm, using soft PZTs C-91 and C-6 (Fuji Ceramics Ltd. Co. Japan), are respectively added to the upper and lower surfaces of an electrode film to make

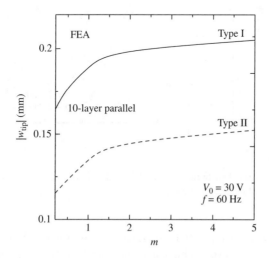

Figure 3.42 Tip deflection versus material volume fraction exponent of the clamped-free FGPM bimorphs

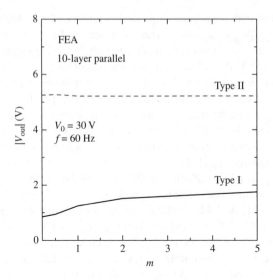

Figure 3.43 Output voltage versus material volume fraction exponent of the clamped-free FGPM bimorphs

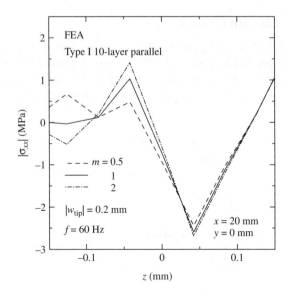

Figure 3.44 Normal stress distribution along the thickness direction for the clamped-free FGPM bimorphs with different volume fraction exponents

the six-layer inward series and parallel bimorphs. The bimorphs have also electrodes on both sides. The thickness of each layer is about $h_k = 0.07$ mm ($k = 1, \ldots, 6$). The clamped–clamped specimen of width $b = 20$ mm and thickness $2h = 0.42$ mm has a span of $a = 40$ mm, and the material volume function exponent m is taken to be 1. The 10-layer parallel bimorph

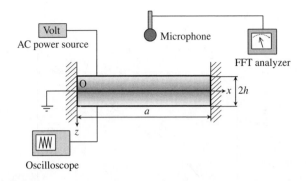

Figure 3.45 A clamped–clamped FGPM bimorph and experimental setup

Table 3.1 Sound pressure level of the clamped–clamped FGPM bimorphs

Six-layer	$V_0 = 10$ V, $f = 400$ Hz L_p (dB)	
$m = 1$	FEA	Test
Type I parallel	49.2	46.5
Type II parallel	47.7	45.9
Type II series	47.7	45.5

($m = 1$) with same dimension is also used. The thickness of each layer is about $h_k = 0.042$ mm ($k = 1, \ldots, 10$). An AC voltage $V_0 \exp(i\omega t)$ is applied to the surface electrode of the upper element, while the center electrode is grounded. We calculate the amplitudes of deflection, sound pressure level, and stress for the FGPM bimorphs under AC electric field and measure the sound pressure level [46]. We also evaluate the self-sensing capability and examine the output voltage amplitude [47]. To calculate the sound pressure level, air density of 1.2 kg/m³ and sound speed of 340 m/s are used in the FEA. Also, the measurement of the sound pressure level is facilitated by a microphone, and the spectrum of a signal is automatically displayed by the use of a fast Fourier transform (FFT) spectrum analyzer. The distance between the microphone position and specimen top surface is about 20 mm. To evaluate the output voltage, one channel of an oscilloscope is connected directly across the lower element.

Table 3.1 shows the sound pressure level L_p of the Type I [C-6/.../C-91]$_s$ six-layer parallel, Type II [C-91/.../C-6]$_s$ six-layer parallel, and inward series bimorphs ($m = 1$) under the AC voltage amplitude $V_0 = 10$ V at frequency $f = 400$ Hz, obtained from the FEA with the domain wall motion effect and test. Comparing the calculation results of the Type II parallel and series bimorphs, no difference is observed. It is shown that the comparison between FEA and test is reasonable. The sound pressure levels for Type I are higher than those for Type II. Although the results are not shown, the sound pressure level increases as the layer number or functionally graded material volume fraction exponent increases. Figure 3.46 shows the FEA and test results for the sound pressure level L_p versus frequency f of the Type I [C-6/.../C-91]$_s$ 10-layer parallel bimorph ($m = 1$) under $V_0 = 10$ V. The sound pressure level

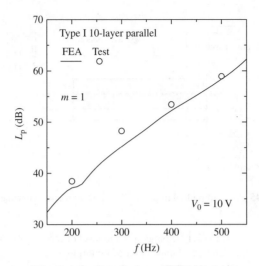

Figure 3.46 Sound pressure level versus frequency of the clamped–clamped FGPM bimorph

Table 3.2 Output voltage of the clamped–clamped
FGPM bimorphs

FEA	$|w_{max}| = 0.5$ μm, $f = 400$ Hz				
Six-layer	$|V_{out}|$ (V)				
$m =$	0.2	0.5	1	2	5
Type I parallel	4.8	4.4	4.1	3.9	3.8
Type II parallel	10	9.6	9.2	8.8	8.5

increases with increasing frequency. Table 3.2 lists the output voltage amplitude $|V_{out}|$ of the Type I and Type II six-layer parallel bimorphs under a maximum central deflection amplitude $|w_{max}| = 0.5$ μm at $f = 400$ Hz for various values of material volume fraction exponent m, obtained from the FEA. The driving voltages of the bimorphs obtained for $|w_{max}| = 0.5$μm are about $V_0 = 41$ V for the Type I and 53 V for the Type II parallel bimorphs ($m = 0.2 \sim 5$) at $f = 400$ Hz, respectively. The output voltage decreases as m increases, and the sensitivity of Type II is larger than that of Type I.

The studies in the area of bending for the FGPM bimorphs will be focused on the following points:

1. The deflection and sound pressure become high as the piezoelectric properties increase toward the midplane of the bimorphs (Type I), that is, this functional grading of materials provides superior actuator characteristics.
2. When the piezoelectric properties decrease toward the midplane of the bimorphs (Type II), the output voltage increases, that is, this functional grading of materials gives superior sensitivity.

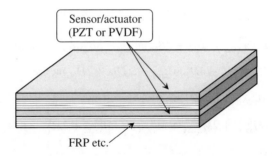

Figure 3.47   A hybrid laminated plate with sensor/actuator

3. Functional grading of piezoelectric materials could effectively reduce the magnitude of internal stresses or obtain an optimal distribution of stresses in sensors and actuators for a given design application.

3.7.3   Laminated Plates

The study of surface-mounted/embedded PZT or PVDF layers in structures has received considerable attention. Figure 3.47 shows an example of such a structure. The distributed piezoelectric sensing layer monitors the structural oscillation due to the direct piezoelectric effect, and the distributed actuator layer suppresses the oscillation via the converse piezoelectric effect [48]. Here, we demonstrate the feasibility of suppressing thermomechanically induced bending of the laminated plates via the converse piezoelectric effect.

Consider a laminated plate of side lengths a, b and thickness $2h$ in a rectangular Cartesian coordinate system (x, y, z) as shown in Fig. 3.23. The coordinate axes x and y are in the middle plane of the hybrid laminate, and the z-axis is normal to this plane. We are here interested in examining the response of a graphite/epoxy laminate formed by adding a PVDF layer. A rectangular laminated plate is simply supported on all edges and is subjected to an applied electric field E_z in addition to the upper and lower surface temperatures Θ_0, Θ_N.

The constitutive relation for the PVDF layer is given by Eq. (3.157). For the graphite/epoxy layer, the constitutive relation is obtained by setting $\bar{e}_{31} = \bar{e}_{32} = \bar{e}_{36} = 0$ in Eq. (3.157). The temperature variation is

$$\Theta(z) = \frac{\Theta_0 + \Theta_N}{2} + \frac{(\Theta_N - \Theta_0)z}{2h} \tag{3.231}$$

From Eqs. (3.186) and (3.187), the basic equations of linear theory for thermopiezoelectric laminates, without inertia and viscoelastic damping, become

$$A_{11}u^0_{x,xx} + 2A_{16}u^0_{x,xy} + A_{66}u^0_{x,yy} + A_{16}u^0_{y,xx} + (A_{12} + A_{66})u^0_{y,xy} + A_{26}u^0_{y,yy}$$
$$- B_{11}w_{,xxx} - 3B_{16}w_{,xxy} - (B_{12} + 2B_{66})w_{,xyy} - B_{26}w_{,yyy}$$
$$- (N^E_{xx,x} + N^E_{xy,y} + N^\Theta_{xx,x} + N^\Theta_{xy,y}) = 0$$

$$A_{16}u^0_{x,xx} + (A_{12} + A_{66})u^0_{x,xy} + A_{26}u^0_{x,yy} + A_{66}u^0_{y,xx} + 2A_{26}u^0_{y,xy} + A_{22}u^0_{y,yy} \tag{3.232}$$
$$- B_{16}w_{,xxx} - (B_{12} + 2B_{66})w_{,xxy} - 3B_{26}w_{,xyy} - B_{22}w_{,yyy}$$
$$- (N^E_{xy,x} + N^E_{yy,y} + N^\Theta_{xy,x} + N^\Theta_{yy,y}) = 0$$

$$B_{11}u^0_{x,xxx} + 3B_{16}u^0_{x,xxy} + (B_{12} + 2B_{66})u^0_{x,xyy} + B_{26}u^0_{x,yyy}$$

$$+ B_{16}u^0_{y,xxx} + (B_{12} + 2B_{66})u^0_{y,xxy} + 3B_{26}u^0_{y,xyy} + B_{22}u^0_{y,yyy}$$

$$- D_{11}w_{,xxxx} - 4D_{16}w_{,xxxy} - 2(D_{12} + 2D_{66})w_{,xxyy}$$

$$- 4D_{26}w_{,xyyy} - D_{22}w_{,yyyy}$$

$$- (M^E_{xx,xx} + 2M^E_{xy,xy} + M^E_{yy,yy}) - (M^\Theta_{xx,xx} + 2M^\Theta_{xy,xy} + M^\Theta_{yy,yy})$$

$$= 0 \tag{3.233}$$

If the laminated plate is supported along its edges in such a manner that only normal in-plane displacements can occur, the boundary conditions (hinge-free normal case) are

$$u^0_y(\pm a/2, y) = w(\pm a/2, y) = N_{xx}(\pm a/2, y) = M_{xx}(\pm a/2, y) = 0$$
$$u^0_x(x, \pm b/2) = w(x, \pm b/2) = N_{yy}(x, \pm b/2) = M_{yy}(x, \pm b/2) = 0 \tag{3.234}$$

Double Fourier series approach is used to obtain the solutions.

In the numerical examples, we consider an eight-layer graphite/epoxy laminate of $a = b = 0.3$ m having fiber orientations $[0°/90°/0°/90°]_s$ and a nine-layer hybrid laminate containing one layer of PVDF. The thickness of each graphite/epoxy layer is $h_k = 0.125$ mm ($k = 1, \ldots, 8$) and the thickness of PVDF layer is $h_9 = 0.25$ mm. Attention is given to the plate with surface temperatures $\Theta_0 = 80$ °C and $\Theta_N = 0$. The material properties of graphite/epoxy are as follows:

$$E^{elas}_{11} = 18.1 \times 10^{10} \text{N/m}^2$$
$$E^{elas}_{22} = 1.03 \times 10^{10} \text{N/m}^2$$
$$G^{elas}_{12} = 0.717 \times 10^{10} \text{N/m}^2$$
$$v^{elas}_{12} = 0.28 \tag{3.235}$$
$$\rho^{elas} = 1580 \text{kg/m}^3$$
$$\alpha^{elas}_{11} = 0.02 \times 10^{-6} \text{K}^{-1}$$
$$\alpha^{elas}_{22} = 22.5 \times 10^{-6} \text{K}^{-1}$$

The material properties of PVDF are

$$E_{11} = 0.2 \times 10^{10} \text{ N/m}^2$$
$$E_{22} = 0.2 \times 10^{10} \text{ N/m}^2$$
$$G_{12} = 0.0752 \times 10^{10} \text{ N/m}^2$$
$$v_{12} = 0.33$$
$$d_{31} = 23 \times 10^{-12} \text{ m/V}$$
$$d_{32} = 23 \times 10^{-12} \text{ m/V} \tag{3.236}$$
$$\rho = 1800 \text{ kg/m}^3$$
$$\alpha_{11} = 120 \times 10^{-6} \text{ K}^{-1}$$
$$\alpha_{22} = 120 \times 10^{-6} \text{ K}^{-1}$$

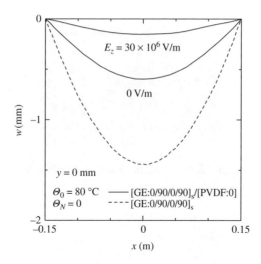

Figure 3.48   Deflection along the length direction for the eight-layer graphite/epoxy laminate and nine-layer hybrid laminate

Figure 3.48 shows the variation of deflection w along the length direction at $y = 0$ mm for the [graphite/epoxy (GE): $0°/90°/0°/90°$]$_s$/[PVDF:$0°$] hybrid laminate under the DC electric fields $E_0 = (E_z)_9 = 0, 30$ MV/m at $\Theta_0 = 80°$C, $\Theta_9 = 0$. Also shown is the result for the [GE: $0°/90°/0°/90°$]$_s$ laminate for comparison. It is found that application of a large electric field to the PVDF layer eliminates the deflection.

3.8   Electromechanical Field Concentrations

Partial electrodes in piezoelectric devices are a source of electric field concentrations, which can result in stress concentrations high enough to fracture the part. In this section, the electromechanical field concentrations in piezoelectric materials and structures are discussed.

3.8.1   *Laminates*

Large force actuators rely on multilayer concept as shown in Fig. 3.24(b). Cracking at the internal electrode tip, as shown in Fig. 3.49, is common cause of failure of many multilayer stacked devices [49]. Here, we focus on the details of the electromechanical field concentrations near the electrode tip in the piezoelectric laminates.

3.8.1.1   Two Dissimilar Piezoelectric Half-Planes

Let us consider two-dimensional piezoelectric materials 1 and 2 as shown in Fig. 3.50. Let $x = x_1, y = x_2$, and $z = x_3$ denote the rectangular Cartesian coordinates. The origin of the coordinate system is located at the center of the electrode. We assume plane strain perpendicular to the y axis. The z axis is assumed to coincide with the poling direction. Consider an electrode of

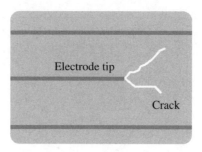

Figure 3.49   Cracking around an electrode tip

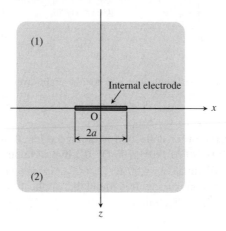

Figure 3.50   Two dissimilar piezoelectric half-planes with an internal electrode

length $2a$ at the interface of two bonded piezoelectric half-planes. The electrode is maintained at a constant voltage V_0. The voltage equals zero at infinity.

The basic equations (3.103), (3.104) without inertia become in-plane strain

$$\sigma_{xx,x} + \sigma_{zx,z} = 0$$
$$\sigma_{xz,x} + \sigma_{zz,z} = 0 \tag{3.237}$$

$$E_{z,x} - E_{x,z} = 0, \quad D_{x,x} + D_{z,z} = 0 \tag{3.238}$$

The constitutive equations (3.120) and (3.121) for piezoelectric ceramics k ($k = 1, 2$) become

$$\left\{\begin{array}{c} \sigma_{xx} \\ \sigma_{zz} \\ \sigma_{zx} \end{array}\right\}_k = \left[\begin{array}{ccc} c_{11} & c_{13} & 0 \\ c_{13} & c_{33} & 0 \\ 0 & 0 & c_{44} \end{array}\right]_k \left\{\begin{array}{c} \varepsilon_{xx} \\ \varepsilon_{zz} \\ 2\varepsilon_{zx} \end{array}\right\}_k - \left[\begin{array}{cc} 0 & e_{31} \\ 0 & e_{33} \\ e_{15} & 0 \end{array}\right]_k \left\{\begin{array}{c} E_x \\ E_z \end{array}\right\}_k \tag{3.239}$$

$$\left\{\begin{array}{c} D_x \\ D_z \end{array}\right\}_k = \left[\begin{array}{ccc} 0 & 0 & e_{15} \\ e_{31} & e_{33} & 0 \end{array}\right]_k \left\{\begin{array}{c} \varepsilon_{xx} \\ \varepsilon_{zz} \\ 2\varepsilon_{zx} \end{array}\right\}_k + \left[\begin{array}{cc} \epsilon_{11} & 0 \\ 0 & \epsilon_{33} \end{array}\right]_k \left\{\begin{array}{c} E_x \\ E_z \end{array}\right\}_k \tag{3.240}$$

Here and in the following, we have dropped the remanent strains and polarizations. The strains are defined in terms of displacements as

$$\varepsilon_{xx} = u_{x,x}$$
$$\varepsilon_{zz} = u_{z,z} \qquad\qquad\qquad (3.241)$$
$$\varepsilon_{zx} = \frac{1}{2}(u_{x,z} + u_{z,x})$$

The electric field components may be written in terms of an electric potential ϕ as

$$E_x = -\phi_{,x}, \quad E_z = -\phi_{,z} \qquad\qquad (3.242)$$

It is clear that Eqs. (3.242) satisfy the first of Eqs. (3.238). Substituting Eqs. (3.239), (3.240) into Eqs. (3.237) and the second of Eqs. (3.238), and considering Eqs. (3.241), (3.242), we obtain the governing equations for piezoelectric ceramics k ($k = 1, 2$) in the plane strain case

$$(c_{11})_k(u_x)_{k,xx} + (c_{44})_k(u_x)_{k,zz} + \{(c_{13})_k + (c_{44})_k\}(u_z)_{k,xz}$$
$$+\{(e_{15})_k + (e_{31})_k\}(\phi)_{k,xz} = 0$$
$$\{(c_{13})_k + (c_{44})_k\}(u_x)_{k,xz} + (c_{44})_k(u_z)_{k,xx} + (c_{33})_k(u_z)_{k,zz}$$
$$+(e_{15})_k(\phi)_{k,xx} + (e_{33})_k(\phi)_{k,zz} = 0 \qquad\qquad (3.243)$$

$$\{(e_{15})_k + (e_{31})_k\}(u_x)_{k,xz} + (e_{15})_k(u_z)_{k,xx} + (e_{33})_k(u_z)_{k,zz}$$
$$- (\epsilon_{11})_k(\phi)_{k,xx} - (\epsilon_{33})_k(\phi)_{k,zz} = 0 \qquad\qquad (3.244)$$

We assume that the mixed boundary conditions are given by

$$\begin{cases} (\phi)_1(x,0) = (\phi)_2(x,0) = V_0 & (0 \le |x| < a) \\ (E_x)_1(x,0) = (E_x)_2(x,0) & (a \le |x| < \infty) \\ (D_z)_1(x,0) = (D_z)_2(x,0) & (a \le |x| < \infty) \end{cases} \qquad (3.245)$$

$$\begin{aligned} (u_x)_1(x,0) &= (u_x)_2(x,0) \\ (u_z)_1(x,0) &= (u_z)_2(x,0) \\ (\sigma_{zx})_1(x,0) &= (\sigma_{zx})_2(x,0) \\ (\sigma_{zz})_1(x,0) &= (\sigma_{zz})_2(x,0) \end{aligned} \qquad\qquad (3.246)$$

Because of the assumed symmetry, it is sufficient to consider the problem for $0 \le x < \infty$ only. Ru [50] considers the loading case where a charge is applied at the electrode, whereas we consider an electric potential load that is more readily achieved in the laboratory [51]. Fourier transforms are used to reduce the mixed boundary value problem to the solution of a pair of dual integral equations. The integral equations are solved exactly. When all material properties of piezoelectric ceramics 1 vanish, the solution reduces to that of a piezoelectric half-plane with a surface electrode as shown in Fig. 3.51. Details can be found in Ref. [52].

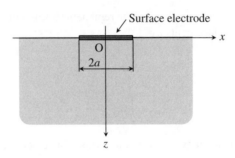

Figure 3.51 A piezoelectric half-plane with a surface electrode

To examine the electromechanical fields near the electrode tip, numerical calculations are carried out. As an example we consider PZT$^+$/PZT$^+$ and PZT$^+$/PZT$^-$ laminates corresponding to the bending and tension actuator models, respectively. The superscripts $+$ and $-$ denote, respectively, the situations for positive and negative poling directions. The present laminates are made of PZT P-7 (Murata Manufacturing Co. Ltd. Japan). The electrode length is $a = 10$ mm and applied voltage is $V_0 = 100$ V. The material properties of P-7 are

$$c_{11} = 13.0 \times 10^{10} \text{ N/m}^2$$
$$c_{12} = 8.3 \times 10^{10} \text{ N/m}^2$$
$$c_{13} = 8.3 \times 10^{10} \text{ N/m}^2$$
$$c_{33} = 11.9 \times 10^{10} \text{ N/m}^2$$
$$c_{44} = 2.5 \times 10^{10} \text{ N/m}^2$$
$$e_{15} = 13.5 \text{ C/m}^2 \tag{3.247}$$
$$e_{31} = -10.3 \text{ C/m}^2$$
$$e_{33} = 14.7 \text{ C/m}^2$$
$$\epsilon_{11} = 171 \times 10^{-10} \text{ C/Vm}$$
$$\epsilon_{33} = 186 \times 10^{-10} \text{ C/Vm}$$
$$E_{\text{c}} = 0.8 \text{ MV/m}$$

In Fig. 3.52, the variation of normal stress σ_{zz} along the x-direction is shown for the P-7/P-7 laminates at $z = 0$ mm. Near the electrode tip, the normal stress for the P-7$^+$/P-7$^+$ laminate is tensile and very high. In the case of the P-7$^+$/P-7$^-$ laminate, the stress ahead of the electrode tip is compressive, while the stress behind the electrode tip is tensile.

3.8.1.2 Multilayer Piezoelectric Actuators

A multilayer piezoelectric actuator, as shown in Fig. 3.24(b), is of particular importance to us. Typically, the actuator consists of tens or hundreds of piezoelectric layers and a coating layer. So, we next consider the piezoelectric actuator with many PZT layers, thin electrodes, and elastic coating layer. In order to discuss the electromechanical fields near the internal electrode, the problem of the multilayer piezoelectric actuator is solved using the unit cell model

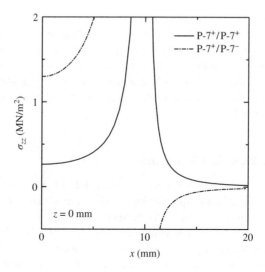

Figure 3.52 Normal stress distribution along the x-direction for the P-7/P-7 laminates

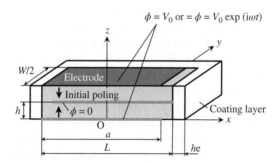

Figure 3.53 Unit cell of multilayer piezoelectric actuators

(two-layer piezoelectric actuator). Figure 3.53 shows the unit cell. A rectangular Cartesian coordinate system (x, y, z) is used, and the origin of the coordinate system coincides with the center of the multilayer actuator. Each PZT layer of length L, width W, and thickness h is sandwiched between the alternating positive and negative thin electrodes of length a and width W, and the layers are electrically in parallel and mechanically in series. An electrically active $2a - L$ region exists in the portions of the actuator where the positive and negative electrodes overlap, while an electrically inactive $L - a$ tab region exists on both sides of the actuator. The thickness of the coating layer h_e is chosen. The poling directions of the upper and lower layers are opposite so that the vertical displacement (z-direction) in adjacent layer can be accumulated. Because of the geometric and loading symmetry, only the half needs to be analyzed. The electrode layers are not incorporated into the model. This unit cell model does not simulate the end effects near the top or bottom of the multilayer piezoelectric actuator and simulates the behavior near the middle of the actuator.

The electric potential on two electrode surfaces $(-L/2 \leq x \leq -L/2 + a, 0 \leq y \leq W/2, z = 0, 2h)$ equals the applied DC voltage, $\phi = V_0$ or AC voltage $\phi = V_0 \exp(i\omega t)$. The electrode surface $(L/2 - a \leq x \leq L/2, 0 \leq y \leq W/2, z = h)$ is connected to the ground, so that $\phi = 0$. The mechanical boundary conditions include the traction-free conditions on the coating layer surfaces at $x = \pm(L/2 + h_e), y = W/2 + h_e$ and the symmetry conditions on the faces at $y = 0, z = 0, 2h$. In addition, the origin is constrained against the displacement in the x-direction, to avoid rigid body motion.

3.8.1.3 Four-Layer Piezoelectric Actuator

Although many studies focus only on a numerical model [53, 54], the emphasis of the work presented by Shindo et al. [55] is on quantifying the electromechanical fields near the electrode tip based on a combined theoretical and experimental approach. FEA and experiment are performed on a four-layer "model" actuator fabricated using soft PZTs C-91 (Fuji Ceramics Ltd. Co. Japan). Figure 3.54 shows a photograph of the specimen with key dimensions. The PZT C-91 layers have dimensions of $30 \times 10 \times 5$ mm, and a 10 mm tab region exists on both sides of the actuator. The total thickness is 20 mm. An external electrode is attached on both sides of the actuator to address the surface and internal electrodes. The actuator does not have a coating layer. Figure 3.55 shows the specimen and setup for the experiment. The specimen is placed on the rigid floor. The high-voltage amplifier used is limited to 1.25 kV/DC. Strain gages are used to measure the strain near the electrode tip. Figure 3.56 shows the half of the finite element model for the actuator. A rectangular Cartesian coordinate system (x, y, z) is used and the origin of the coordinate system coincides with the center of the actuator. The length, width, and thickness of the model are $L = 30$ mm, $W/2 = 5$ mm, and $2h = 10$ mm, respectively. The electrode length is $a = 20$ mm. As mentioned earlier, the electric potential on two electrode surfaces $(-L/2 \leq x \leq -L/2 + a, 0 \leq y \leq W/2, z = 0, 2h)$ equals the applied DC voltage $\phi = V_0$. The electric potential on the electrode surface $(L/2 - a \leq x \leq L/2, 0 \leq y \leq W/2, z = h)$ is $\phi = 0$. The mechanical boundary conditions include the traction-free conditions on the actuator surfaces at $x = \pm L/2, y = W/2, z = 2h$ and the symmetry conditions on the faces at $y = 0, z = 0$.

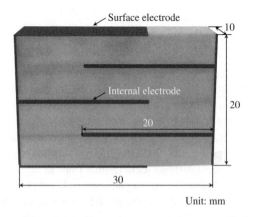

Figure 3.54 A four-layer piezoelectric actuator

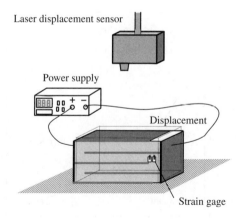

Figure 3.55 A four-layer actuator and experimental setup

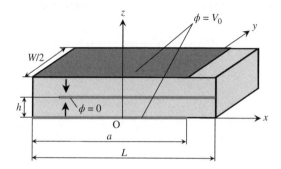

Figure 3.56 Finite element model of the four-layer piezoelectric actuator

Figure 3.57 shows the normal strain ε_{zz} as a function of the distance x at the face of the actuator (at $y = 5$ mm plane) under DC electric field $E_0 = V_0/h = 0.16$ MV/m for the distance from interface $z = 0, 0.8$ mm. It is shown that a portion near the electrode tip is highly strained. The comparison between prediction and measurement is reasonable at several locations and therefore provides confidence in the results from the FEA. Figure 3.58 provides the normal strain ε_{zz} versus DC electric field E_0 at the face of the actuator (at $y = 5$ mm plane) for $x = 5$ mm and $z = 0.8$ mm. A negative electric field increases compressive strain, and a monotonically increasing negative electric field causes polarization switching. The switching in a local region leads to a decrease in compressive strain.

3.8.1.4 Four-Layer Partially Poled Piezoelectric Actuator

We consider the partially poled piezoelectric actuators [56]. If unpoled piezoelectric materials or films are coated with electrode, laminated, and then poled, the resulting polarization of the multilayer actuators will inevitably be partial. Consider the unit cell shown in Fig. 3.56 again. Each element is taken to represent a single grain, and piezoelectric layers of crystalline

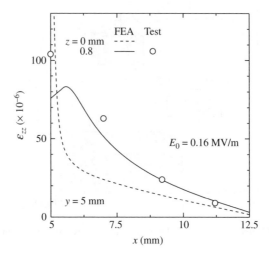

Figure 3.57 Normal strain distribution along the x-direction for the four-layer piezoelectric actuator

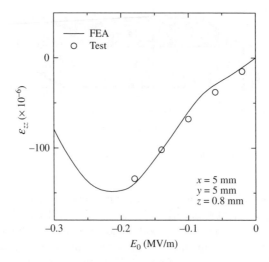

Figure 3.58 Normal strain versus DC electric field of the four-layer piezoelectric actuator

grains with an idealized microstructure are assumed [57], where each grain has a random polarization. The response is a mixture of each grain's properties, and the resulting material possesses no overall polarization. Figure 3.59 shows a typical model with a fine microstructure. Grain polarizations are taken to be along just six orientations (black, white, and several gray level images), corresponding to the two directions along the x, y, and z axes. Other directions are not considered for simplicity. This model is equivalent to an unpoled piezoelectric actuator.

The partially poled piezoelectric actuator model is created through the high electric field. The electric potential on two electrode surfaces $(-L/2 \leq x \leq -L/2 + a, 0 \leq y \leq W/2, z = 0, 2h)$ equals the applied DC voltage $\phi = V_0$. The electric potential on the electrode surface

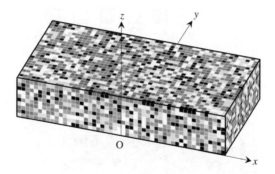

Figure 3.59 Typical model of the four-layer unpoled piezoelectric actuator

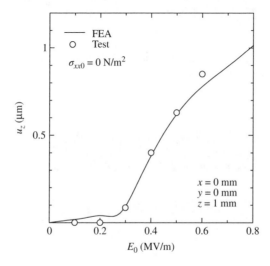

Figure 3.60 Displacement versus DC electric field of the four-layer unpoled piezoelectric actuator

$(L/2 - a \leq x \leq L/2, 0 \leq y \leq W/2, z = h)$ is $\phi = 0$. We consider two following mechanical boundary conditions at both ends $(x = \pm L/2)$ of the actuator: (i) each end is stress free and (ii) uniformly distributed compressive load (compressive stress $-\sigma_{xx0}$) is applied at the end of the actuator along the x-direction. Other conditions are the traction-free conditions on the actuator surfaces at $y = W/2, z = 2h$ and the symmetry conditions on the faces at $y = 0, z = 0$.

In order to validate the predictions, we fabricate the four-layer unpoled PZT C-91 actuator. The dimensions of the specimen are $L = 5$ mm, $W = 5$ mm, and $2h = 1$ mm. The electrode length is $a = 4.5$ mm. The specimen is placed on the rigid floor, and voltage is applied using a power supply for voltages up to 1.25 kV/DC. To measure the magnitude of the deformation, a laser displacement sensor is used as shown in Fig. 3.55.

Figure 3.60 shows the prediction and measurement of displacement u_z versus DC electric field $E_0 = V_0/h$ at $x = 0$ mm, $y = 0$ mm, and $z = 1$ mm for the unpoled actuator under stress-free condition. The open circle denotes the half of the measured value. The displacement

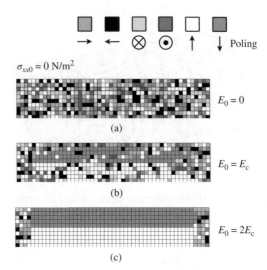

Figure 3.61 Images of poling for the four-layer piezoelectric actuator under $\sigma_{xx0} = 0$ N/m² at (a) $E_0 = 0$, (b) $E_0 = E_c$, and (c) $E_0 = 2E_c$

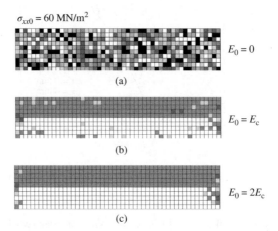

Figure 3.62 Images of poling for the four-layer piezoelectric actuator under $\sigma_{xx0} = 60$ MN/m² at (a) $E_0 = 0$, (b) $E_0 = E_c$, and (c) $E_0 = 2E_c$

of the unpoled actuator increases suddenly when the electric field reaches about 0.25 MV/m, whereas the coercive electric field is $E_c = 0.35$ MV/m. We see that the trend is sufficiently similar between FEA and test. Figure 3.61 shows the distribution of poled area for the actuator under stress-free condition at (a) $E_0 = 0$, (b) $E_0 = E_c = 0.35$ MV/m, and (c) $E_0 = 2E_c = 0.7$ MV/m. At the electric field of $2E_c$, the active region of the actuator is poled. Figure 3.62 shows the similar area for the actuator under compressive stress $\sigma_{xx0} = 60$ MN/m². Under the coercive electric field E_c, there are small unpoled areas. It is interesting to note that the unpoled area diminishes at the electric field of $2E_c$, compared to the case under stress-free condition.

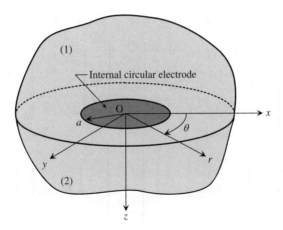

Figure 3.63 Two dissimilar piezoelectric half-spaces with an internal circular electrode

Experimental results [58] showed that compressive stress applied transverse to the electric field increases the potential for domain alignment during poling. Increase of the domain alignment degree leads to increased polarization. Our model describes this phenomenon quantitatively.

3.8.2 Disk Composites

Some designs are in existence for piezoelectric devices with circular electrodes. Bonded piezo-electric and elastic disks with a circular electrode embedded at the interface are also used as sounders. In this section, we discuss a few examples for the analysis and test of the electrome-chanical field concentrations near the circular electrode.

3.8.2.1 Two Dissimilar Piezoelectric Half-Spaces with Circular Electrode

We consider two piezoelectric half-spaces with a circular electrode at the interface as shown in Fig. 3.63. Let $x = x_1, y = x_2$, and $z = x_3$ denote the rectangular Cartesian coordinates. The origin of the coordinates coincides with the center of the circular electrode. The z axis is assumed to coincide with the poling direction. The problem is formulated using a cylindrical coordinate system (r,θ,z). Let a piezoelectric material 1 occupy the upper half-space, $z < 0$, and a piezoelectric material 2 occupy the lower half-space, $z > 0$. The electrode is maintained at a constant voltage V_0, and the voltage equals zero at infinity.

If the circular electrode is subjected to a constant voltage that is independent of θ, then the system possesses axial symmetry about the z-axis. The basic equations for piezoelectric ceramics without inertia [59, 60] are

$$
\begin{aligned}
\sigma_{rr,r} + \sigma_{zr,z} + \frac{\sigma_{rr} - \sigma_{\theta\theta}}{r} &= 0 \\
\sigma_{rz,r} + \sigma_{zz,z} + \frac{\sigma_{rz}}{r} &= 0
\end{aligned}
\tag{3.248}
$$

$$
E_{z,r} - E_{r,z} = 0, \quad D_{r,r} + \frac{D_r}{r} + D_{z,z} = 0
\tag{3.249}
$$

The constitutive equations for piezoelectric ceramics k ($k = 1, 2$) become

$$
\begin{Bmatrix} \sigma_{rr} \\ \sigma_{\theta\theta} \\ \sigma_{zz} \\ \sigma_{zr} \end{Bmatrix}_k = \begin{bmatrix} c_{11} & c_{12} & c_{13} & 0 \\ c_{12} & c_{11} & c_{13} & 0 \\ c_{13} & c_{13} & c_{33} & 0 \\ 0 & 0 & 0 & c_{44} \end{bmatrix}_k \begin{Bmatrix} \varepsilon_{rr} \\ \varepsilon_{\theta\theta} \\ \varepsilon_{zz} \\ 2\varepsilon_{zr} \end{Bmatrix}_k - \begin{bmatrix} 0 & e_{31} \\ 0 & e_{31} \\ 0 & e_{33} \\ e_{15} & 0 \end{bmatrix}_k \begin{Bmatrix} E_r \\ E_z \end{Bmatrix}_k
\tag{3.250}
$$

$$
\begin{Bmatrix} D_r \\ D_z \end{Bmatrix}_k = \begin{bmatrix} 0 & 0 & 0 & e_{15} \\ e_{31} & e_{31} & e_{33} & 0 \end{bmatrix}_k \begin{Bmatrix} \varepsilon_{rr} \\ \varepsilon_{\theta\theta} \\ \varepsilon_{zz} \\ 2\varepsilon_{zr} \end{Bmatrix}_k + \begin{bmatrix} \epsilon_{11} & 0 \\ 0 & \epsilon_{33} \end{bmatrix}_k \begin{Bmatrix} E_r \\ E_z \end{Bmatrix}_k
\tag{3.251}
$$

Here and in the following, we have dropped the remanent strains and polarizations. The strain–displacement relations can be written in the cylindrical coordinate system as

$$
\varepsilon_{rr} = u_{r,r}
$$

$$
\varepsilon_{\theta\theta} = \frac{u_r}{r}
$$

$$
\varepsilon_{zz} = u_{z,z}
\tag{3.252}
$$

$$
\varepsilon_{zr} = \frac{1}{2}(u_{r,z} + u_{z,r})
$$

The electric field components may be written as

$$
E_r = -\phi_{,r}, \quad E_z = -\phi_{,z}
\tag{3.253}
$$

Substituting Eqs. (3.250), (3.251) into Eqs. (3.248) and the second of Eqs. (3.249), and considering Eqs. (3.252), (3.253), we obtain the governing equations for piezoelectric ceramics k ($k = 1, 2$) in the axisymmetric case

$$
(c_{11})_k \left\{ (u_r)_{k,rr} + \frac{(u_r)_{k,r}}{r} - \frac{(u_r)_k}{r^2} \right\} + (c_{44})_k (u_r)_{k,zz}
$$
$$
+ \{(c_{13})_k + (c_{44})_k\}(u_z)_{k,rz} + \{(e_{31})_k + (e_{15})_k\}(\phi)_{k,rz} = 0
$$

$$
\tag{3.254}
$$

$$
\{(c_{13})_k + (c_{44})_k\} \left\{ (u_r)_{k,rz} + \frac{(u_r)_{k,z}}{r} \right\} + (c_{44})_k \left\{ (u_z)_{k,rr} + \frac{(u_z)_{k,r}}{r} \right\}
$$
$$
+ (c_{33})_k(u_z)_{k,zz} + (e_{15})_k \left\{ (\phi)_{k,rr} + \frac{(\phi)_{k,r}}{r} \right\} + (e_{33})_k(\phi)_{k,zz} = 0
$$

$$
\{(e_{31})_k + (e_{15})_k\} \left\{ (u_r)_{k,rz} + \frac{(u_r)_{k,z}}{r} \right\} + (e_{15})_k \left\{ (u_z)_{k,rr} + \frac{(u_z)_{k,r}}{r} \right\}
$$
$$
+ (e_{33})_k(u_z)_{k,zz} - (\epsilon_{11})_k \left\{ (\phi)_{k,rr} + \frac{(\phi)_{k,r}}{r} \right\} - (\epsilon_{33})_k(\phi)_{k,zz} = 0
\tag{3.255}
$$

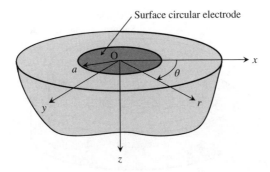

Figure 3.64 A piezoelectric half-space with a surface circular electrode

The mixed boundary conditions are given by

$$\begin{cases} (\phi)_1(r,0) = (\phi)_2(r,0) = V_0 & (0 \leq r < a) \\ (E_r)_1(r,0) = (E_r)_2(r,0) & (a \leq r < \infty) \\ (D_z)_1(r,0) = (D_z)_2(r,0) & (a \leq r < \infty) \end{cases} \tag{3.256}$$

$$\begin{aligned} (u_r)_1(r,0) &= (u_r)_2(r,0) \\ (u_z)_1(r,0) &= (u_z)_2(r,0) \\ (\sigma_{zr})_1(r,0) &= (\sigma_{zr})_2(r,0) \\ (\sigma_{zz})_1(r,0) &= (\sigma_{zz})_2(r,0) \end{aligned} \tag{3.257}$$

Hankel transforms are employed to reduce the electroelastic problem to the solution of a pair of dual integral equations. Then the closed form solutions of these dual integral equations are obtained [61]. When all material properties of piezoelectric ceramics 1 vanish, the solution reduces to that of a piezoelectric half-space with a surface circular electrode as shown in Fig. 3.64. Details can be found in Ref. [62].

3.8.2.2 Two Dissimilar Piezoelectric Disks with Circular Electrodes

Consider the piezoelectric disk 1 with radius b_1 and thickness h_1. The piezoelectric disk 1 is bonded to the upper surface of the piezoelectric disk 2 with radius b_2 and thickness h_2 as shown in Fig. 3.65. The origin of the coordinates (r,θ,z) coincides with the center of the interface considered as $z = 0, 0 \leq r \leq b_1$. The z axis is assumed to coincide with the poling direction. Three parallel circular electrodes of radius a lie in the planes $z = -h_1, 0, h_2$. Let the voltage applied to the internal electrode be denoted by V_0. The surface electrodes are grounded.

The electrical loading conditions are given by

$$\begin{cases} (\phi)_1(r,-h_1) = 0 & (0 \leq r < a) \\ (D_z)_1(r,-h_1) = 0 & (a \leq r \leq b_1) \end{cases} \tag{3.258}$$

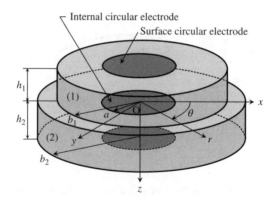

Figure 3.65   A piezoelectric disk composite with circular electrodes

$$\begin{cases} (\phi)_1(r,0) = V_0 & (0 \leq r < a) \\ (E_r)_1(r,0) = (E_r)_2(r,0) & (a \leq r \leq b_1) \\ (D_z)_1(r,0) = (D_z)_2(r,0) & (a \leq r \leq b_1) \\ (D_z)_2(r,0) = 0 & (b_1 < r \leq b_2) \end{cases} \tag{3.259}$$

$$\begin{cases} (\phi)_2(r,h_2) = 0 & (0 \leq r < a) \\ (D_z)_2(r,h_2) = 0 & (a \leq r \leq b_2) \end{cases} \tag{3.260}$$

The axisymmetric model is generated, and FEA is performed. Consider the case where the piezoelectric disk C-91 of radius $b_1 = 15$ mm and thickness $h_1 = 1$ mm is bonded to the upper surface of a C-91 or brass disk of radius $b_2 = 25$ mm and thickness $h_2 = 5$ mm. The electrode radius is fixed at $a = 10$ mm. Young's modulus E^{elas} and Poisson's ratio v^{elas} of brass are

$$E^{\text{elas}} = 10.06 \times 10^{10} \text{ N/m}^2$$
$$v^{\text{elas}} = 0.35 \tag{3.261}$$

Figure 3.66 displays the computed radial strain ε_{rr} against DC electric field $E_0 = V_0/h_1$ for bonded C-91 and brass disk at $r = 11.5$ mm, $z = 0$ mm. The FEA and test data for C-91 disk ($h_2 = 0$ mm) as shown in Fig. 3.67 [62] are also plotted. As the electric field E_0 is reduced from zero initially, the compressive strain increases. Local polarization switching can cause an unexpected decrease in compressive strain near the circular electrode tip and the sudden increase in tensile strain during switching. The stain near the circular electrode tip of the piezoelectric disk composite is smaller than that of the piezoelectric disk. Although the results are not shown here, the strain of C-91/C-91 disk composite is smaller than that of C-91/brass disk composite.

3.8.3   Fiber Composites

The development of piezoelectric fiber composites helps to overcome some of the limitations of conventional PZT ceramics, especially brittleness, lack of reliability, difficulty in

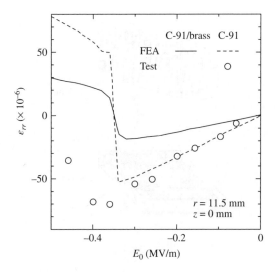

Figure 3.66 Strain versus DC electric field of the piezoelectric disk composite

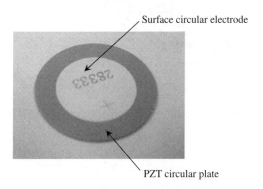

Figure 3.67 A piezoelectric circular plate with circular electrodes

their attachment to curved structures, and low flexibility. There are several types of piezoelectric composites, namely 1–3 composites by Smart Materials Corp., active fiber composites (AFCs) developed by MIT, and macrofiber composites (MFCs) constructed at NASA Langley Research Center, as shown in Fig. 3.68. Shindo and Togawa [63] discussed the multiple scattering of antiplane shear waves in piezoelectric fibrous composites with slip at the interfaces. The focus of the discussion presented here is on the electromechanical field concentrations in the piezoelectric fiber composites.

3.8.3.1 1–3 piezocomposites

Here, the focus is on the 1–3 piezocomposites. Figure 3.69 shows the geometry of the finite element study for the 1–3 piezocomposites. Let $x = x_1, y = x_2$, and $z = x_3$ denote the rectangular Cartesian coordinates. The z axis is assumed to coincide with the poling direction. We

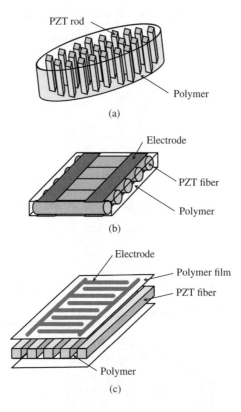

Figure 3.68   Sketch of (a) 1–3 composite, (b) AFC, and (c) MFC

consider two classes of 1–3 piezocomposites. $4N^2$ square PZT rods of side a and height h (Fig. 3.69(a)) or circular PZT rods of diameter d and height h (Fig. 3.69(b)) are embedded in a polymer matrix and arranged in a square array, where N is half of the number of rods aligned in the x- or y-direction. So, the height of the 1–3 piezocomposites is h, and the widths of the square- and circular-type are $2N(a + a_g)$ and $2N(d + d_g)$, respectively; where a_g and d_g are the gaps between the PZT rods. Repeating units of the composites with $4N^2$ square and circular PZT rods occupy the region $(0 \leq x \leq a + a_g, 0 \leq y \leq a + a_g, 0 \leq z \leq h/2)$ and $(0 \leq x \leq d + d_g, 0 \leq y \leq d + d_g, 0 \leq z \leq h/2)$, and the volume fractions V_f of the square- and circular-type 1–3 piezocomposites are $a^2/(a + a_g)^2$ and $\pi d^2/4(d + d_g)^2$, respectively.

Electrodes lie in the top and bottom ends of each PZT rod. We consider two electrical loading conditions in FEA. That is, the electric potential on the top electrode surfaces equals the constant DC voltage V_0 [64] and AC voltage $V_0 \exp(i\omega t)$ [65]. The bottom electrode surfaces are connected to the ground, so that the electric potential is zero. In the case under AC voltage, a calculation of the ratio of the AC voltage $V(t)$ of the 1–3 piezocomposite to an alternating current $I(t)$ leads to the impedance expression:

$$Z = \frac{V(t)}{I(t)} = |Z|e^{i\varphi} \tag{3.262}$$

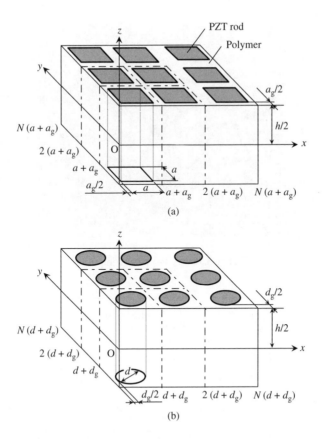

Figure 3.69 Finite element model of the 1–3 piezocomposite with (a) square and (b) circular PZT rods

where $|Z|$ is the impedance magnitude and φ is the phase difference between the voltage and current. The alternating current $I(t)$ for the unit of the composites is obtained as

$$I(t) = i\omega \int_A D_z(x, y, h/2, t)\mathrm{d}A \qquad (3.263)$$

where A is cross-sectional area of PZT rod.

In order to validate the predictions, we fabricate the 1–3 piezocomposite of $N = 2$ using 16 square PZT rods C-6 and epoxy resin P80B10, as shown in Fig. 3.70. Solid PZT ceramics are diced with a diamond saw and then back-filled with epoxy. Electrodes are attached on whole areas of the top and bottom surfaces of the 1–3 piezocomposite. The dimension of the composite is 32 mm × 32 mm × 27 mm. The piezoelectric rods have side $a = 7$ mm and height $h = 27$ mm, and the gap is $a_g = 1$ mm. So the volume fraction is about $V_f = 0.77$. Young's modulus E^m, Poisson's ratio v^m, and mass density ρ^m (the superscript m means epoxy matrix) of P80B10 are

$$E^m = 0.065 \times 10^{10} \text{ N/m}^2$$
$$v^m = 0.35 \qquad (3.264)$$
$$\rho^m = 1100 \text{ kg/m}^3$$

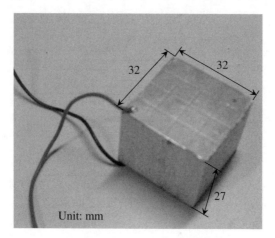

Figure 3.70   A 1–3 piezocomposite

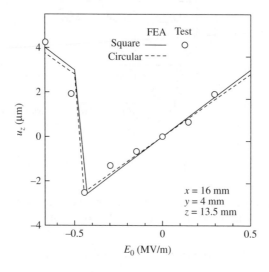

Figure 3.71   Displacement versus DC electric field of the 1–3 piezocomposites

A voltage is applied by using a power supply for voltages up to 125 kV/DC, and the magnitude of the deformation is measured using a digital microscope. Then, resonant frequencies are measured by an Agilent 4294A impedance/phase analyzer.

We present some results for $N = 2$ and $V_f = 0.77$. First, consider the 1–3 piezocomposites under DC electric field. Figure 3.71 shows the displacement u_z versus DC electric field $E_0 = -V_0/h$ for the square-type 1–3 piezocomposite at the measured point (at $x = 16$ mm, $y = 4$ mm, $z = 13.5$ mm). The line represents the FEA result while the open circle denotes the half of the measured value. Also shown is the FEA result for the circular-type composite. Nonlinear behavior is observed under negative electric field due to the polarization switching. An applied electric field of approximately -0.445 MV/m (below the coercive electric

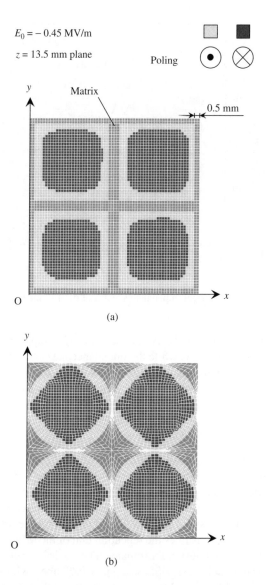

$E_0 = -0.45$ MV/m

$z = 13.5$ mm plane

Poling

(a)

(b)

Figure 3.72 Images of polarization switching for the 1–3 piezocomposites under $E_0 = -0.45$ MV/m

field $E_c = 0.45$ MV/m) causes polarization switching in the portion of the composite. The polarization switching can cause an unexpected decrease in the negative displacement. As the magnitude of positive electric field increases, the positive displacement increases. The trend is similar between analyses and experiments. Figure 3.72 shows the polarization switching zones at the top surface ($z = 13.5$ mm) of the composites under $E_0 = -E_c = -0.45$ MV/m. Next, we consider the 1–3 piezocomposites under AC electric field. The impedance-frequency spectra of the square-type 1–3 piezocomposite under AC electric field amplitude $E_0 = -V_0/h = 0.1$ MV/m are plotted in Fig. 3.73, in which both the FEA and measured data are shown. Also

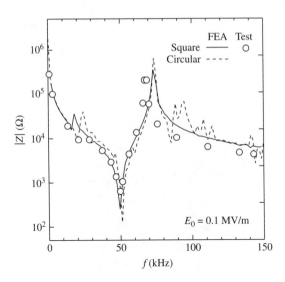

Figure 3.73   Electrical impedance spectra of the 1–3 piezocomposites

shown is the FEA result for the circular-type composite. The impedance minimum peak corresponds to the resonance frequency, while the impedance maximum corresponds to antiresonance frequency, and the fundamental resonances for square- and circular-type composites are about 49 and 51 Hz, respectively. It can be seen that the trend is sufficiently similar between analyses and tests.

3.8.3.2   MFCs

Here, we focus on the MFCs. Figure 3.74(a) shows the lay-up and geometry of the MFC. The MFC is comprised of very thin unpoled PZT fibers (width $2w_p$, thickness $2h_p$) that are aligned in an epoxy matrix and sandwiched between two sets of copper interdigitated electrode (IDE) (width $2w_e$, thickness h_e) patterns. The IDE patterns are printed on a Kapton film (thickness h_k). Figure 3.74(b) illustrates the repeating unit of the MFC. Let $x = x_1, y = x_2$, and $z = x_3$ denote the rectangular Cartesian coordinates. The z axis is assumed to coincide with the PZT fiber direction. The model consists of rectangular PZT fiber (width w_p, length L) embedded in an epoxy matrix and Kapton film (width W, length L). Electrodes 1, 3 of area $W \times w_e$ and electrode 2 of area $W \times 2w_e$ are also incorporated into the model. The total thickness of the model is $H = h_p + h_e + h_k$. We consider the following electrical boundary conditions: the electric potentials on the interface between PZT fiber and electrode 1 ($0 \leq x \leq w_p, y = h_p$, $0 \leq z \leq w_e$) and the interface between PZT fiber and electrode 3 ($0 \leq x \leq w_p, y = h_p, L - w_e \leq z \leq L$) equal the applied voltage, $\phi = V_0$, and the interface between PZT fiber and electrode 2 ($0 \leq x \leq w_p, y = h_p, L/2 - w_e \leq z \leq L/2 + w_e$) is connected to the ground, so that $\phi = 0$. The applied electric fields E_0 of the regions ($0 \leq x \leq W, 0 \leq y \leq H, 0 \leq z \leq L/2$) and ($0 \leq x \leq W$, $0 \leq y \leq H, L/2 \leq z \leq L$) can be approximately estimated to be $V_0/(L/2)$ and $-V_0/(L/2)$, respectively, and the denominator ($L/2$) means the IDE spacing. The mechanical boundary conditions include the traction-free condition on the top surface at $y = H$ and the symmetry

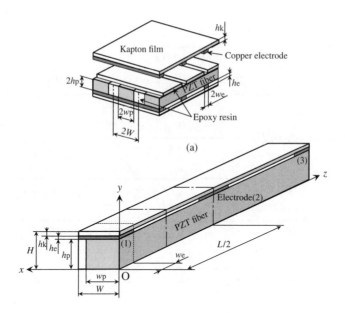

Figure 3.74   Schematic drawing of (a) lay-up and geometry and (b) repeating unit of MFC

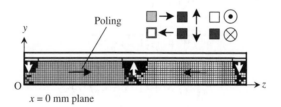

Figure 3.75   Image of poling for the partially poled PZT fiber

conditions on the $x = 0, y = 0, z = 0$ and $x = W, z = L$ faces. Under these conditions, the partially poled PZT fiber model is created, as shown in Fig. 3.75 [66].

The strain ε_{zz} of the unit without Kapton film and copper/epoxy layer ($0 \leq x \leq W, 0 \leq y \leq h_{\mathrm{p}}, 0 \leq z \leq L/2$) is related to the electric field E_0 by the expression [67]:

$$\varepsilon_{zz} = d_{33e}E_0 \tag{3.265}$$

Here, the effective longitudinal piezoelectric coefficient d_{33e} is obtained using the micro-electromechanics models [68]. Assuming that $90°$ switching would not occur, the effective coefficient for the unit for the fully poled PZT fiber is given by

$$d_{33e} = \begin{cases} e_{31e}s_{13e} + e_{32e}s_{23e} + e_{33e}s_{33e} & (E_0 > -E_c) \\ -(e_{31e}s_{13e} + e_{32e}s_{23e} + e_{33e}s_{33e}) & (E_0 \leq -E_c) \end{cases} \tag{3.266}$$

where

$$e_{31e} = \frac{c_{11}^m e_{31} w_p/W}{c_{11}^m w_p/W + c_{11}(1 - w_p/W)}$$

$$e_{32e} = \frac{c_{11}^m e_{31e} w_p/W + (c_{11} - c_{12} + c_{12}^m)e_{31e}(1 - w_p/W)}{c_{11}^m}$$

$$e_{33e} = e_{33} w_p/W + \frac{(c_{12}^m - c_{13})e_{31e}(1 - w_p/W)}{c_{11}^m}$$

$$s_{13e} = \frac{C_{12}C_{23} - C_{13}C_{22}}{(C_{11}C_{22} - C_{12}^2)C_{33} - C_{11}C_{23}^2 + 2C_{12}C_{13}C_{23} - C_{13}^2 C_{22}}$$

$$s_{23e} = \frac{C_{12}C_{13} - C_{11}C_{23}}{(C_{11}C_{22} - C_{12}^2)C_{33} - C_{11}C_{23}^2 + 2C_{12}C_{13}C_{23} - C_{13}^2 C_{22}}$$

$$s_{33e} = \frac{C_{11}C_{22} - C_{22}^2}{(C_{11}C_{22} - C_{12}^2)C_{33} - C_{11}C_{23}^2 + 2C_{12}C_{13}C_{23} - C_{13}^2 C_{22}}$$

$$(3.267)$$

and

$$c_{11}^m = \frac{E^m(1 - v^m)}{(1 + v^m)(1 - 2v^m)}$$

$$c_{12}^m = \frac{E^m v^m}{(1 + v^m)(1 - 2v^m)}$$

$$C_{11} = \frac{c_{11}c_{11}^m}{c_{11}^m w_p/W + c_{11}(1 - w_p/W)}$$

$$C_{12} = \frac{c_{12}c_{11}^m w_p/W + c_{11}c_{12}^m(1 - w_p/W)}{c_{11}^m w_p/W + c_{11}(1 - w_p/W)}$$

$$C_{13} = \frac{c_{13}c_{11}^m w_p/W + c_{11}c_{12}^m(1 - w_p/W)}{c_{11}^m w_p/W + c_{11}(1 - w_p/W)}$$

$$C_{22} = c_{11}w_p/W + c_{11}^m(1 - w_p/W) + \frac{C_{12}^2}{C_{11}}$$
$$- \left\{ \frac{c_{12}^2 w_p/W}{c_{11}} + \frac{c_{12}^{m2}(1 - w_p/W)}{c_{11}^m} \right\}$$

$$(3.268)$$

$$C_{23} = c_{13}w_p/W + c_{12}^m(1 - w_p/W) + \frac{C_{12}C_{13}}{C_{11}}$$
$$- \left\{ \frac{c_{12}c_{13}w_p/W}{c_{11}} + \frac{c_{12}^{m2}(1 - w_p/W)}{c_{11}^m} \right\}$$

$$C_{33} = c_{33}w_p/W + c_{11}^m(1 - w_p/W) + \frac{C_{13}^2}{C_{11}}$$
$$- \left\{ \frac{c_{13}^2 w_p/W}{c_{11}} + \frac{c_{12}^{m2}(1 - w_p/W)}{c_{11}^m} \right\}$$

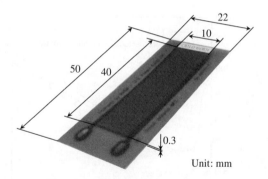

Figure 3.76 A M-4010-P1-type MFC

In order to validate the predictions, we measure the strain of the commercial MFC as shown in Fig. 3.76. Consider a M-4010-P1-type MFC (Smart Material Corp., Sarasota, FL) with active area of about 40 mm × 10 mm. The overall dimensions are about 50 mm × 22 mm × 0.3 mm. The MFC is comprised of very thin PZT 5A fibers that are unidirectionally aligned with the epoxy matrix. The material properties of PZT 5A and epoxy matrix are

$$c_{11} = 12.0 \times 10^{10} \text{ N/m}^2$$
$$c_{12} = 7.51 \times 10^{10} \text{ N/m}^2$$
$$c_{13} = 7.51 \times 10^{10} \text{ N/m}^2$$
$$c_{33} = 21.1 \times 10^{10} \text{ N/m}^2$$
$$c_{44} = 2.1 \times 10^{10} \text{ N/m}^2$$
$$e_{15} = 11.8 \text{ C/m}^2 \tag{3.269}$$
$$e_{31} = -3.05 \text{ C/m}^2$$
$$e_{33} = 14.7 \text{ C/m}^2$$
$$\epsilon_{11} = 81.1 \times 10^{-10} \text{ C/Vm}$$
$$\epsilon_{33} = 73.5 \times 10^{-10} \text{ C/Vm}$$
$$E_c = 1.5 \text{ MV/m}$$

$$E^m = 0.29 \times 10^{10} \text{ N/m}^2$$
$$v^m = 0.3 \tag{3.270}$$

Also, Young's modulus E^{elas} and Poisson's ratio v^{elas} of copper electrode and Kapton film are

$$E^{\text{elas}} = \begin{cases} 11.7 \times 10^{10} \text{ N/m}^2 \text{ (copper electrode)} \\ 0.28 \times 10^{10} \text{ N/m}^2 \text{ (Kapton film)} \end{cases}$$
$$v^{\text{elas}} = \begin{cases} 0.31 \text{ (copper electrode)} \\ 0.3 \text{ (Kapton film)} \end{cases} \tag{3.271}$$

Strain gages are bonded symmetrically at the center of the active area on both sides of the MFC. Voltage is applied using a power supply for voltages up to 1.25 kV/DC in order to generate the

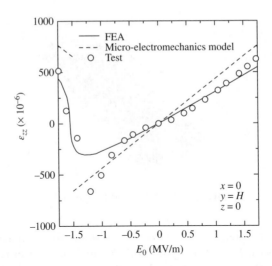

Figure 3.77   Strain versus DC electric field of the MFC

strain versus electric field curve. The MFC is then cut and observed by using a digital micro-scope to obtain realistic geometrical parameters. The main values are $w_p = 184.9$ μm, $W = 206.95$ μm, $h_p = 103.2$ μm, $h_e = 18$ μm, $h_k = 25$ μm, $w_e = 63.5$ μm, and $L/2 = 533.4$ μm. So the MFC has about 24 PZT fibers.

Figure 3.77 shows the strain ε_{zz} versus DC electric field E_0 for the MFC at $x = 0, y = H$, $z = 0$. The solid and dashed lines represent the strain ε_{zz} obtained from the FEA and micro-electromechanics model, respectively. The open circle denotes the test data. The FEA results show that as the electric field E_0 is increased from zero initially, the tensile strain increases linearly. On the other hand, the compressive strain increases as the electric field is lowered from zero; then, the compressive strain reaches the maximum at about $E_0 = -1.25$ MV/m as localized polarization switching occurs. The compressive strain then decreases, and the tensile strain rises. It is noted that there is a good agreement between the FEA and test. Although the strain obtained from the microelectromechanics model is larger than that from the FEA due to neglecting the Kapton film and copper/epoxy layer, the microelectromechanics model may serve to predict roughly the strain response of the MFCs.

3.8.4   MEMS Mirrors

PZT thick or thin film is very well suited for resonant microactuators in MEMS applications, and MEMS technology is for creating micromirrors. Here, we discuss the electromechanical field concentrations near electrodes in piezoelectric thick films for MEMS mirrors.

We consider a MEMS mirror configuration with four suspended unimorph PZT actuator beams and a square mirror plate supported by hinges, as shown in Fig. 3.78. Let $x = x_1, y = x_2$, and $z = x_3$ denote the rectangular Cartesian coordinates. The origin of the coordinate system is located at the top center of the mirror plate, and the z axis is assumed to coincide with the

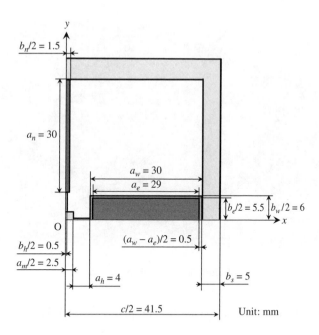

Figure 3.78 Schematic drawing of mirror device

poling direction. Two wide PZT thick films (length $a_w = 30$ mm, width $b_w = 12$ mm, thickness $h = 95$ μm) with partial top electrode (length $a_e = 29$ mm, width $b_e = 11$ mm, thickness $h_e = 5$ μm) and whole bottom electrode (length a_w, width b_w, thickness h_e) and two narrow PZT thick films (length $a_n = a_w$, width $b_n = 3$ mm, thickness h) with whole top and bottom electrodes (length a_n, width b_n, thickness h_e) are arranged on four elastic substrates (thickness $h_s = 120$ μm). The inactive width of the wide PZT thick films is $(a_w - a_e)/2 = (b_w - b_e)/2 = 0.5$ mm. Side length and thickness of the square mirror plate are $a_m = 5$ mm and h_s, and length, width, and thickness of the hinge are $a_h = 4$ mm, $b_h = 1$ mm, and h_s, respectively. Surrounding substrate has width $b_s = 5$ mm and thickness h_s, and total width of the structure is $c = 83$ mm.

We perform the FEA to predict the resonant frequency, mirror tilt angle, and electromechanical fields in the piezoelectric mirrors under electric fields. For simplicity, we ignore the effects of electrode layers and surrounding substrate. The edges at $x = y = \pm(c/2 - b_s)$ are clamped. In the case of wide PZT thick films, the films are partially poled. The electric potential on the top electrode surface $(a_m/2 + a_h + (a_w - a_e)/2 \leq x \leq a_m/2 + a_h + (a_w + a_e)/2, |y| \leq b_e/2, z = h)$ equals the constant DC voltage V_0 [69] and AC voltage $V_0 \exp(i\omega t)$ [70]. The bottom electrode surface $(a_m/2 + a_h \leq x \leq a_m/2 + a_h + a_w, |y| \leq b_w/2, z = 0)$ is connected to the ground, so that the electric potential is zero.

In order to validate the predictions, we fabricate the mirror device with four unimorph actuator beams using PZT C-91 thick films, Fe-42Ni substrate and Ag electrodes, as shown in Fig. 3.79. Fe-42Ni plate is cut as shown in Fig. 3.78. A mirror, four hinges, four substrates, and surrounding substrate are one-component (Fe-42Ni), and four bottom electrodes are coated onto Fe-42Ni substrates. C-91 thick films are then bonded, and four top electrodes are coated

Figure 3.79   A mirror device

on the surface of C-91 thick films. The material properties of C-91 thick film are

$$
\begin{aligned}
s_{11} &= 7.46 \times 10^{-12} \ \text{m}^2/\text{N} \\
s_{12} &= -2.1 \times 10^{-12} \ \text{m}^2/\text{N} \\
s_{13} &= -2.1 \times 10^{-12} \ \text{m}^2/\text{N} \\
s_{33} &= 10.4 \times 10^{-12} \ \text{m}^2/\text{N} \\
s_{44} &= 17.2 \times 10^{-12} \ \text{m}^2/\text{N} \\
d_{15} &= 943 \times 10^{-12} \ \text{m/V} \\
d_{31} &= -377 \times 10^{-12} \ \text{m/V} \\
d_{33} &= 645 \times 10^{-12} \ \text{m/V} \\
\epsilon_{11}^{T} &= 743 \times 10^{-10} \ \text{C/Vm} \\
\epsilon_{33}^{T} &= 906 \times 10^{-10} \ \text{C/Vm} \\
\rho &= 7800 \ \text{kg/m}^3 \\
E_{c} &= 0.7 \ \text{MV/m}
\end{aligned}
\tag{3.272}
$$

Some of the aforementioned properties are characterized by combined finite element simulations and electromechanical tests. A voltage is applied to a top electrode of the wide PZT
thick film, while a bottom electrode is grounded. The mirror tilt angle θ, defined in Fig. 3.80,
under DC electric field is measured using a displacement sensor. The sound pressure under
AC electric field is also measured by a microphone, and the resonant frequency is determined.
The distance between the microphone position and specimen top surface is about 10 mm.

Figure 3.81 shows the mirror tilt angle θ versus DC electric field $E_0 = -V_0/h$ for the mirror
device. The line represents the FEA result while the open circle denotes the measured value.
Nonlinear behavior is observed under negative electric fields due to the polarization switching,
and the trend is similar between analyses and tests. Table 3.3 lists the fundamental resonance

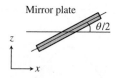

Figure 3.80 Mirror tilt angle

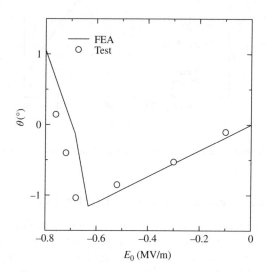

Figure 3.81 Tilt angle versus DC electric field of the mirror device

Table 3.3 Fundamental resonance frequency of the mirror device

f_r (Hz)	
FEA	Test
10.4	10.6

frequency f_r obtained from the FEA and test for the mirror device. The finite element model quantitatively predicts the resonant frequency, that is, experimental results show that our model is valid. Figure 3.82 shows the variation of normal stress σ_{xx} along the length direction at the top surface ($z = h = 95$ μm plane) and $y = 5$ mm of the wide PZT thick film under $|\theta| = 30°$ at $f = 10.4$ kHz. If the PZT thick films are operated near the resonance frequency, the stresses in the PZT thick films are very high near the electrode tip in the neighborhood of the clamped end. The results are useful in designing MEMS mirrors and in reducing the problems of fracture and so on that frequently occur during service.

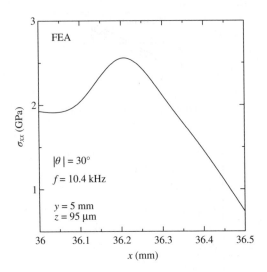

Figure 3.82   Normal stress distribution along the length direction for PZT thick film

3.9   Cryogenic and High-Temperature Electromechanical Responses

PZT ceramics in special electronic devices such as structural health monitoring systems of liquid rocket engines and microvalves for space applications are under extreme temperatures (e.g., cryogenic conditions). PZT ceramics are also used in active fuel injectors in severe environments [71, 72]. In the application of the multilayer stacked actuators to hydrogen fuel injectors, the actuators are operated at cryogenic and high temperatures. In this section, we first evaluate the temperature-dependent piezoelectric properties of PZT ceramics and discuss the cryogenic response of PZT stacked actuators under electric fields. Next, we propose a simple phenomenological model of depolarization and examine the electromechanical response of PZT stacked actuators at high temperatures.

3.9.1   Cryogenic Electromechanical Response

Consider the phase diagram of PZT ceramics as shown in Fig. 3.20. The MPB between the tetragonal and rhombohedral/monoclinic phases is the origin of the unusually high piezoelectric response of PZT ceramics, and this MPB is numerically predicted [73]. For simplicity, we ignore the orthorhombic phase and the octahedral tilt transition, which differentiates the high-temperature (HT) and low-temperature (LT) rhombohedral phases. An energy function for the solid solution between the two end-members $PbTiO_3$ and $PbZrO_3$ is given by Bell and Furman [74]

$$\Delta G_{\mathrm{PZT}} = X G_{\mathrm{PT}}(p_i) + (1 - X) G_{\mathrm{PZ}}(q_i) + G_C(p_i, q_i) \tag{3.273}$$

where G_{PT} and G_{PZ} are identical to a Landau–Devonshire potential up to the sixth order, G_C represents the coupling energy, p_i and q_i ($i = 1,2,3$) denote the contribution of the polarization from the $PbTiO_3$ and $PbZrO_3$, respectively, and

$$
\begin{aligned}
G_{PT}(p_i) &= 3.74 \times 10^5 (T - T_{PT})(p_1^2 + p_2^2 + p_3^2) - 7.9 \times 10^7 (p_1^4 + p_2^4 + p_3^4) \\
&\quad + 7.5 \times 10^8 (p_1^2 p_2^2 + p_2^2 p_3^2 + p_3^2 p_1^2) + 2.61 \times 10^8 (p_1^6 + p_2^6 + p_3^6) \\
&\quad + 6.3 \times 10^8 \{ p_1^4 (p_2^2 + p_3^2) + p_2^4 (p_3^2 + p_1^2) + p_3^4 (p_1^2 + p_2^2) \} \\
&\quad - 3.66 \times 10^9 p_1^2 p_2^2 p_3^2 \\
G_{PZ}(q_i) &= 2.82 \times 10^5 (T - T_{PZ})(q_1^2 + q_2^2 + q_3^2) + 5.12 \times 10^8 (q_1^4 + q_2^4 + q_3^4) \\
&\quad - 6.5 \times 10^8 (q_1^2 q_2^2 + q_2^2 q_3^2 + q_3^2 q_1^2) + 5.93 \times 10^8 (q_1^6 + q_2^6 + q_3^6) \\
&\quad + 2 \times 10^9 \{ q_1^4 (q_2^2 + q_3^2) + q_2^4 (q_3^2 + q_1^2) + q_3^4 (q_1^2 + q_2^2) \} \\
&\quad - 9.5 \times 10^9 q_1^2 q_2^2 q_3^2 \\
G_C(p_i, q_i) &= \gamma_{200}(p_1^2 q_1^2 + p_2^2 q_2^2 + p_3^2 q_3^2) \\
&\quad + \gamma_{220} \{ p_1^2 (q_2^2 + q_3^2) + p_2^2 (q_3^2 + q_1^2) + p_3^2 (q_1^2 + q_2^2) \}
\end{aligned}
\tag{3.274}
$$

In Eqs. (3.274), $T_{PT} = 766$ K and $T_{PZ} = 503$ K are the Curie temperatures of $PbTiO_3$ and $PbZrO_3$, respectively, and γ_{200} and γ_{220} are unknown coefficients. The equilibrium thermodynamic state can be determined via minimization of ΔG_{PZT} with respect to p_i and q_i. For each temperature T and mole fraction X, we obtain the local minima in ΔG_{PZT} systematically for the following phases:

Cubic

$$
p_1 = p_2 = p_3 = 0, \quad q_1 = q_2 = q_3 = 0
\tag{3.275}
$$

Tetragonal

$$
p_1 = p_2 = 0, p_3 \neq 0, \quad q_1 = q_2 = 0, q_3 \neq 0
\tag{3.276}
$$

Rhombohedral

$$
p_1 = p_2 = p_3 \neq 0, \quad q_1 = q_2 = q_3 \neq 0
\tag{3.277}
$$

Monoclinic

$$
p_1 = p_2 \neq 0, p_3 \neq 0, \quad q_1 = q_2 \neq 0, q_3 \neq 0
\tag{3.278}
$$

We then compare the energies of the minima to define the stable state. In the simulation, we assume $\gamma_{200} = 6.0 \times 10^8$ and $\gamma_{220} = 1.2 \times 10^8$. Figure 3.83 shows the analytical PZT phase diagrams. We also plot the experimental phase diagrams of PZT. The solid triangle, open square, and open circle are the results from Jaffe et al. [18], Pandey et al. [75], and Noheda et al. [19], respectively. The analytical result reasonably agrees with the experimental data and shows that the MPB is located at about 44 mol%Ti at RT and there is an apparent shift of about 13 mol% in the MPB location on cooling to 0 K. Since the piezoelectric coefficient d_{33} of PZT shows a significant decrease of about 85 % due to a shift of 12 mol% of Zr at RT [76], we propose the following equation for temperature-dependent piezoelectric coefficient:

$$
\bar{d}_{ijk} = \begin{cases} (1.8 \times 10^{-4} T + 0.95) d_{ijk} & 192 \leq T \\ (5.5 \times 10^{-6} T^2 + 2.8 \times 10^{-3} T + 0.10) d_{ijk} & 0 < T \leq 192 \end{cases}
\tag{3.279}
$$

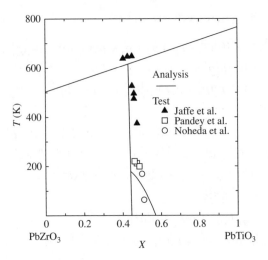

Figure 3.83 Analytical and experimental phase diagrams of PZT

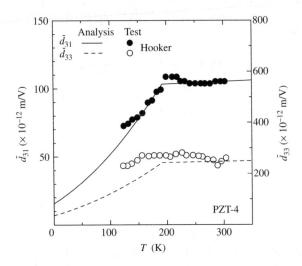

Figure 3.84 Piezoelectric coefficients of PZT-4 from room to cryogenic temperatures

Figure 3.84 shows our calculated piezoelectric coefficients $\bar{d}_{31} = \bar{d}_{311} = \bar{d}_{322}$ and $\bar{d}_{33} = \bar{d}_{333}$ versus temperature T for PZT-4, together with the experimental data of Hooker [77]. The piezoelectric coefficients d_{31} and d_{33} for PZT-4 at RT are 105×10^{-12} and 248×10^{-12} m/V, respectively. The proposed temperature-dependent piezoelectric coefficients are reasonably good.

In order to discuss the cryogenic response of PZT stacked actuators under electric fields, we consider the unit cell as shown in Fig. 3.53 again. Electrical loading and mechanical boundary conditions have already been mentioned in Section 3.8.1(b). In the analysis, we need the temperature-dependent coercive electric field $\bar{E}_c$. First-principles free-energy calculations for ferroelectric perovskites show that the domain wall energy increases linearly about 50% as T decreases from RT to 260 K [78]. Since higher domain wall energy leads to higher coercive electric field, we assume the following temperature-dependent coercive electric field:

$$\overline{E}_c = (4.84 - 0.0129T)E_c \qquad (3.280)$$

To validate the predictions, we prepare the piezoelectric stacked actuator using 300 poled PZT N-10 layers (NEC/Tokin Co. Ltd., Japan) of length $L = 5.2$ mm, width $W = 5.2$ mm, and thickness $h = 0.1$ mm. The electrode length is $a = L = 5.2$ mm (fully electrode), and the piezoelectric stacked actuator is coated with epoxy layer of thickness $h_e = 0.5$ mm. The total dimensions of the actuator are side widths of 6.2 mm and length of 40.5 mm as shown in Fig. 3.85. Let $x = x_1, y = x_2$, and $z = x_3$ denote the rectangular Cartesian coordinates. The origin coincides with the center of the actuator, and the z axis is assumed to coincide with the

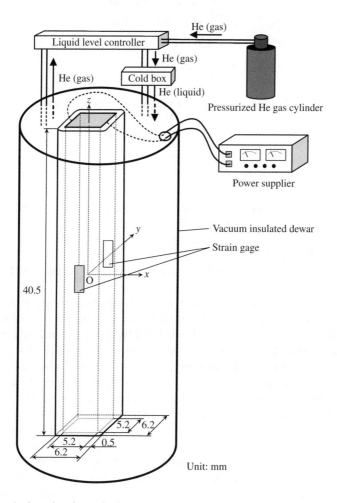

Figure 3.85   A piezoelectric stacked actuator and experimental setup at cryogenic temperatures

poling direction. Two strain gages are attached at the center of the $y = \pm 3.1$ mm planes of the actuator. The material properties of N-10 are

$$s_{11} = 14.8 \times 10^{-12} \text{ m}^2/\text{N}$$

$$s_{12} = -5.1 \times 10^{-12} \text{ m}^2/\text{N}$$

$$s_{13} = -5.8 \times 10^{-12} \text{ m}^2/\text{N}$$

$$s_{33} = 18.1 \times 10^{-12} \text{ m}^2/\text{N}$$

$$s_{44} = 44.9 \times 10^{-12} \text{ m}^2/\text{N}$$

$$d_{15} = 930 \times 10^{-12} \text{ m/V}$$

$$d_{31} = -287 \times 10^{-12} \text{ m/V} \tag{3.281}$$

$$d_{33} = 635 \times 10^{-12} \text{ m/V}$$

$$\epsilon_{11}^T = 443 \times 10^{-10} \text{ C/Vm}$$

$$\epsilon_{33}^T = 481 \times 10^{-10} \text{ C/Vm}$$

$$\rho = 8000 \text{ kg/m}^3$$

$$E_c = 0.36 \text{ MV/m}$$

Young's modulus E^{elas}, Poisson's ratio v^{elas}, and mass density ρ^{elas} of epoxy coating layer are

$$E^{\text{elas}} = 0.335 \times 10^{10} \text{ N/m}^2$$

$$v^{\text{elas}} = 0.214 \tag{3.282}$$

$$\rho^{\text{elas}} = 1100 \text{ kg/m}^3$$

The piezoelectric stacked actuator is placed on the rigid body in a vacuum insulated dewar, DC voltage [73] or AC voltage at a frequency of 400 Hz [79] is applied using a power supplier, and the magnitude of strain is measured. To control the temperature of the actuator, we use an automated helium (He) refill system (TRG-300, Taiyo Toyo Sanso Co. Ltd., Japan). Low-temperature environments are achieved by immersing the actuator in He gas, and temperature is measured with a thermocouple.

Figure 3.86 shows the predicted normal strain ε_{zz} versus temperature T at $x = 0$ mm, $y = 3.1$ mm, and $z = 0$ mm for the piezoelectric stacked actuator with $a = L = 5.2$ mm (fully electrodes) under DC electric field $E_0 = V_0/h = 0.1$ MV/m. The measured data are also shown. The electric field induced stain decreases with decreasing temperature due to a shift in the MPB, and we observe a reasonable agreement between our predictions and measurements. Figure 3.87 shows the predicted and measured normal strain ε_{zz} versus E_0 at $x = 0$ mm, $y = 3.1$ mm, and $z = 0$ mm for the piezoelectric stacked actuator with $a = 5.2$ mm (fully electrodes) at $T = 196$ and 298 K. As the electric field is cycled, the butterfly loop is repeated, and there is a reasonable agreement between the predictions and measurements. Figure 3.88 shows the normal strain ε_{zz} versus AC electric field amplitude $E_0 = V_0/h$ at $x = 0$ mm, $y = 3.1$ mm, and $z = 0$ mm for the piezoelectric stacked actuator with $a = 5.2$ mm (fully electrodes) at frequency $f = 400$ Hz and temperature $T = 20$ K (liquid hydrogen temperature). The predicted result for the piezoelectric stacked actuator with $a = 5.0$ mm (partially electrodes) is also shown. It is interesting to note that a small difference is observed in the electric field induced strains for

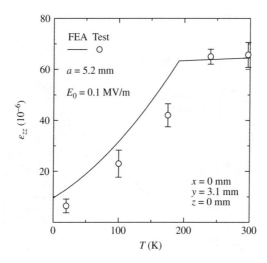

Figure 3.86 Normal strain versus temperature of the piezoelectric stacked actuator (cryogenic temperature range)

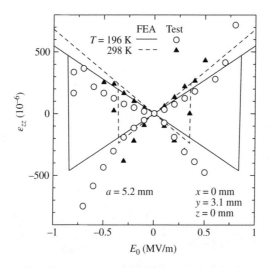

Figure 3.87 Normal strain versus DC electric field of the piezoelectric stacked actuator

$a = 5.0$ and 5.2 mm. Agreement between analyses with domain wall motion effect and experiments is fair. Figure 3.89 shows the normal stress distribution σ_{zz} and electric field distribution E_z as a function of x at $y = 0$ mm and $z = 0.025$ mm for the piezoelectric stacked actuators with $a = 5.0$ mm (partially electrodes) and $a = 5.2$ mm (fully electrodes) under electric field amplitude $E_0 = 1.65$ MV/m at $f = 400$ Hz and 20 K. The coercive electric field at 20 K is about 1.65 MV/m. For the piezoelectric stacked actuator with partially electrodes, high normal stress, and electric field occur near the electrode tip ($x = 2.4$ mm).

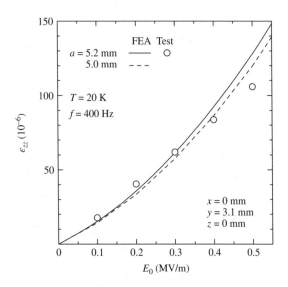

Figure 3.88 Normal strain versus AC electric field of the piezoelectric stacked actuator at liquid hydrogen temperature

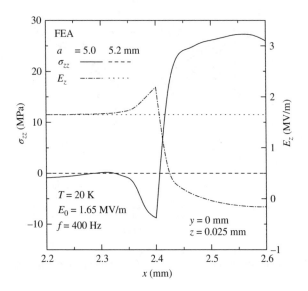

Figure 3.89 Normal strain and electric field distributions along the x-direction for the piezoelectric stacked actuator at liquid hydrogen temperature

The results of the stress and electric field distributions show that if the piezoelectric stacked actuator with partially electrodes is applied to hydrogen fuel injector components, the failure or dielectric breakdown is expected near the electrode tip. The fuel injector designers need to be aware of electromechanical behavior at cryogenic temperatures.

3.9.2 High-Temperature Electromechanical Response

When PZT ceramics are heated, the internal kinetic energy increases. At a certain temperature, called the Curie temperature T_c, the crystal structure changes to cubic as shown in Fig. 3.20, the dipole alignment is lost, and all the piezoelectric activity disappears [80]. Generally, the operating temperatures of PZT ceramics have been limited to $T_c/2$.

Consider the unit cell of piezoelectric stacked actuator as shown in Fig. 3.53 again. We assume that when the temperature T increases from RT, the directions of the polarizations begin to rotate at $T_c/2$ and the rotation angle α becomes $\pi/2$ at T_c. Figure 3.90 shows the model of the depolarization. The angle $\alpha = 0$ from RT to $T_c/2$ corresponds to the direction of first poling axis (z-direction). Also, assume that the piezoelectric properties decrease linearly as T increases from $T_c/2$ to T_c, and the elastic and dielectric properties remain unchanged after the depolarization occurs. The temperature-dependent direct piezoelectric coefficient $\bar{d}_{ikl}$ is in place of Eq. (3.136)

$$\bar{d}_{ikl} = \{d_{33}n_{Pi}n_{Pk}n_{Pl} + d_{31}(n_{Pi}\delta_{kl} - n_{Pi}n_{Pk}n_{Pl})$$

$$+ \frac{1}{2}d_{15}(\delta_{ik}n_{Pl} - 2n_{Pi}n_{Pk}n_{Pl} + \delta_{il}n_{Pk})\} \qquad (3.283)$$

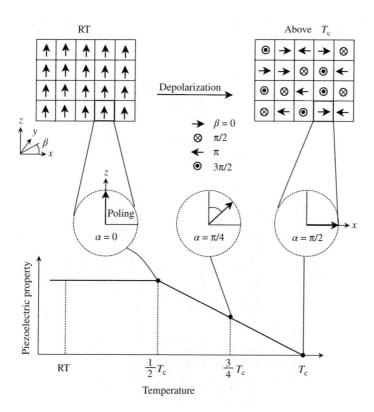

Figure 3.90 Image of depolarization

Here, we define the rotation angle of poling in each element to be temperature-dependent function $\alpha(T)$, and the unit vector in Eq. (3.283) is given by

$$
\begin{aligned}
n_{P1} &= \sin \alpha(T) \cos \beta \\
n_{P2} &= \sin \alpha(T) \sin \beta \\
n_{P3} &= \cos \alpha(T)
\end{aligned}
\tag{3.284}
$$

A random number generator is used to create only the angle β, with value of 0, $\pi/2$, π, or $3\pi/2$ as shown in Fig. 3.90.

To validate the predictions, we use the piezoelectric stacked actuator mentioned in Section 3.9.1. Figure 3.91 shows a specimen and experimental setup. Let $x = x_1, y = x_2$, and $z = x_3$ denote the rectangular Cartesian coordinates. The origin coincides with the center of the actuator, and the z axis is assumed to coincide with the poling direction. Two strain gages are attached at the center of the $y = \pm 3.1$ mm planes of the actuator. The Curie temperature of N-10 is $T_c = 145°C$. So, the temperature-dependent rotation angle of the poling direction for N-10 is assumed to be [81]

$$
\alpha(T) = \begin{cases}
0 & (25\,°C \leq T \leq 73\,°C) \\
\dfrac{\pi}{2}(1.38 \times 10^{-2} T - 1) & (73\,°C < T \leq 145\,°C)
\end{cases}
\tag{3.285}
$$

The actuator is placed on the rigid body in a constant temperature oven (DVS402, Yamato Scientific Co. Ltd., Japan), and DC voltage is applied by using a power supplier. Two strain

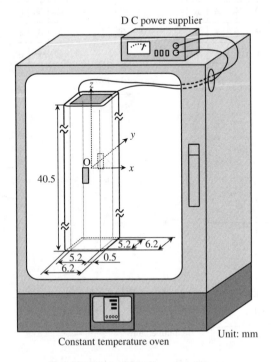

Figure 3.91   A piezoelectric stacked actuator and experimental setup at high temperatures

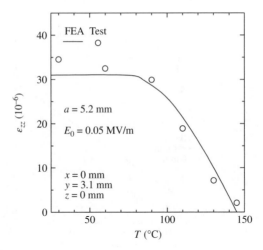

Figure 3.92   Normal strain versus temperature of the piezoelectric stacked actuator (high temperature range)

gages are attached at the center of the $y = \pm3.1$ mm planes, and the magnitude of strain is measured.

Figure 3.92 shows the predicted normal strain ε_{zz} versus temperature T at $x = 0$ mm, $y = 3.1$ mm, and $z = 0$ mm for the piezoelectric stacked actuator with $a = L = 5.2$ mm (fully electrodes) under DC electric field $E_0 = V_0/h = 0.05$ MV/m. The measured data are also shown. The electric field-induced strain decreases due to the depolarization as the temperature increases from about 75°C. There is a good agreement between the predictions and measurements. Figure 3.93 shows the predicted and measured normal strain ε_{zz} versus E_0 at $x = 0$ mm, $y = 3.1$ mm, and $z = 0$ mm for the piezoelectric stacked actuator with $a = 5.2$ mm (fully electrodes) at $T = 100$ and 125°C. At $T = 125$°C, the predicted result for the piezoelectric stacked actuator with $a = 5.0$ mm (partially electrodes) is also shown. The results show that the slope of the strain versus the electric field curve decreases with increasing temperature, and small difference is observed in the electric field induced strains for $a = 5.0$ and 5.2 mm. Although the results are not shown here, high normal stress and electric field occur near the electrode tip for the piezoelectric stacked actuator with partially electrodes, similarly to the case at cryogenic temperatures. It is noticed that our model may be useful in designing high-performance fuel injectors for high-temperature environments.

3.10   Electric Fracture and Fatigue

Electromechanical field concentrations at defects or inhomogeneities in piezoelectric material systems and structures can contribute to critical crack growth and subsequent mechanical failure or dielectric breakdown. The concentrations also tend to develop near the electrode tip as mentioned in Section 3.8, resulting in localized electromechanical failure. Since most of the piezoelectric materials used in practice are ceramics such as PZT, they are susceptible to brittle fracture that can lead to catastrophic failure. Moreover, PZT ceramics experience fatigue cracking and degradation when subjected to cyclic electromechanical loading. Therefore, to

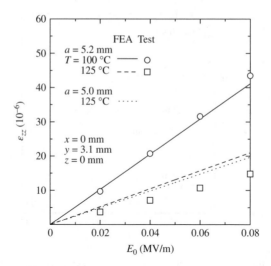

Figure 3.93 Normal strain versus DC electric field of the piezoelectric stacked actuator at high temperatures

prevent failure during service and to secure the structural integrity of piezoelectric devices, understanding of fracture and fatigue of piezoelectric material systems and structures is of great importance. In this section, we overview the theoretical and experimental observations in the fracture and fatigue behavior of piezoelectric materials under DC and AC electric fields.

3.10.1   Fracture Mechanics Parameters

The stress intensity factor or energy release rate approach of linear elastic fracture mechanics has proved to be successful in predicting the unstable fracture of brittle materials. In the case of piezoelectric materials, however, we need to generalize the fracture mechanics concepts so that the electromechanical coupling phenomenon can be properly taken into account. An extensive body of theoretical and experimental data on fracture behavior for a wide variety of piezoelectric ceramics over the last two decades was summarized in books [82, 83] and review papers [84, 85]. Here, we present the fracture mechanics parameters of piezoelectric materials.

3.10.1.1   Crack Face Boundary Conditions

When studying piezoelectric crack problems, we need to use the electrical boundary conditions across the crack surface. With reference to the theoretical studies of the piezoelectric crack problems, the researchers differ in opinions on the electrical boundary conditions at the crack surfaces. As the permittivity of the air or the medium between the crack surfaces is very small as compared to that of the piezoelectric material, most of the works have assumed the permittivity in the medium inside the crack to be zero (the so-called impermeable condition), that is,

$$D_n^+ = D_n^- = 0 \qquad (3.286)$$

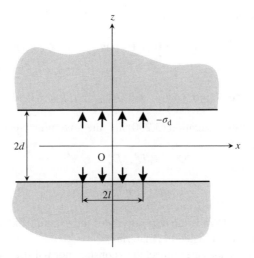

Figure 3.94   Two piezoelectric half-planes

where D_n is the normal component of the electric displacement vector at the crack surfaces, and superscripts $+$ and $-$ denote the upper and lower crack surfaces, respectively. Across the crack surfaces, having no charge density, we need to satisfy the relevant boundary conditions [86], which are the continuity of the tangential component of the electric field intensity vector E_t at the crack surfaces

$$E_t^+ = E_t^-$$
(3.287)

and the continuity of the normal component of the electric displacement vector D_n

$$D_n^+ = D_n^-$$
(3.288)

In order to examine the effect of the electrical surface conditions on the electromechanical fields, we consider two piezoelectric half-planes as shown in Fig. 3.94. Let $x = x_1, y = x_2$, and $z = x_3$ denote the rectangular Cartesian coordinates. We assume plane strain perpendicular to the y axis. The z axis is assumed to coincide with the poling direction. Two piezoelectric half-planes are placed at a distance of $2d$ apart and occupy the region $(-\infty < x < \infty, |z| \geq d)$. We suppose that the surfaces of the half-planes are subjected to a force distributed over the segments of length $2l$ with constant normal stress σ_d.

The constitutive equations are in place of Equations (3.239) and (3.240)

$$\begin{Bmatrix} \sigma_{xx} \\ \sigma_{zz} \\ \sigma_{zx} \end{Bmatrix} = \begin{bmatrix} c_{11} & c_{13} & 0 \\ c_{13} & c_{33} & 0 \\ 0 & 0 & c_{44} \end{bmatrix} \begin{Bmatrix} \varepsilon_{xx} \\ \varepsilon_{zz} \\ 2\varepsilon_{zx} \end{Bmatrix} - \begin{bmatrix} 0 & e_{31} \\ 0 & e_{33} \\ e_{15} & 0 \end{bmatrix} \begin{Bmatrix} E_x \\ E_z \end{Bmatrix}$$
(3.289)

$$\begin{Bmatrix} D_x \\ D_z \end{Bmatrix} = \begin{bmatrix} 0 & 0 & e_{15} \\ e_{31} & e_{33} & 0 \end{bmatrix} \begin{Bmatrix} \varepsilon_{xx} \\ \varepsilon_{zz} \\ 2\varepsilon_{zx} \end{Bmatrix} + \begin{bmatrix} \epsilon_{11} & 0 \\ 0 & \epsilon_{33} \end{bmatrix} \begin{Bmatrix} E_x \\ E_z \end{Bmatrix}$$
(3.290)

Also, the governing equations are in place of Eqs. (3.243) and (3.244)

$$c_{11}u_{x,xx} + c_{44}u_{x,zz} + (c_{13} + c_{44})u_{z,xz} + (e_{15} + e_{31})\phi_{,xz} = 0$$
$$(c_{13} + c_{44})u_{x,xz} + c_{44}u_{z,xx} + c_{33}u_{z,zz} + e_{15}\phi_{,xx} + e_{33}\phi_{,zz} = 0$$

(3.291)

$$(e_{15} + e_{31})u_{x,xz} + e_{15}u_{z,xx} + e_{33}u_{z,zz} - \epsilon_{11}\phi_{,xx} - \epsilon_{33}\phi_{,zz} = 0$$

(3.292)

In a vacuum, the constitutive equation (3.290) and the governing equation (3.292) become

$$D_x^e = \epsilon_0 E_x^e, \quad D_z^e = \epsilon_0 E_z^e$$

(3.293)

$$\phi_{,xx}^e + \phi_{,zz}^e = 0$$

(3.294)

where $\epsilon_0 = 8.85 \times 10^{-12}$ C/Vm is the permittivity of the vacuum and the superscript e stands for the electric field quantity outside the material.

The piezoelectric half-planes are subjected to constant normal stress σ_d at $z = \pm d$, $-l \leq x \leq l$. Hence, the boundary conditions are given by

$$\sigma_{zx}(x, \pm d) = 0 \quad (0 < |x| < \infty)$$

(3.295)

$$\sigma_{zz}(x, \pm d) = \begin{cases} -\sigma_d & (0 \leq |x| \leq l) \\ 0 & (l < |x| < \infty) \end{cases}$$

(3.296)

We consider next two possible cases of electrical boundary conditions on $z = \pm d$.
Impermeability:

$$D_z(x, \pm d) = 0 \quad (0 \leq |x| < \infty)$$

(3.297)

Electrical continuity:

$$E_x(x, \pm d) = E_x^e(x, \pm d) \quad (0 \leq |x| < \infty)$$

(3.298)

$$D_z(x, \pm d) = D_z^e(x, \pm d) \quad (0 \leq |x| < \infty)$$

(3.299)

Solutions for the electromechanical fields are obtained by using Fourier transforms. To simulate a crack, the distance d is taken to be very small. This is equivalent to setting $d \rightarrow 0$. In the limit $\epsilon_0 \rightarrow 0$, the solutions for the electrical continuity of Eqs. (3.298) and (3.299) become those for the impermeability of Eq. (3.297). However, the solutions for the impermeability do not tend to those for the electrical continuity as $d \rightarrow 0$ [87]. In order to observe this trend more clearly, we show some numerical examples for PZT C-91. $\sigma_d = 10$ MN/m^2 and $l = 10$ mm are chosen. Figure 3.95 shows the normal strain ε_{zz} versus distance d at $x = l = 10$ mm and $z = d$. Figure 3.96 shows the similar results for the electric field E_z. The results illustrate how rapidly the applicability of the impermeable assumption deteriorates. Therefore, the electric boundary condition given by Eq. (3.297) is not appropriate for a slit crack in piezoelectric materials.

Similarly, the problem of an elliptic hole in an infinite piezoelectric material was examined by Sosa and Khutoryansky [88]. An infinite piezoelectric material containing an elliptic hole with major and minor axes $2a$ and $2b$ was considered as shown in Fig. 3.97, and the cavity was assumed to be filled with a homogeneous gas of ϵ_0. The major axis was normal to the

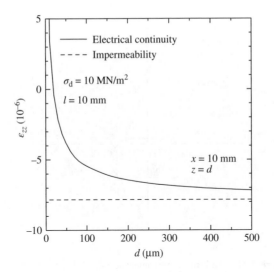

Figure 3.95   Normal strain versus distance of the two piezoelectric half-planes

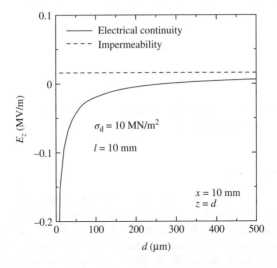

Figure 3.96   Electric field versus distance of the two piezoelectric half-planes

poling direction. Expressions for the electromechanical fields inside and outside the cavity were derived in closed form, and the following were found: (1) If $b/a \gg 10^{-4}$, the models for the impermeability and electrical continuity provide virtually the same results. (2) If ϵ_0 is retained, in the limit of $b/a \rightarrow 0$, the dramatic differences for the fields are provided by both models. It was concluded that the condition of electric impermeability at the boundary of the cavity may result in erroneous conclusions for the case of very slender ellipses or sharp cracks. The fracture mechanics parameters for the aforementioned two models will be presented later.

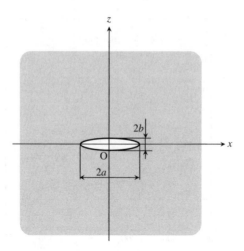

Figure 3.97   An infinite piezoelectric material with an elliptic hole

The electric permeability of the air in a crack gap was considered by Hao and Shen [89]. The conditions are

$$D_n^+ = D_n^- \tag{3.300}$$

$$D_n^+(u_n^+ - u_n^-) = -\epsilon_0(\phi^+ - \phi^-) \tag{3.301}$$

where u_n is the normal component of the displacement vector at the crack surfaces. This open crack model will also be discussed later.

3.10.1.2   Plane Strain Crack

We consider an infinite piezoelectric material containing a crack of length $2a$ as shown in Fig. 3.98. Let $x = x_1, y = x_2$, and $z = x_3$ denote the rectangular Cartesian coordinates. We assume plane strain perpendicular to the y axis. The z axis is assumed to coincide with the poling direction. The crack is assumed with faces normal (Case 1) or parallel (Case 2) to the polarization axis. A uniform tensile strain ε_∞ is applied in the z-direction (Case 1) or x-direction (Case 2) at infinity. Usually, the piezoelectric material is driven by a poling direction electric field and so we consider the piezoelectric material subjected to a uniform electric field E_0 parallel to the poling direction. The electric field normal to the poling is not important to investigate the plane strain piezoelectric crack behavior because the constitutive relations of Eq. (3.289) do not exhibit coupling between the electric field E_x and the normal stress σ_{xx} or σ_{zz}. The electric potentials at $z = \pm$ const. have equal magnitude but opposite sign. Only the first quadrant with appropriate boundary conditions needs to be analyzed due to symmetry.

We first consider Case 1. The boundary conditions at $z = 0$ can be obtained as

$$\sigma_{zx}(x,0) = 0 \quad (0 \le x < \infty) \tag{3.302}$$

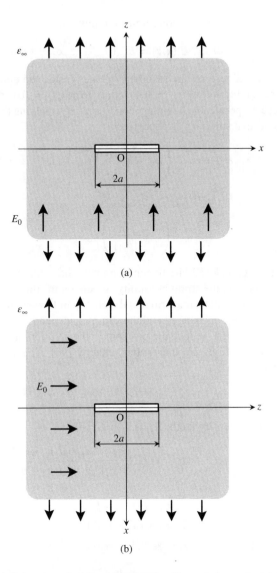

Figure 3.98   An infinite piezoelectric material with a plane strain crack: (a) Case 1, (b) Case 2

$$\sigma_{zz}(x,0) = 0 \quad (0 \le x < a)$$
$$u_z(x,0) = 0 \quad (a \le x < \infty)$$

$$(3.303)$$

$$E_x(x,0) = E_x^e(x,0) \quad (0 \le x < a)$$
$$\phi(x,0) = 0 \quad (a \le x < \infty)$$

$$(3.304)$$

$$D_z(x,0) = D_z^e(x,0) \, (0 \le x < a)$$

$$(3.305)$$

The mechanical and electrical loading conditions at infinity are

$$\varepsilon_{zz}(x,z) = \varepsilon_\infty, \quad E_z(x,z) = E_0 \quad (0 \le x < \infty, z \to \infty) \tag{3.306}$$

The electric potential is all zero on the symmetry planes inside the crack and ahead of the crack, so the boundary conditions of Eqs. (3.304) reduce to $\phi(x,0) = 0 (0 \le x < \infty)$. Equations (3.304) and (3.305) are the permeable boundary conditions. By applying the loading conditions (3.306), the far-field normal stress σ_∞ is expressed as

$$\sigma_\infty = \sigma_0 - \left\{ e_{33} - \left(\frac{c_{13}}{c_{11}} \right) e_{31} \right\} E_0 \tag{3.307}$$

where

$$\sigma_0 = \left(c_{33} - \frac{c_{13}^2}{c_{11}} \right) \varepsilon_\infty \tag{3.308}$$

Note that σ_0 is the stress for a closed-circuit condition with the potential forced to remain zero ($E_0 = 0$) and depends only on the strain at infinity. When a uniform strain ε_∞ is applied and fixed at infinity, the stress σ_0 will be uniform. On the other hand, when the stress σ_∞ is applied and fixed at infinity, σ_∞ is left unchanged and the strain ε_∞ depends on E_0.

Fourier transforms are used to reduce the problem to the solution of a pair of dual integral equations. The integral equations are then solved exactly [90]. The mode I stress intensity factor K_I is obtained as

$$K_I = \lim_{x \to a^+} \{2\pi(x-a)\}^{1/2}\sigma_{zz}(x,0) = \sigma_\infty(\pi a)^{1/2} \tag{3.309}$$

The electric displacement intensity factor K_D is also given by

$$K_D = \lim_{x \to a^+} \{2\pi(x-a)\}^{1/2}D_z(x,0) = \frac{d_1 h_1 + d_2 h_2 + d_3 h_3}{d_1 g_1 + d_2 g_2 + d_3 g_3} K_I \tag{3.310}$$

where

$$d_1 = \gamma_1(b_2 f_3 - b_3 f_2)$$
$$d_2 = \gamma_2(b_3 f_1 - b_1 f_3) \tag{3.311}$$
$$d_3 = \gamma_3(b_1 f_2 - b_2 f_1)$$

$$\begin{aligned} g_j &= c_{13}a_j - c_{33} + e_{33}b_j \\ h_j &= e_{31}a_j - e_{33} - \epsilon_{33}b_j \end{aligned} \quad (j = 1,2,3) \tag{3.312}$$

and

$$f_j = c_{44}(a_j\gamma_j^2 + 1) - e_{15}b_j$$

$$a_j = \frac{(c_{33}\gamma_j^2 - c_{44})(e_{31} + e_{15}) - (c_{13} + c_{44})(e_{33}\gamma_j^2 - e_{15})}{(c_{44}\gamma_j^2 - c_{11})(e_{33}\gamma_j^2 - e_{15}) + (c_{13} + c_{44})(e_{31} + e_{15})\gamma_j^2}$$

$$b_j = \frac{(c_{44}\gamma_j^2 - c_{11})a_j + (c_{13} + c_{44})}{e_{31} + e_{15}}$$

$$(j = 1,2,3) \tag{3.313}$$

γ_j^2 $(j = 1, 2, 3)$ are the roots of the following characteristic equations:

$$a_0\gamma^6 + b_0\gamma^4 + c_0\gamma^2 + d_0 = 0 \tag{3.314}$$

where

$$
\begin{aligned}
a_0 &= c_{44}(c_{33}\epsilon_{33} + e_{33}^2) \\
b_0 &= -2c_{44}e_{15}e_{33} - c_{11}e_{33}^2 - c_{33}(c_{44}\epsilon_{11} + c_{11}\epsilon_{33}) + (c_{13} + c_{44})^2\epsilon_{33} \\
&\quad + 2e_{33}(c_{13} + c_{44})(e_{31} + e_{15}) - (c_{44})^2\epsilon_{33} - c_{33}(e_{31} + e_{15})^2 \\
c_0 &= 2c_{11}e_{15}e_{33} + c_{44}e_{15}^2 + c_{11}(c_{33}\epsilon_{11} + c_{44}\epsilon_{33}) - (c_{13} + c_{44})^2\epsilon_{11} \\
&\quad -2(c_{13} + c_{44})(e_{31} + e_{15})e_{15} + (c_{44})^2\epsilon_{11} + c_{44}(e_{31} + e_{15})^2 \\
d_0 &= -c_{11}(c_{44}\epsilon_{11} + e_{15}^2)
\end{aligned}
\tag{3.315}
$$

The displacements and electric potential near the crack tip are given by

$$u_x = \frac{K_I}{F}\left(\frac{r_1}{\pi}\right)^{1/2}\sum_{j=1}^{3}a_jd_j\{(\cos^2\theta_1 + \gamma_j^2\sin^2\theta_1)^{1/2} + \cos\theta_1\}^{1/2}$$

$$\tag{3.316}$$

$$u_z = -\frac{K_I}{F}\left(\frac{r_1}{\pi}\right)^{1/2}\sum_{j=1}^{3}\frac{d_j}{\gamma_j}\{(\cos^2\theta_1 + \gamma_j^2\sin^2\theta_1)^{1/2} - \cos\theta_1\}^{1/2}$$

$$\phi = \frac{K_I}{F}\left(\frac{r_1}{\pi}\right)^{1/2}\sum_{j=1}^{3}\frac{b_jd_j}{\gamma_j}\{(\cos^2\theta_1 + \gamma_j^2\sin^2\theta_1)^{1/2} - \cos\theta_1\}^{1/2} \tag{3.317}$$

where the polar coordinates r_1 and θ_1 are defined by $r_1 = \{(x-a)^2 + z^2\}^{1/2}, \theta_1 = \tan^{-1}(z/(x-a))$, and

$$F = \sum_{j=1}^{3}d_jg_j \tag{3.318}$$

The singular parts of strains, stresses, electric field intensities, and electric displacements are expressed as

$$\epsilon_{xx} = \frac{K_I}{2F(\pi r_1)^{1/2}}\sum_{j=1}^{3}a_jd_jR_j^c(\theta_1)$$

$$\epsilon_{zx} = -\frac{K_I}{4F(\pi r_1)^{1/2}}\sum_{j=1}^{3}\frac{(a_j\gamma_j^2 + 1)d_j}{\gamma_j}R_j^s(\theta_1) \tag{3.319}$$

$$\epsilon_{zz} = -\frac{K_I}{2F(\pi r_1)^{1/2}}\sum_{j=1}^{3}d_jR_j^c(\theta_1)$$

$$\sigma_{xx} = \frac{K_{\mathrm{I}}}{2F(\pi r_1)^{1/2}} \sum_{j=1}^{3} d_j M_j R_j^c(\theta_1)$$

$$\sigma_{zx} = -\frac{K_{\mathrm{I}}}{2F(\pi r_1)^{1/2}} \sum_{j=1}^{3} \frac{d_j f_j}{\gamma_j} R_j^s(\theta_1) \tag{3.320}$$

$$\sigma_{zz} = \frac{K_{\mathrm{I}}}{2F(\pi r_1)^{1/2}} \sum_{j=1}^{3} d_j g_j R_j^c(\theta_1)$$

$$E_x = -\frac{K_{\mathrm{I}}}{2F(\pi r_1)^{1/2}} \sum_{j=1}^{3} \frac{b_j d_j}{\gamma_j} R_j^s(\theta_1)$$

$$\tag{3.321}$$

$$E_z = -\frac{K_{\mathrm{I}}}{2F(\pi r_1)^{1/2}} \sum_{j=1}^{3} b_j d_j R_j^c(\theta_1)$$

$$D_x = -\frac{K_{\mathrm{I}}}{2F(\pi r_1)^{1/2}} \sum_{j=1}^{3} \frac{d_j N_j}{\gamma_j} R_j^s(\theta_1)$$

$$\tag{3.322}$$

$$D_z = \frac{K_{\mathrm{I}}}{2F(\pi r_1)^{1/2}} \sum_{j=1}^{3} d_j h_j R_j^c(\theta_1)$$

where

$$\begin{aligned} M_j &= c_{11}a_j - c_{13} + e_{31}b_j \\ N_j &= e_{15}(a_j\gamma_j^2 + 1) + \epsilon_{11}b_j \end{aligned} \qquad (j = 1, 2, 3) \tag{3.323}$$

$$R_j^c(\theta_1) = \left\{ \frac{(\cos^2\theta_1 + \gamma_j^2 \sin^2\theta_1)^{1/2} + \cos\theta_1}{\cos^2\theta_1 + \gamma_j^2 \sin^2\theta_1} \right\}^{1/2}$$

$$R_j^s(\theta_1) = -\left\{ \frac{(\cos^2\theta_1 + \gamma_j^2 \sin^2\theta_1)^{1/2} - \cos\theta_1}{\cos^2\theta_1 + \gamma_j^2 \sin^2\theta_1} \right\}^{1/2} \qquad (j = 1, 2, 3) \tag{3.324}$$

The energy release rate G can be obtained by using the concept of crack closure energy [91] as

$$G = \lim_{\Delta a \to 0} \frac{1}{\Delta a} \int_0^{\Delta a} \{\sigma_{zz}(x)u_z(\Delta a - x) + \sigma_{zx}(x)u_x(\Delta a - x)$$

$$+ D_z(x)\phi(\Delta a - x)\} dx \tag{3.325}$$

where Δa is the assumed crack extension. The energy release rate G is also obtained from the following J-integral:

$$J = \int_{\Gamma_0} \{Hn_x - (\sigma_{xx}u_{x,x} + \sigma_{zx}u_{z,x})n_x - (\sigma_{zx}u_{x,x} + \sigma_{zz}u_{z,x})n_z$$

$$+ D_x E_x n_x + D_z E_x n_z\} d\Gamma \tag{3.326}$$

where Γ_0 is a small contour closing a crack tip and n_x, n_z are the components of the outer unit vector $\mathbf{n}$ normal to Γ_0. The electrical enthalpy density H is

$$H = \frac{1}{2}(c_{11}\varepsilon_{xx}^2 + c_{33}\varepsilon_{zz}^2 + 2c_{13}\varepsilon_{xx}\varepsilon_{zz} + 4c_{44}\varepsilon_{zx}^2)$$

$$-\frac{1}{2}(\epsilon_{11}E_x^2 + \epsilon_{33}E_z^2) - \{2e_{15}\varepsilon_{zx}E_x + (e_{31}\varepsilon_{xx} + e_{33}\varepsilon_{zz})E_z\} \tag{3.327}$$

Writing the energy release rate expression for the permeable crack model in terms of the stress intensity factor [90], we obtain

$$G = J = \frac{1}{2F^2}\left(-F\sum_{j=1}^{3}\frac{d_j}{\gamma_j} + \sum_{k=1}^{3}d_k h_k \sum_{j=1}^{3}\frac{b_j d_j}{\gamma_j}\right)K_I^2 \tag{3.328}$$

The energy density fracture criterion can be developed by referring to the amount of energy stored in a volume element ahead of the crack [92]. For the piezoelectric material, the energy stored in the volume element dV [93] is

$$dW = \left\{\frac{1}{2}(\sigma_{xx}\varepsilon_{xx} + \sigma_{xz}\varepsilon_{xz} + \sigma_{zx}\varepsilon_{zx} + \sigma_{zz}\varepsilon_{zz}) + \frac{1}{2}(D_x E_x + D_z E_z)\right\} dV \tag{3.329}$$

The energy density S is given by

$$S = r_1 \frac{dW}{dV} \tag{3.330}$$

The energy density criterion can be used to predict the path of the mixed-mode crack initiation and propagation. The energy density for the permeable crack model [94] becomes

$$S = \{a_M(\theta_1) + a_E(\theta_1)\}K_I^2 \tag{3.331}$$

where

$$a_M(\theta_1) = \frac{1}{8\pi F^2}\left\{-\sum_{j=1}^{3}d_j g_j R_j^c(\theta_1)\sum_{j=1}^{3}d_j R_j^c(\theta_1)\right.$$

$$+ \sum_{j=1}^{3}d_j M_j R_j^c(\theta_1)\sum_{j=1}^{3}a_j d_j R_j^c(\theta_1) + \sum_{j=1}^{3}\frac{d_j f_j}{\gamma_j}R_j^s(\theta_1)\sum_{j=1}^{3}\frac{(a_j\gamma_j^2 + 1)d_j}{\gamma_j}R_j^s(\theta_1)\right\}$$

$$a_E(\theta_1) = \frac{1}{8\pi F^2}\left[-\sum_{j=1}^{3}d_j h_j R_j^c(\theta_1)\sum_{j=1}^{3}b_j d_j R_j^c(\theta_1)\right.$$

$$\left. + \sum_{j=1}^{3}\frac{d_j N_j}{\gamma_j}R_j^s(\theta_1)\sum_{j=1}^{3}\frac{b_j d_j}{\gamma_j}R_j^s(\theta_1)\right] \tag{3.332}$$

The stress and electric displacement intensity factors, energy release rate, and energy density for the permeable crack model depend only on the stress σ_∞, and the electric loading dependence on these parameters is different for the two mechanical loading conditions, that is, under applied strain and under applied stress. For some practical applications, the piezoelectric material seems to be operated with the strain fixed at some values. For a general remote boundary condition prescribed by fixed strain, the stress σ_∞ is determined by Eq. (3.307) with σ_0 and E_0. The stress and electric displacement intensity factors, energy release rate, and energy density for the permeable crack model depend on the electric field E_0, since the stress σ_∞ increases or decreases depending on the magnitude and direction of the electric field E_0. However, if the stress is applied and fixed, the value of the stress σ_∞ is left unchanged. Consequently, the stress and electric displacement intensity factors, energy release rate, and energy density for the permeable crack model are independent of the electric field E_0.

The aforementioned phenomena are not observed in the impermeable or open crack model of the piezoelectric material. A solution for the impermeable and open crack models in the infinite piezoelectric material is outlined in the next two paragraphs.

The crack surface electrical boundary condition for the impermeable crack model is

$$D_z(x,0) = 0 \quad (0 \le x < a)$$
$$\phi(x,0) = 0 \quad (a \le x < \infty) \tag{3.333}$$

The energy release rate G for the impermeable crack model [95] is given by

$$
\begin{aligned}
G = J = -\frac{1}{2(F^{\mathrm{I}})^2} & \left\{ \left(F^{\mathrm{I}} \sum_{j=1}^{3} \frac{s_j}{\gamma_j} - \sum_{k=1}^{3} h_k s_k \sum_{j=1}^{3} \frac{b_j s_j}{\gamma_j} \right) K_{\mathrm{I}}^2 \right. \\
& + \left(-F^{\mathrm{I}} \sum_{j=1}^{3} \frac{t_j}{\gamma_j} + \sum_{k=1}^{3} h_k s_k \sum_{j=1}^{3} \frac{b_j t_j}{\gamma_j} + \sum_{k=1}^{3} h_k t_k \sum_{j=1}^{3} \frac{b_j s_j}{\gamma_j} \right) K_{\mathrm{I}} K_{\mathrm{D}} \\
& \left. - \left(\sum_{k=1}^{3} h_k t_k \sum_{j=1}^{3} \frac{b_j t_j}{\gamma_j} \right) K_{\mathrm{D}}^2 \right\}
\end{aligned}
\tag{3.334}
$$

where

$$F^{\mathrm{I}} = F_{11}F_{22} - F_{12}F_{21} \tag{3.335}$$

$$
\begin{aligned}
s_j &= d_j F_{22} - l_j F_{21} \\
t_j &= d_j F_{12} - l_j F_{11}
\end{aligned} \quad (j = 1, 2, 3)
\tag{3.336}
$$

and

$$
\begin{aligned}
l_1 &= \gamma_1 (f_2 - f_3) \\
l_2 &= \gamma_2 (f_3 - f_1) \\
l_3 &= \gamma_3 (f_1 - f_2)
\end{aligned}
\tag{3.337}
$$

$$F_{11} = \frac{1}{b_1(f_2 - f_3) + b_2(f_3 - f_1) + b_3(f_1 - f_2)} \sum_{j=1}^{3} d_j g_j$$

$$F_{12} = \frac{1}{b_1(f_2 - f_3) + b_2(f_3 - f_1) + b_3(f_1 - f_2)} \sum_{j=1}^{3} g_j l_j$$

$$F_{21} = \frac{1}{b_1(f_2 - f_3) + b_2(f_3 - f_1) + b_3(f_1 - f_2)} \sum_{j=1}^{3} d_j h_j \qquad (3.338)$$

$$F_{22} = \frac{1}{b_1(f_2 - f_3) + b_2(f_3 - f_1) + b_3(f_1 - f_2)} \sum_{j=1}^{3} h_j l_j$$

In Eq. (3.334), K_I is given by Eq. (3.309) and K_D is obtained as

$$K_D = D_\infty(\pi a)^{1/2} \qquad (3.339)$$

where the far-field electric displacement D_∞ is given by

$$D_\infty = \frac{c_{13}e_{31} - c_{11}e_{33}}{c_{13}^2 - c_{33}c_{11}} \sigma_0 + \left(\frac{e_{31}^2}{c_{11}} + \epsilon_{33} \right) E_0 \qquad (3.340)$$

Also, the energy density S for the impermeable crack model [93] is

$$S = \frac{1}{8\pi(F^I)^2}[\{\beta_{M1}(\theta_1) + \beta_{E1}(\theta_1)\}K_I^2 + \{\beta_{M2}(\theta_1) + \beta_{E2}(\theta_1)\}K_I K_D$$

$$+ \{\beta_{M3}(\theta_1) + \beta_{E3}(\theta_1)\}K_D^2] \qquad (3.341)$$

where

$$\beta_{M1}(\theta_1) = -\sum_{k=1}^{3} g_k s_k R_k^c(\theta_1) \sum_{j=1}^{3} s_j R_j^c(\theta_1) + \sum_{k=1}^{3} s_k M_k R_k^c(\theta_1) \sum_{j=1}^{3} a_j s_j R_j^c(\theta_1)$$

$$+ 2\sum_{k=1}^{3} \frac{f_k s_k}{\gamma_k} R_k^s(\theta_1) \sum_{j=1}^{3} \frac{(a_j \gamma_j^2 + 1)s_j}{\gamma_j} R_j^s(\theta_1)$$

$$\beta_{M2}(\theta_1) = \sum_{k=1}^{3} g_k s_k R_k^c(\theta) \sum_{j=1}^{3} t_j R_j^c(\theta_1) + \sum_{k=1}^{3} g_k t_k R_k^c(\theta_1) \sum_{j=1}^{3} s_j R_j^c(\theta_1)$$

$$- \sum_{k=1}^{3} s_k M_k R_k^c(\theta_1) \sum_{j=1}^{3} a_j t_j R_j^c(\theta_1) - \sum_{k=1}^{3} t_k M_k R_k^c(\theta_1) \sum_{j=1}^{3} a_j s_j R_j^c(\theta_1)$$

$$- 2 \sum_{k=1}^{3} \frac{f_k s_k}{\gamma_k} R_k^s(\theta_1) \sum_{j=1}^{3} \frac{(a_j \gamma_j^2 + 1)t_j}{\gamma_j} R_j^s(\theta_1)$$

$$- 2 \sum_{k=1}^{3} \frac{f_k t_k}{\gamma_k} R_k^s(\theta_1) \sum_{j=1}^{3} \frac{(a_j \gamma_j^2 + 1)s_j}{\gamma_j} R_j^s(\theta_1)$$

$$\beta_{M3}(\theta_1) = - \sum_{k=1}^{3} g_k t_k R_k^c(\theta_1) \sum_{j=1}^{3} t_j R_j^c(\theta_1) + \sum_{k=1}^{3} t_k M_k R_k^c(\theta_1) \sum_{j=1}^{3} a_j t_j R_j^c(\theta_1)$$

$$+ 2 \sum_{k=1}^{3} \frac{f_k t_k}{\gamma_k} R_k^s(\theta_1) \sum_{j=1}^{3} \frac{(a_j \gamma_j^2 + 1)t_j}{\gamma_j} R_j^s(\theta_1)$$

$$\beta_{E1}(\theta_1) = - \sum_{k=1}^{3} h_k s_k R_k^c(\theta_1) \sum_{j=1}^{3} b_j s_j R_j^c(\theta_1) + \sum_{k=1}^{3} \frac{s_k N_k}{\gamma_k} R_k^s(\theta_1) \sum_{j=1}^{3} \frac{b_j s_j}{\gamma_j} R_j^s(\theta_1)$$

$$\beta_{E2}(\theta_1) = \sum_{k=1}^{3} h_k s_k R_k^c(\theta_1) \sum_{j=1}^{3} b_j t_j R_j^c(\theta_1) + \sum_{k=1}^{3} h_k t_k R_k^c(\theta_1) \sum_{j=1}^{3} b_j s_j R_j^c(\theta_1)$$

$$- \sum_{k=1}^{3} \frac{s_k N_k}{\gamma_k} R_k^s(\theta_1) \sum_{j=1}^{3} \frac{b_j t_j}{\gamma_j} R_j^s(\theta_1) - \sum_{k=1}^{3} \frac{t_k N_k}{\gamma_k} R_k^s(\theta_1) \sum_{j=1}^{3} \frac{b_j s_j}{\gamma_j} R_j^s(\theta_1)$$

$$\beta_{E3}(\theta_1) = - \sum_{k=1}^{3} h_k t_k R_k^c(\theta_1) \sum_{j=1}^{3} b_j t_j R_j^c(\theta_1) + \sum_{k=1}^{3} \frac{t_k N_k}{\gamma_k} R_k^s(\theta_1) \sum_{j=1}^{3} \frac{b_j t_j}{\gamma_j} R_j^s(\theta_1) \qquad (3.342)$$

The crack surface electrical boundary condition for the open crack model becomes

$$\begin{aligned}
D_z^+ &= D_z^- & (0 \le x < a) \\
D_z^+(u_z^+ - u_z^-) &= \epsilon_0(\phi^- - \phi^+) & (0 \le x < a) \\
\phi(x, 0) &= 0 & (a \le x < \infty)
\end{aligned} \qquad (3.343)$$

The energy release rate G [96] and energy density S for the open crack model are given by Eq. (3.334) and Eq. (3.341), respectively, with

$$K_D = (D_\infty - D_0)(\pi a)^{1/2} \qquad (3.344)$$

where

$$D_0 = -\epsilon_0 \frac{F_{21}\sigma_\infty + F_{11}(D_0 - D_\infty)}{F_{22}\sigma_\infty + F_{12}(D_0 - D_\infty)} \qquad (3.345)$$

Next, we consider Case 2 as shown in Fig. 3.98. The crack surface boundary and loading conditions can be written as

$$\sigma_{xz}(0, z) = 0 \quad (0 \le z < \infty) \qquad (3.346)$$

$$\begin{aligned}
\sigma_{xx}(0, z) &= 0 & (0 \le z < a) \\
u_x(0, z) &= 0 & (a \le z < \infty)
\end{aligned} \qquad (3.347)$$

$$E_z(0, z) = E_z^e(0, z) \quad (0 \le z < a)$$
$$\phi_{,x}(0, z) = 0 \quad\quad (a \le z < \infty) \quad\quad (3.348)$$

$$D_x(0, z) = D_x^e(0, z) \ (0 \le z < a) \quad\quad (3.349)$$

$$\varepsilon_{xx}(x, z) = \varepsilon_\infty \quad (0 \le z < \infty, x \to \infty)$$
$$E_z(x, z) = E_0 \quad (0 \le x < \infty, z \to \infty) \quad\quad (3.350)$$

The stress intensity factor K_I is given by

$$K_I = \lim_{z \to a^+} \{2\pi(z - a)\}^{1/2} \sigma_{xx}(0, z) \quad\quad (3.351)$$

Also, the energy release rate and energy density can easily be obtained.

Shindo et al. [86] made a comparison of the stress intensity factor without electric field between Case 1 and Case 2. A piezoelectric strip with a permeable central crack parallel to the edges of the strip was considered as shown in Fig. 3.99. It was found that since the value of the stress intensity factor for Case 1 is always larger than that for Case 2, the crack propagates easily normal to the poling direction. The indentation fracture tests on the piezoelectric ceramics [97] showed that the indentation cracks are longer normal to the poling direction and shorter parallel to the poling as shown in Fig. 3.100. The theoretical results are in agreement with the experimental results. For piezoelectric materials, a crack normal to the poling (Case 1) is critical, not only without electric field but also with electric field. For Case 1, Narita and Shindo [98] obtained a crack growth rate equation for a permeable central crack in the piezoelectric

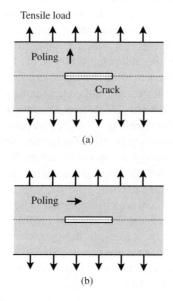

Figure 3.99 Sketch of piezoelectric strip with a central crack: (a) Case 1, (b) Case 2

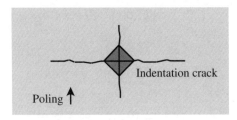

Figure 3.100   Indentation crack in piezoelectric ceramics

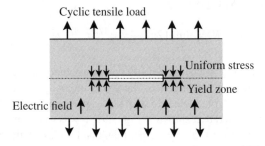

Figure 3.101   Sketch of piezoelectric strip containing a central crack with two postulated yield zones

strip under cyclic tensile load and monotonic electric field as shown in Fig. 3.101, by using the yield strip model and plastic accumulated displacement criterion for small crack growth, and discussed the effect of electric field on the crack growth rate. On the other hand, Shindo et al. [90] considered a piezoelectric strip with a permeable central crack normal to the edges of the strip as shown in Fig. 3.102(a) and discussed the effect of electric field on the stress intensity factor and energy release rate. Lin et al. [94] made a comparison of the energy release rate and energy density criteria using the example of a piezoelectric strip bonded between two half-spaces of a different elastic material with a crack normal to the interfaces as shown in Fig. 3102(b). They summarized the following findings: (1) For the permeable crack model, no difference is found in the effect of electric field on crack propagation for the criteria (the stress intensity factor, energy release rate, and energy density), and (2) If the impermeable crack model is used, different criteria give different results for crack propagation in piezoelectric materials. Similar results were found for the piezoelectric strip bonded between two piezoelectric half-spaces with different polarization (central active piezoelectric transformer) as shown in Fig. 3.102(c) [99]. No consensus is reached on the fracture criteria for the impermeable crack model, and therefore, the fracture criteria for the impermeable crack model are unreliable and may yield misleading results.

There are some publications concerning the dynamic behavior of a permeable crack in the piezoelectric material. Shindo and Ozawa [100] investigated the scattering of normally incident plane harmonic waves by a crack in the piezoelectric material and discussed the dynamic stress and electric field intensity factors. Shindo et al. [101] also determined the transient dynamic stress intensity factor and energy release rate for the cracked piezoelectric material under normal impact. Although a number of papers on the dynamic behavior of an impermeable crack have also been appeared, questions remain regarding the crack surface boundary and loading conditions.

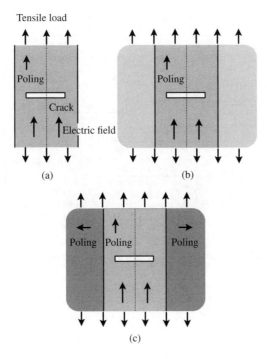

Figure 3.102 Sketch of (a) cracked piezoelectric strip, (b) cracked piezoelectric strip bonded between two elastic half-planes, and (c) cracked piezoelectric strip bonded between two piezoelectric half-planes

3.10.1.3 Penny-Shaped Crack

Consider a penny-shaped crack of radius a embedded in an infinite piezoelectric material. Let $x = x_1, y = x_2$, and $z = x_3$ denote the rectangular Cartesian coordinates. The z axis is assumed to coincide with the poling direction. Referring to Fig. 3.103, the crack is located in the plane $z = 0$ of the piezoelectric material and is centered at the origin. The formulation can be expressed most conveniently in terms of a cylindrical coordinate system (r,θ,z). The material is subjected to a far-field uniform tensile strain $\varepsilon_{zz} = \varepsilon_\infty$ and electric field $E_z = E_0$. The electric potentials at $z = \pm$ const. have equal magnitude but opposite sign.

The constitutive equations are in place of Equations (3.250) and (3.251)

$$
\begin{Bmatrix} \sigma_{rr} \\ \sigma_{\theta\theta} \\ \sigma_{zz} \\ \sigma_{zr} \end{Bmatrix} = \begin{bmatrix} c_{11} & c_{12} & c_{13} & 0 \\ c_{12} & c_{11} & c_{13} & 0 \\ c_{13} & c_{13} & c_{33} & 0 \\ 0 & 0 & 0 & c_{44} \end{bmatrix} \begin{Bmatrix} \varepsilon_{rr} \\ \varepsilon_{\theta\theta} \\ \varepsilon_{zz} \\ 2\varepsilon_{zr} \end{Bmatrix} - \begin{bmatrix} 0 & e_{31} \\ 0 & e_{31} \\ 0 & e_{33} \\ e_{15} & 0 \end{bmatrix} \begin{Bmatrix} E_r \\ E_z \end{Bmatrix} \tag{3.352}
$$

$$
\begin{Bmatrix} D_r \\ D_z \end{Bmatrix} = \begin{bmatrix} 0 & 0 & 0 & e_{15} \\ e_{31} & e_{31} & e_{33} & 0 \end{bmatrix} \begin{Bmatrix} \varepsilon_{rr} \\ \varepsilon_{\theta\theta} \\ \varepsilon_{zz} \\ 2\varepsilon_{zr} \end{Bmatrix} + \begin{bmatrix} \epsilon_{11} & 0 \\ 0 & \epsilon_{33} \end{bmatrix} \begin{Bmatrix} E_r \\ E_z \end{Bmatrix} \tag{3.353}
$$

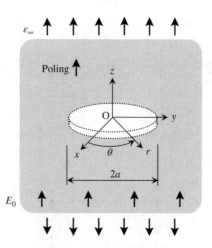

Figure 3.103   An infinite piezoelectric material with a penny-shaped crack

Also, the governing equations are in place of (3.254) and (3.255)

$$c_{11}\left(u_{r,rr} + \frac{u_{r,r}}{r} - \frac{u_r}{r^2}\right) + c_{44}u_{r,zz} + (c_{13} + c_{44})u_{z,rz} + (e_{31} + e_{15})\phi_{,rz} = 0$$

$$(c_{13} + c_{44})\left(u_{r,rz} + \frac{u_{r,z}}{r}\right) + c_{44}\left(u_{z,rr} + \frac{u_{z,r}}{r}\right) + c_{33}u_{z,zz} \tag{3.354}$$

$$+ e_{15}\left(\phi_{,rr} + \frac{\phi_{,r}}{r}\right) + e_{33}\phi_{,zz} = 0$$

$$(e_{31} + e_{15})\left(u_{r,rz} + \frac{u_{r,z}}{r}\right) + e_{15}\left(u_{z,rr} + \frac{u_{z,r}}{r}\right) + e_{33}u_{z,zz}$$

$$- \epsilon_{11}\left(\phi_{,rr} + \frac{\phi_{,r}}{r}\right) - \epsilon_{33}\phi_{,zz} = 0 \tag{3.355}$$

In a vacuum, the constitutive equation (3.353) and the governing equation (3.355) become

$$D_r^e = \epsilon_0 E_r^e, \quad D_z^e = \epsilon_0 E_z^e \tag{3.356}$$

$$\phi_{,rr}^e + \frac{\phi_{,r}^e}{r} + \phi_{,zz}^e = 0 \tag{3.357}$$

Referring to the semi-infinite region $0 \le r < \infty, 0 \le \theta \le 2\pi, 0 \le z < \infty$, the boundary conditions can be expressed in the form

$$\sigma_{zr}(r,0) = 0 \quad (0 \le r < \infty) \tag{3.358}$$

$$\sigma_{zz}(r,0) = 0 \quad (0 \leq r < a)$$
$$u_z(r,0) = 0 \quad (a \leq r < \infty)$$
(3.359)

$$E_r(r,0) = E_r^e(r,0) \quad (0 \leq r < a)$$
$$\phi(r,0) = 0 \quad (a \leq r < \infty)$$
(3.360)

$$D_z(r,0) = D_z^e(r,0) \, (0 \leq r < a)$$
(3.361)

$$\varepsilon_{zz}(r,z) = \varepsilon_\infty, \quad E_z(r,z) = E_0 \quad (0 \leq r < \infty, z \to \infty)$$
(3.362)

The far-field normal stress σ_∞ can be obtained as

$$\sigma_\infty = \sigma_0 - \left\{ \frac{(c_{11} + c_{12})e_{33} - 2c_{13}e_{31}}{c_{11} + c_{12}} \right\} E_0$$
(3.363)

where σ_0 is given by Eq. (3.308).

Hankel transforms are used to reduce the problem to the solution of a pair of dual integral equations. The integral equations are then solved exactly [102]. The stress intensity factor k_1 is given by

$$k_1 = \lim_{r \to a^+} \{2(r-a)\}^{1/2} \sigma_{zz}(r,0) = \frac{2}{\pi} \sigma_\infty a^{1/2}$$
(3.364)

The energy release rate and energy density for the permeable crack model are obtained as

$$G = \frac{\pi}{2F^2} \left(-F \sum_{j=1}^{3} \frac{d_j}{\gamma_j} + \sum_{k=1}^{3} h_k d_k \sum_{j=1}^{3} \frac{b_j d_j}{\gamma_j} \right) k_1^2$$
(3.365)

$$S = \pi\{a_M(\theta_1) + a_E(\theta_1)\} k_1^2$$
(3.366)

The stress intensity factor, energy release rate, and energy density for the penny-shaped crack depend on the electric field E_0, similar to the plane strain crack.

Narita et al. [103] performed the analysis of a penny-shaped crack in a piezoelectric cylinder as shown in Fig. 3.104(a) and discussed the effect of electric field on the stress intensity factor, energy release rate, and energy density. Lin et al. [104] also investigated the electromechanical response of a penny-shaped crack in a piezoelectric fiber embedded in an elastic matrix as shown in Fig. 3.104(b).

3.10.1.4 Antiplane Shear Crack

We consider a crack located in the interior of an infinite piezoelectric material as shown in Fig. 3.105. Let $x = x_1, y = x_2$, and $z = x_3$ denote the rectangular Cartesian coordinates. The z axis is assumed to coincide with the poling direction. When the crack front is parallel to

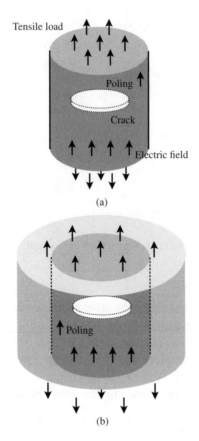

(a)

(b)

Figure 3.104   Sketch of (a) cracked piezoelectric cylinder and (b) cracked piezoelectric fiber embedded in an elastic matrix

the poling direction (z-axis), the in-plane electrical fields couple only with the out-of-plane mechanical fields. This particular configuration is referred to as antiplane case. Hence, we consider the case where the material is subjected to a far-field uniform shear strain $2\varepsilon_{yz} = \gamma_\infty$ and electric field $E_y = E_0$. Shear actuators often offer many advantages over extension actuators. The electric potentials at $y = \pm$ const. have equal magnitude but opposite sign. Due to symmetry of the problem, only the first quadrant with appropriate boundary conditions needs to be analyzed.

For the out-of-plane displacement and in-plane electric fields, the constitutive equations (3.120) and (3.121) become

$$
\left\{ \begin{array}{c} \sigma_{xz} \\ \sigma_{yz} \end{array} \right\} = \left[\begin{array}{cc} c_{44} & 0 \\ 0 & c_{44} \end{array} \right] \left\{ \begin{array}{c} 2\varepsilon_{xz} \\ 2\varepsilon_{yz} \end{array} \right\} - \left[\begin{array}{cc} e_{15} & 0 \\ 0 & e_{15} \end{array} \right] \left\{ \begin{array}{c} E_x \\ E_y \end{array} \right\}
\tag{3.367}
$$

$$
\left\{ \begin{array}{c} D_x \\ D_y \end{array} \right\} = \left[\begin{array}{cc} e_{15} & 0 \\ 0 & e_{15} \end{array} \right] \left\{ \begin{array}{c} 2\varepsilon_{xz} \\ 2\varepsilon_{yz} \end{array} \right\} + \left[\begin{array}{cc} \epsilon_{11} & 0 \\ 0 & \epsilon_{11} \end{array} \right] \left\{ \begin{array}{c} E_x \\ E_y \end{array} \right\}
\tag{3.368}
$$

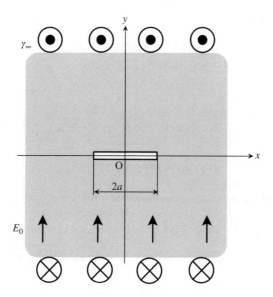

Figure 3.105   An infinite piezoelectric material with an antiplane shear crack

Here and in the following, we have dropped the remanent strains and polarizations. The strains are defined in terms of displacement as

$$\varepsilon_{xz} = \frac{1}{2}u_{z,x}, \quad \varepsilon_{yz} = \frac{1}{2}u_{z,y} \tag{3.369}$$

The electric field components may be written in terms of electric potential as

$$E_x = -\phi_{,x}, \quad E_y = -\phi_{,y} \tag{3.370}$$

The basic equations (3.103), (3.104) without inertia become in antiplane shear

$$\sigma_{xz,x} + \sigma_{yz,y} = 0 \tag{3.371}$$

$$E_{x,y} - E_{y,x} = 0, \quad D_{x,x} + D_{y,y} = 0 \tag{3.372}$$

Substituting Eqs. (3.367), (3.368) into Eqs. (3.371) and the second of Eqs. (3.372), and considering Eqs. (3.369), (3.370), the governing equations become

$$c_{44}(u_{z,xx} + u_{z,yy}) + e_{15}(\phi_{,xx} + \phi_{,yy}) = 0 \tag{3.373}$$

$$e_{15}(u_{z,xx} + u_{z,yy}) - \epsilon_{11}(\phi_{,xx} + \phi_{,yy}) = 0 \tag{3.374}$$

In a vacuum, the constitutive equation (3.368) and the governing equation (3.374) become

$$D_x^e = \epsilon_0 E_x^e, \quad D_y^e = \epsilon_0 E_y^e \tag{3.375}$$

$$\phi_{,xx}^e + \phi_{,yy}^e = 0 \tag{3.376}$$

The boundary conditions are

$$\sigma_{yz}(x,0) = 0 \quad (0 \leq x < a)$$
$$u_z(x,0) = 0 \quad (a \leq x < \infty) \tag{3.377}$$

$$E_x(x,0) = E_x^e(x,0) \quad (0 \leq x < a)$$
$$\phi(x,0) = 0 \quad (a \leq x < \infty) \tag{3.378}$$

$$D_y(x,0) = D_y^e(x,0) \ (0 \leq x < a) \tag{3.379}$$

The mechanical and electrical loading conditions at infinity are

$$2\varepsilon_{yz}(x,y) = \gamma_\infty, \quad E_y(x,y) = E_0 \ (0 \leq x < \infty, y \to \infty) \tag{3.380}$$

By applying the loading conditions (3.380), the far-field shear stress τ_∞ is expressed as

$$\tau_\infty = \tau_0 - e_{15}E_0 \tag{3.381}$$

where

$$\tau_0 = c_{44}\gamma_\infty \tag{3.382}$$

Fourier transforms are used to reduce the problem to the solution of a pair of dual integral equations. The integral equations are then solved exactly [105]. The mode III stress intensity factor K_{III} is obtained as

$$K_{\text{III}} = \lim_{x \to a^+} \{2\pi(x-a)\}^{1/2}\sigma_{yz}(x,0) = \tau_\infty(\pi a)^{1/2} \tag{3.383}$$

Writing the energy release rate expression for the permeable crack model in terms of the stress intensity factor, we obtain

$$G = \frac{1}{2c_{44}}K_{\text{III}}^2 \tag{3.384}$$

Shindo et al. discussed the effect of electric field on the stress intensity factor and energy release rate of the permeable crack in a piezoelectric strip parallel [106] and normal [105] to the edges of the strip under antiplane shear as shown in Fig. 3.106(a) and (b). Narita and Shindo [107] also considered the antiplane problem of bonded piezoelectric and orthotropic layers containing an interface crack as shown in Fig. 3.106(c).

For the dynamic crack problems, Narita and Shindo [108] applied the dynamic theory of antiplane piezoelectricity to solve the problem of a permeable crack subjected to horizontally polarized shear waves in an arbitrary direction as shown in Fig. 3.107(a). Narita and Shindo [109] also considered the scattering of horizontally polarized shear waves by a permeable crack in a composite laminate containing a piezoelectric layer as shown in Fig. 3.107(b). Moreover, Shindo et al. [110] analyzed the problem of antiplane shear waves scattered from multiple circular-arc interface cracks between a piezoelectric fiber and its surrounding polymer matrix as shown in Fig. 3.107(c).

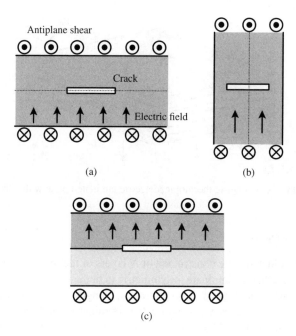

Figure 3.106 Sketch of (a) piezoelectric strip with a crack parallel to the edges of the strip, (b) piezoelectric strip with a crack normal to the edges of the strip, and (c) piezoelectric laminate with an interface crack

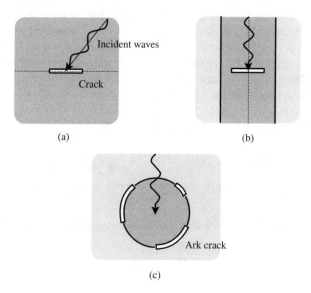

Figure 3.107 Sketch of (a) cracked piezoelectric material, (b) cracked piezoelectric laminate, and (c) arc-shaped interface cracks between a piezoelectric fiber and a polymer matrix

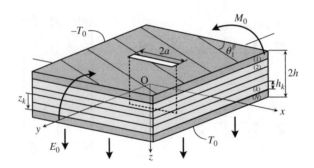

Figure 3.108   A symmetric thermopiezoelectric laminated plate with a through crack

3.10.1.5   Through Crack

Let us consider a laminated plate constructed of N layers of the thermopiezoelectric materials and FRPs. We assume that the laminate is symmetric in both geometry and material properties about the middle surface and has a through crack of length $2a$ as shown in Fig. 3.108. The principal material coordinates are x_1, x_2, x_3. Let the coordinate axes x and y be such that they are in the middle plane of the hybrid laminate and the $z = x_3$ axis is normal to this plane. θ_1^p is the angle between the lamina x-axis and lamina principal x_1-axis. Here, we consider the case where the laminae of the symmetric laminate have their major principal material directions alternating at $\theta_1^p = 0°$ and $90°$ (regular symmetric cross-ply laminate). The total thickness is $2h$ and the kth layer has thickness $h_k = z_k - z_{z-1}(k = 1, \ldots, N)$, where $z_0 = -h$ and $z_N = h$. The crack is located on the line $-a < x < a, y = 0$, and the cracked hybrid laminate is bent by uniform moments of intensity $M_{yy} = M_0$ at infinity and is subjected to an applied electric field $E_z = E_0$ in addition to the upper and lower surface temperatures $\Theta_0 = -T_0, \Theta_N = T_0$. It is assumed that the electric field resulting from variations in stress and temperature is insignificant compared with the applied electric field. Because of the assumed symmetry in geometry and loading, it is sufficient to consider the problem for $0 \le x < \infty, 0 \le y < \infty$ only.

The constitutive relation is given by Eq. (3.157). From Eq. (3.191), the governing equations without inertia and viscoelastic damping are simplified to

$$D_{11}w_{,xxxx} + 2(D_{12} + 2D_{66})w_{,xxyy} + D_{22}w_{,yyyy}$$
$$+ M^E_{xx,xx} + M^E_{yy,yy} + M^\Theta_{xx,xx} + M^\Theta_{yy,yy} = 0 \qquad (3.385)$$

From Eq. (3.231), the plate is subjected to the following linear temperature variation:

$$\Theta(z) = \frac{T_0 z}{h} \qquad (3.386)$$

For a stress-free crack surface, the thin-plate theory involves two conditions [111] and requires the vanishing of the bending moment M_{yy} and equivalent shear V_y given by

$$V_y = Q_y + M_{xy,x} \qquad (3.387)$$

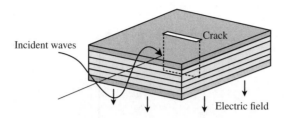

Figure 3.109   Sketch of symmetric piezoelectric laminated plate with a through crack

The boundary conditions can be written as

$$V_y(x,0) = 0 \qquad (0 \leq x < \infty) \qquad (3.388)$$

$$\begin{cases} M_{yy}(x,0) = 0 & (0 \leq x < a) \\ u_y(x,0) = 0 & (a \leq x < \infty) \end{cases} \qquad (3.389)$$

Fourier transforms are used to reduce the problem to the solution of a pair of dual integral equations. The integral equations are then solved exactly [112]. The moment intensity factor K_I is obtained as

$$K_\mathrm{I} = \lim_{x \to a^+} \{2\pi(x-a)\}^{1/2} M_{yy}(x,0) = M_0 \left(1 - \frac{M_{yy}^E + M_{yy}^\Theta}{M_0} \right) (\pi a)^{1/2} \qquad (3.390)$$

Shindo et al. [113] studied the scattering of time harmonic flexural waves by a through crack in a symmetric piezoelectric laminated plate as shown in Fig. 3.109. They discussed the effect of electric field on the dynamic moment intensity factor.

3.10.2   Cracked Rectangular Piezoelectric Material

For piezoelectric materials, the mode I energy release rate has been found to be a useful fracture mechanics parameter. It has also been found that electrical crack surface boundary conditions strongly affect the electric field effect characteristics of the electromechanical behavior and fracture mechanics parameters such as energy release rate [114]. There are three commonly used electrical boundary conditions, namely, permeable crack model (Eqs. (3.304) and (3.305)), impermeable crack model (Eq. (3.333)), and open crack model (Eq. (3.343)), and the condition remains a debating issue. Although the impermeable and open crack models may provide the mathematical solutions of the crack in piezoelectric material, it is still questionable to use these models as mentioned earlier. Recently, Landis [115] proposed a nonlinear electrically discharging crack model in the piezoelectric material and discussed the changes in the energy release rate due to fictional value of a critical electric field level for discharge E_d within the crack gap. Here, we highlight the effect of electric field on the mode I energy release rate for the rectangular piezoelectric material with a crack. Various boundary conditions are treated.

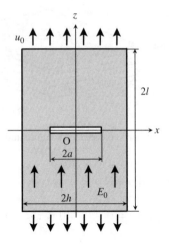

Figure 3.110 A rectangular piezoelectric material with a central crack normal to the poling and electric field

3.10.2.1 Central Crack Normal to the Poling and Electric Field

Consider a rectangular piezoelectric material of width $2h$ and length $2l$ containing a central crack of length $2a$ as shown in Fig. 3.110. Let $x = x_1, y = x_2$, and $z = x_3$ denote the rectangular Cartesian coordinates. We assume plane strain perpendicular to the y axis. The z axis is assumed to coincide with the poling direction. The material is loaded by a uniform displacement u_0 with a uniform electric field E_0 in the z-direction. The electric potentials of the same magnitude but opposite sign are applied at $z = \pm l$. Because of the assumed symmetry in geometry and loading, it is sufficient to consider the problem for $0 \le x \le h, 0 \le z \le l$ only.

We consider the permeable, impermeable, and open crack models. Mechanical boundary conditions at $z = 0$ are given by Eqs. (3.302) and (3.303), where the regions $(0 \le x < \infty)$ and $(a \le x < \infty)$ are replaced by $(0 \le x \le h)$ and $(a \le x \le h)$, respectively. Also, the boundary conditions for the permeable, impermeable, and open crack models are Eqs. (3.304), (3.305), (3.333), and (3.343), respectively, where $(a \le x < \infty)$ is replaced by $(a \le x \le h)$. For the permeable crack model, the electrical conditions at $z = 0$ reduce to $\phi(x, 0) = 0$ $(0 \le x \le h)$. The loading conditions may be stated as follows:

$$u_z(x, l) = u_0, \quad \phi(x, l) = -E_0 l \quad (0 \le x \le h) \tag{3.391}$$

$$\sigma_{xx}(h, z) = 0, \quad \sigma_{xz}(h, z) = 0$$
$$D_x(h, z) = 0 \tag{3.392}$$

The strain ε_{zz} for the uncracked piezoelectric material is given by

$$\varepsilon_{zz}(x, z) = \varepsilon_0 = \frac{u_0}{l} \tag{3.393}$$

By applying the loading conditions (3.391) and (3.392), the normal stress σ_0^{unc} at $z = l$ for the uncracked material is given by

$$\sigma_0^{\mathrm{unc}} = \sigma_0 - \left\{ e_{33} - \left(\frac{c_{13}}{c_{11}} \right) e_{31} \right\} E_0 \tag{3.394}$$

where

$$\sigma_0 = \left(c_{33} - \frac{c_{13}^2}{c_{11}} \right) \varepsilon_0 \tag{3.395}$$

For the rectangular piezoelectric material, finite element method is effective for the energy release rate calculation. The method can include nonlinear effects such as polarization switching. Due to the polarization switching, the piezoelectric material is often nonhomogeneous. The piezoelectric properties vary from one location to the other, and the variations are either continuous or discontinuous. An extension of J-integral to multiphase materials was proposed by Weichert and Schulz [116]. The energy release rate G can be obtained from the following J-integral [117]:

$$\begin{aligned} G = J = \int_{\Gamma_0} &\{ H n_x - (\sigma_{xx} u_{x,x} + \sigma_{zx} u_{z,x}) n_x - (\sigma_{zx} u_{x,x} + \sigma_{zz} u_{z,x}) n_z \\ &+ D_x E_x n_x + D_z E_x n_z \} d\Gamma \\ - \int_{\Gamma_p} &\{ H n_x - (\sigma_{xx} u_{x,x} + \sigma_{zx} u_{z,x}) n_x - (\sigma_{zx} u_{x,x} + \sigma_{zz} u_{z,x}) n_z \\ &+ D_x E_x n_x + D_z E_x n_z \} d\Gamma \end{aligned} \tag{3.396}$$

where Γ_0 is a small contour closing a crack tip, and Γ_p is a path embracing that part of phase boundary that is enclosed by Γ_0 as shown in Fig. 3.111.

FEA is performed for the permeable, open, and impermeable crack models, and the effects of electric field and polarization switching on the fracture mechanics parameters such as energy release rate are discussed for the rectangular piezoelectric material. The calculation of the energy release rate for the open crack model is more complicated than those for the permeable

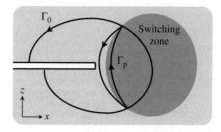

Figure 3.111   Typical contours Γ_0 and Γ_p for the evaluation of path-independent line integral

and impermeable crack models. The open crack model calculation starts with $\phi = 0$ on the crack surface [118]. The crack opening displacement and electric displacement on the crack surface are calculated, and the resulting potential difference is applied to the crack surface. The electromechanical fields are again solved, leading to new crack opening displacement and electric displacement on the crack surface. When this is accomplished, the potential difference is applied once more to the crack surface. Such a procedure is repeated until the evolution of the objective solutions shows no improvements. If the crack gap can store energy [119], nonzero contributions to the energy release rate arise from the contour segments along the crack surface. The values of the energy release rate for the open crack model can be obtained by computing contour integrations and then subtracting the electrical enthalpy density of the crack gap $H^e = -\epsilon_0\{(E_x^e)^2 + (E_z^e)^2\}/2$ times the crack opening displacement $2u_z^+$ evaluated at the intersection x^Γ of the contour with the crack surfaces [115]. So, Eq. (3.396) becomes

$$G = \int_{\Gamma_0} \{Hn_x - (\sigma_{xx}u_{x,x} + \sigma_{zx}u_{z,x})n_x - (\sigma_{zx}u_{x,x} + \sigma_{zz}u_{z,x})n_z$$

$$+ D_x E_x n_x + D_z E_x n_z\}d\Gamma$$

$$- \int_{\Gamma_p} \{Hn_x - (\sigma_{xx}u_{x,x} + \sigma_{zx}u_{z,x})n_x - (\sigma_{zx}u_{x,x} + \sigma_{zz}u_{z,x})n_z$$

$$+ D_x E_x n_x + D_z E_x n_z\}d\Gamma - 2H^e(x^\Gamma)u_z^+(x^\Gamma) \tag{3.397}$$

It is required that the contour Γ_0 intersects the upper and lower crack surfaces at the same position x^Γ as shown in Fig. 3.112. Note that the electrical traction $H^e + E_z^e D_z^e$ on the crack surfaces is accounted for in the calculation of Eq. (3.397).

Here, we show some numerical examples for rectangular piezoelectric material PZT C-91 of $2l = 20$ mm and $2h = 20$ mm. Table 3.4 presents the energy releases rates for the permeable, impermeable, and open crack models under displacement $u_0 = 0.5$ μm and electric field $E_0 = -0.1$ MV/m for $2a = 2$ mm. For the calculation of the energy release rate, four contours are defined in the finite element mesh as shown in Fig. 3.112, and the average values are presented. The values in parentheses are the contribution from the crack interior (see Eq. (3.397)). It is noted that for the open crack model, the contribution $2H^e(x^\Gamma)u_z^+(x^\Gamma)$ from the crack interior is negligibly small under a practical loading condition. We wish to add that the effect of electrical discharge within the crack is not accounted for in the calculation of the energy release rate for

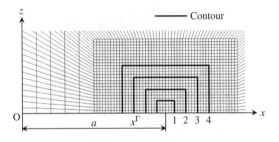

Figure 3.112   Finite element mesh and paths of *J*-integral

Table 3.4 Comparison of energy release rates for four crack models

G (J/m^2)			
Permeable	Impermeable	Open	Discharging
1.678	1.053	1.530 (-4.48×10^{-2})	1.676 (-6.09×10^{-6})

Values in parentheses are the results of $2H^e(x^\Gamma)u_z^+(x^\Gamma)$ in Eq. (3.397)

the open crack model. Landis [115] proposed a nonlinear electrically discharging crack model. He assumed that the crack gap behaves in a linear dielectric manner when the electric field within the crack gap is below the critical electric field level for discharge E_d. The solution for the discharging crack model is also presented in this table. It is assumed that the critical electric field level for discharge equals to a dielectric breakdown strength of air, that is, $E_d = 3$ MV/m. Similar results can be found in Ref. [120], although the four contours are not different from Fig. 3.112. For piezoelectric cracks, FEA predicts that

1. The value of $2H^e(x^\Gamma)u_z^+(x^\Gamma)$ for the discharging crack model is much smaller than that for the open crack model.
2. Using the standard air breakdown strength as the critical electric field level for discharge within the crack gap, the energy release rates predicted by the permeable and discharging crack models are not significantly different.

Figure 3.113 shows the contribution $2H^e(x^\Gamma)u_z^+(x^\Gamma)$ as a function of the critical electric field level for discharge E_d for the discharging crack model under $u_0 = 0.5$ μm and $E_0 = -0.1$ MV/m for $2a = 2$ mm. The contribution from the crack interior increases with increasing discharging level and approaches the value of the open crack model. Although the data are not

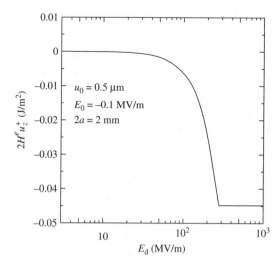

Figure 3.113 Contribution from the crack interior versus critical electric field level for discharge

shown here, higher displacement and electric field lead to higher contribution from the crack interior. If unrealistic critical electric field level for discharge is used to calculate the energy release rate for the discharging crack model, meaningless result is obtained.

Figure 3.114 shows the dependence of the energy release rate G on the electric field E_0 of the permeable crack model under displacement $u_0 = 0.5$ μm for $2a = 2, 4$ mm. Positive electric field decreases the value of G, while negative electric field has an opposite effect. The energy release rate for $2a = 4$ mm is higher than that for $2a = 2$ mm. Figure 3.115 summarizes

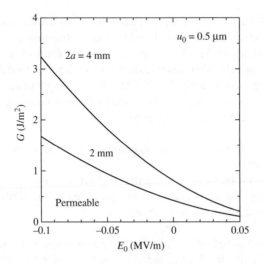

Figure 3.114 Energy release rate versus electric field of the rectangular piezoelectric material under applied displacement (permeable crack model)

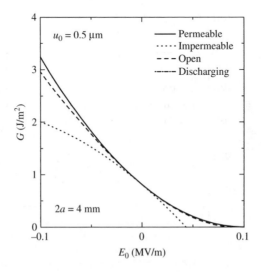

Figure 3.115 Energy release rate versus electric field of the rectangular piezoelectric material under applied displacement (permeable, impermeable, open, and discharging crack models)

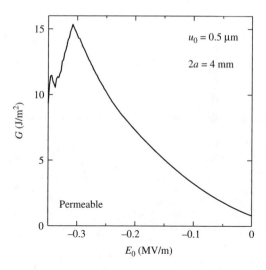

Figure 3.116   Energy release rate versus negative electric field of the rectangular piezoelectric material under applied displacement (permeable crack model)

the variation of G with E_0 of the permeable, impermeable, open, and discharging crack models under $u_0 = 0.5\,\mu m$ for $2a = 4\,mm$. Note that small differences are observed between the permeable (solid line) and discharging (dot-dashed line) crack models under various electric fields. For the impermeable and open crack models, a negative energy release rate can be induced with positive electric field. According to the fracture mechanics interpretation, a negative energy release rate would correspond to a crack that could absorb energy due to crack propagation. Since this would exclude the fracture in the piezoelectric material under the electric field, the energy release rates for the impermeable and open crack models have questionable physical significance. Hence, it is clear that the impermeable and open crack models are not appropriate for a slit crack in the piezoelectric material. Figure 3.116 displays the variation of G versus negative E_0 of the permeable crack model under $u_0 = 0.5\,\mu m$ for $2a = 4\,mm$. When the negative electric field increases, the energy release rate increases. A monotonically increasing negative electric field causes polarization switching. After the electric field reaches about $-0.3\,MV/m$ (below the coercive electric field $E_c = 0.35\,MV/m$), the polarization switching in a local region leads to an unexpected decrease in the energy release rate.

Figure 3.117 shows the G versus E_0 of the permeable and impermeable crack models under stress $\sigma_0^{unc} = 3\,MPa$. The stress $\sigma_0^{unc} = 3\,MPa$ corresponds to the uniform displacement $u_0 = 0.5\,\mu m$ for the uncracked piezoelectric material without the electric field. The results for the positive electric field under applied stress are different from those under applied displacement, and the energy release rate for the permeable crack model is independent of the positive electric field. The behavior of the energy release rate under the negative electric field is complicated because of the polarization switching phenomena. For the impermeable crack model, a negative energy release rate is produced when both positive and negative electric fields are larger. Figure 3.118 shows similar results under stress $\sigma_0^{unc} = 10\,MPa$. When the negative electric field increases, the energy release rate for the impermeable crack model decreases and then increases because of the localized polarization switching. The energy release rate for the

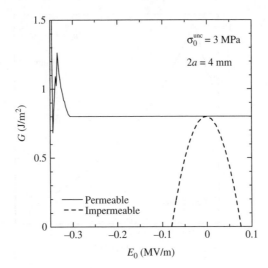

Figure 3.117   Energy release rate versus electric field of the rectangular piezoelectric material under applied stress (permeable and impermeable crack models)

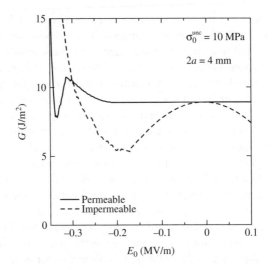

Figure 3.118   Energy release rate versus electric field of the rectangular piezoelectric material under high applied stress (permeable and impermeable crack models)

impermeable crack model does not become negative, but the piezoelectric material will fail under $\sigma_0^{unc} = 10$ MPa and the result may be meaningless due to unrealistic loading condition (high applied stress). The same can be said about the open crack model (not shown here). Although the fracture mechanics parameters such as energy release rate can easily be computed numerically by the FEA, unrealistic situations may be obtained. We have to be careful with loading conditions.

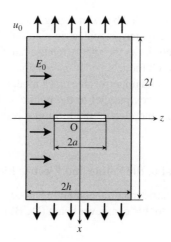

Figure 3.119 A rectangular piezoelectric material with a central crack parallel to the poling and electric field

3.10.2.2 Central Crack Parallel to the Poling and Electric Field

Consider a rectangular piezoelectric material of width $2h$ and length $2l$ containing a central crack of length $2a$ as shown in Fig. 3.119. Let $x = x_1, y = x_2$, and $z = x_3$ denote the rectangular Cartesian coordinates. We assume plane strain perpendicular to the y axis. The z axis is assumed to coincide with the poling direction. The material is loaded by the uniform displacement u_0 in the x-direction with the uniform electric field E_0 in the z-direction. Because of the assumed symmetry in geometry and loading, it is sufficient to consider the problem for $0 \le x \le l, 0 \le z \le h$ only.

Mechanical boundary conditions at $x = 0$ are given by Eqs. (3.346) and (3.347), where the regions $(0 \le z < \infty)$ and $(a \le z < \infty)$ are replaced by $(0 \le z \le h)$ and $(a \le z \le h)$, respectively. Also, the boundary conditions for the permeable crack model are Eqs. (3.348) and (3.349), where $(a \le z < \infty)$ is replaced by $(a \le z \le h)$. For the permeable crack model, the electrical conditions at $x = 0$ reduce to $\phi_{,x}(0, z) = 0$ $(0 \le z \le h)$. The loading conditions may be expressed in the form

$$u_x(l, z) = u_0 \quad (0 \le z \le h)$$
$$\phi(x, h) = -E_0 h \quad (0 \le x \le l) \tag{3.398}$$

$$\sigma_{zz}(x, h) = 0, \quad \sigma_{zx}(x, h) = 0$$
$$D_z(x, h) = 0 \tag{3.399}$$

The $z = 0$ plane is the electrical symmetry plane, that is, $\phi(x, 0) = 0$ $(0 \le x \le l)$. The strain ε_{xx} for the uncracked piezoelectric material is given by

$$\varepsilon_{xx}(x, z) = \varepsilon_0 = \frac{u_0}{l} \tag{3.400}$$

The energy release rate G is obtained by exchanging x and z in Eq. (3.397). The numerical results are summarized for rectangular piezoelectric material PZT C-91 of $2l = 20\,\text{mm}$ and $2h = 20\,\text{mm}$ in Fig. 3.120, which shows the energy release rate G versus electric field E_0 of the permeable and impermeable crack models under $u_0 = 5\,\mu\text{m}$ for $2a = 4\,\text{mm}$. No difference is observed between the permeable and impermeable crack models. The electric field dependence of the energy release rate for a crack parallel to the poling and electric field is smaller than for a crack normal to the poling and electric field. Similar results can be found at cryogenic temperatures [121].

3.10.2.3   Edge Crack Normal to the Poling and Electric Field

Consider a rectangular piezoelectric material containing an edge crack as shown in Fig. 3.121. Here, we treat an example of a double cantilever beam (DCB) piezoelectric material of length

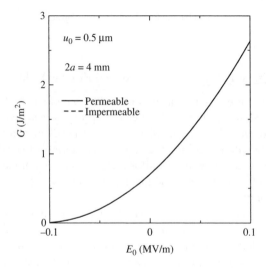

Figure 3.120   Energy release rate versus electric field of the rectangular piezoelectric material under applied displacement (permeable and impermeable crack models)

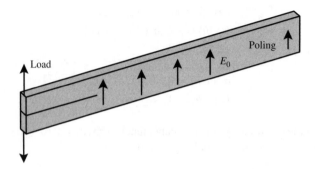

Figure 3.121   A rectangular piezoelectric material with an edge crack normal to the poling and electric field

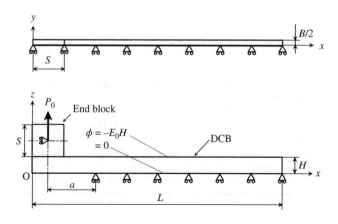

Figure 3.122   Geometry and boundary conditions of the DCB

L, width B, and thickness $2H$, and Fig. 3.122 illustrates the geometry of the DCB with end blocks of side length S. Let $x = x_1, y = x_2$, and $z = x_3$ denote the rectangular Cartesian coordinates. The z axis is assumed to coincide with the poling direction. The edge crack length is a. The end block is treated as an extremely stiff elastic body (nearly five times of PZT stiffness) and is assumed to be perfectly bonded on the DCB surface. The DCB is subjected to a uniformly distributed load P_0/B at the center of the end block ($x = S/2, 0 \leq y \leq B/2, z = H + S/2$) and a uniform electric field E_0 in the z-direction. The electric potentials of the same magnitude but opposite sign are applied at $z = \pm H$. Because of symmetry, only a quarter of the DCB is considered.

We consider the permeable crack model with stress-free crack surfaces. The crack surface electrical boundary conditions can be written as

$$
\begin{aligned}
E_x(x, y, 0) &= E_x^e(x, y, 0) \quad &(0 \leq x < S/2 + a, 0 \leq y \leq B/2) \\
E_y(x, y, 0) &= E_y^e(x, y, 0) \quad &(0 \leq x < S/2 + a, 0 \leq y \leq B/2) \\
\phi(x, y, 0) &= 0 \quad &(S/2 + a \leq x \leq L, 0 \leq y \leq B/2)
\end{aligned}
\tag{3.401}
$$

$$
D_z(x, y, 0) = D_z^e(x, y, 0) \quad (0 \leq x < S/2 + a, 0 \leq y \leq B/2)
\tag{3.402}
$$

For the permeable crack model, the electrical conditions at $z = 0$ reduce to $\phi(x, y, 0) = 0$ ($0 \leq x \leq L, 0 \leq y \leq B/2$). The electrical loading condition is

$$
\phi(x, y, H) = -E_0 H \quad (0 \leq x \leq L, 0 \leq y \leq B/2)
\tag{3.403}
$$

We also consider the discharging, open, and impermeable crack models.

A 3D finite element model is necessary to accurately obtain the energy release rate. We show some numerical examples for PZT C-91 DCB specimen of $L = 40, B = 1.5, 2H = 5$ mm, and $S = 5$ mm. Table 3.5 presents the energy releases rates at the mid-width ($y = 0$ mm) for the permeable, impermeable, open, and discharging crack models under load $P_0 = 1.5$ N and electric field $E_0 = -0.1$ MV/m for $a = 7.5$ mm. The results obtained by the four contours as shown in

Table 3.5 Comparison of energy release rates at the midwidth for four crack models

		G (J/m^2)		
	Permeable	Impermeable	Open	Discharging
Contour 1	0.969	−0.241	0.534	0.961
Contour 2	0.962	−0.247	0.484	0.952
Contour 3	0.962	−0.249	0.444	0.950
Contour 4	0.960	−0.254	0.404	0.946
Avg.	0.963	−0.248	0.466	0.952

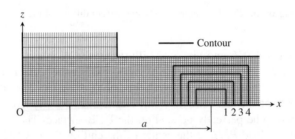

Figure 3.123 Finite element mesh of the DCB with end block and paths of J-integral

Fig. 3.123 and the average values (Avg.) are shown. Although the data are not shown here, the contribution from the crack interior is small for the open and discharging crack models. Little differences are observed between the permeable and discharging crack models.

Figure 3.124 displays the variation of the energy release rate G with the electric field E_0 of the permeable, impermeable, open, and discharging crack models under $P_0 = 1.5$ N for $a = 7.5$ mm. For the impermeable and open crack models, a negative energy release rate can be induced with positive and negative electric fields. If unrealistic loading condition is applied to the DCB specimen, the energy release rates for the impermeable and open crack models do not become negative, similar to the case for the rectangular piezoelectric material with a central crack. Figure 3.125 shows the G versus E_0 of the impermeable crack model under $P_0 = 5$ N for $a = 7.5$ mm. This is an unphysical result because the DCB will fail under $P_0 = 5$ N. Although the graph is not shown, similar unphysical result can be obtained for the open crack model.

Shindo et al. [122] performed the plane strain FEA for the rectangular piezoelectric material with an edge crack normal to the poling and electric field. They predicted the dielectric breakdown regions near the crack tip and concluded that under a high positive electric field below a dielectric breakdown strength, 10 MV/m, of PZT ceramics [123], dielectric breakdown occurs near the impermeable crack tip. However, the breakdown did not occur near the permeable crack tip as shown in Fig. 3.126. Different crack models give different results for the dielectric breakdown, and the unreliable impermeable and open crack models yield meaningless results.

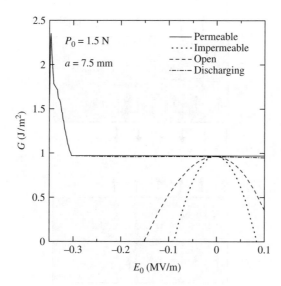

Figure 3.124 Energy release rate versus electric field of the DCB piezoelectric material under applied load (permeable, impermeable, open, and discharging crack models)

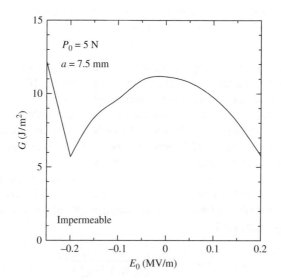

Figure 3.125 Energy release rate versus electric field of the DCB piezoelectric material under high applied load (impermeable crack model)

3.10.3 Indentation Fracture Test

The most important parameter for fracture mechanics is the fracture toughness K_{IC}. Usually, two experimental techniques, the indentation fracture (IF) and single-edge precracked beam (SEPB) methods, have been used to evaluate the fracture toughness in brittle materials. In the

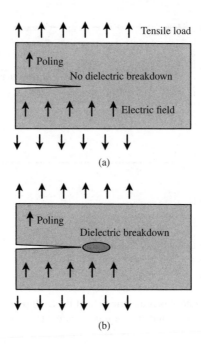

Figure 3.126 Images of dielectric breakdown region near (a) permeable and (b) impermeable cracks induced by tensile load and high positive electric field

case of the IF method, the cracks propagating from the corner of a surface indent are assumed to be directly related to the fracture toughness. Here, we describe the IF test and FEA conducted on piezoelectric ceramics under electromechanical loading [97].

We cut $5 \times 5 \times 15$ mm samples from some commercially available PZT ceramics and make Vickers indentations with loads $P = 4.9, 9.8$, and 19.6 N for 15 s under a uniform electric field E_0 parallel to the poling. Figure 3.127 shows the specimen and setup for the experiment, and a_n and a_p are the indentation crack lengths. The subscripts n and p stand for the normal and parallel to the poling, respectively. To generate the electric fields, we use a power supplier.

The IF toughness values K_{Cn} and K_{Cp} of the piezoelectric ceramics can be calculated in the first approximation using the equation of the IF toughness K_C of fine ceramics [124] corrected in the following way [125, 126]:

$$K_{Cn} = 0.018 \frac{1}{(s_{33}H_v)^{1/2}} \frac{P}{a_n^{3/2}}$$

$$K_{Cp} = 0.018 \frac{1}{(s_{11}H_v)^{1/2}} \frac{P}{a_p^{3/2}} \tag{3.404}$$

where H_v is hardness. Zhang and Raj [127] estimated an error of no more than 10% in the measurement of the fracture toughness by the Vickers indentation method in the PZT ceramics.

We now report some results for PZT P-7 under $P = 9.8$ N. The IF test results show that average values of seven indentation crack lengths normal to the poling for open- and closed-circuit condition are $a_n = 119.3$ and 106.0 μm, respectively. The indentation crack for the open-circuit

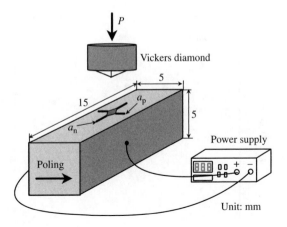

Figure 3.127 IF test setup

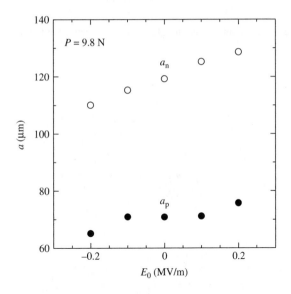

Figure 3.128 Crack length versus electric field of the IF specimen

condition is longer than that for the closed-circuit condition. The different behavior between the two conditions was also shown theoretically. For example, Shindo et al. [90] indicated that the stress intensity factor for the open-circuit condition is larger than that for the closed-circuit condition. Figure 3.128 shows the indentation crack lengths a_n, a_p under various electric fields E_0. It is found that the indentation cracks are longer normal to the poling and shorter parallel to the poling. This is illustrated in Fig. 3.100. The positive electric field assists the crack growth. Figure 3.129 shows the apparent fracture toughnesses K_{Cn}, K_{Cp} under various electric fields E_0. The apparent fracture toughness is not very much affected by the electric field.

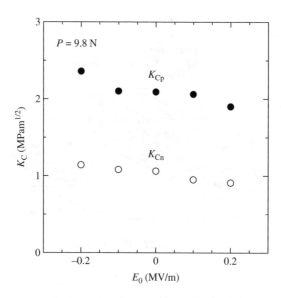

Figure 3.129   Apparent fracture toughness versus electric field of the IF specimen

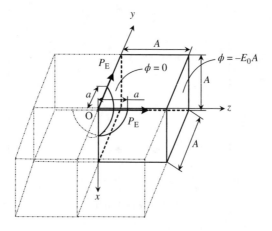

Figure 3.130   Finite element model of the IF test

We perform the FEA to demonstrate the fracture mechanics parameters such as energy release rate by using the simplest model. Figure 3.130 shows the 3D finite element model for two point-force loaded half-penny-shaped cracks in the piezoelectric material of side lengths $2A$ and thickness A. Let $x = x_1, y = x_2$, and $z = x_3$ denote the rectangular Cartesian coordinates. The z axis is assumed to coincide with the poling direction. Two point forces P_E are applied on the center of the permeable or impermeable crack of radius a in a large piezoelectric material $(A/a = 5.0)$, and the electric potential $\phi = -E_0 A$ is also added on the edge $z = A$. The electric potential on the edge $z = -A$ is $\phi = E_0 A$. For the permeable crack model, the

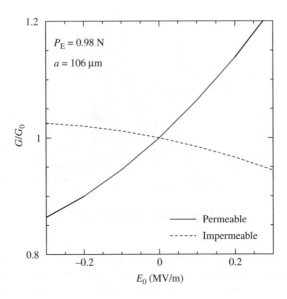

Figure 3.131 Energy release rate versus electric field of the IF specimen

$z = 0$ plane is equivalent to the ground, that is, $\phi = 0$. Because of the double symmetry of the specimen and loading, only one-quarter of the specimen is modeled.

Results are computed giving the values of the energy release rate G at maximum depth point of the crack normal to the poling direction for the residual force $P_E = 0.1P = 0.98$ N derived from the indentation plastic zone. The ratio G/G_0 for the permeable crack model and the corresponding ratio for the impermeable crack model are plotted versus electric field E_0 in Fig. 3.131 for $a = 106$ μm, where G_0 is the energy release rate under no electric field. For the permeable crack model, the energy release rate is higher for positive electric field and lower for negative electric field. Hence, the positive electric field will tend to enhance the crack growth and the negative electric field will tend to slow the crack growth. In the impermeable case, an opposite effect can be found.

3.10.4 Modified Small Punch Test

In the past decade, the small punch (SP) test had been used successfully to characterize the ductility and fracture resistance of metals and ceramics with specimens measuring only 0.5 mm in thickness [128, 129]. Here, we describe analytical and experimental efforts in bending fracture of piezoelectric ceramics by using a modified small punch (MSP) test technique.

We select PZT P-7 for the MSP test [130] and slice small thin-plate specimens of 10 × 10 × 0.5 mm. Poling is done along the axis of the 0.5 mm dimension. We conduct the MSP tests using a 10 kN screw-driven test machine and a power supply. Figure 3.132 shows the specimen, punch, and specimen holder designed for the MSP test. The specimen holder consists of an upper and a lower die. Load P is applied and recorded as a function of machine crosshead displacement δ, and fracture load is measured for each set of specimens for various uniform electric fields E_0. Machine crosshead speed is 0.2 mm/min. For $E_0 = 0, \pm 0.4, \pm 0.8$,

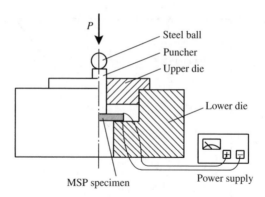

Figure 3.132   MSP test setup

and ± 1.0 MV/m, four or five tests are performed. After the test, the microstructures of the fracture surfaces are examined using a confocal scanning laser microscope (CSLM). Electrodes are removed from the fractured specimens, and the surfaces of the plates are also observed by an optical microscope.

We perform the FEA to calculate the MSP energy, maximum strain energy density, and so on [131]. Figure 3.133 shows the three-dimensional model of the MSP specimen with load and boundary conditions. The rectangular Cartesian coordinates $x = x_1, y = x_2$, and $z = x_3$ are used with the z axis coinciding with the poling direction. The contact between the specimen and the lower die is modeled using contact elements. A mechanical load is produced by the application of either a prescribed force P or a prescribed displacement δ along the z-direction. For electrical load, the electric potential $\phi = -E_0 h$ is added on the surface $z = h = 0.5$ mm; h is the specimen thickness. The surface $z = 0$ is grounded, that is, $\phi = 0$. Because of the double symmetry of the specimen and loading, only one-quarter of the specimen is modeled. The switching criterion of Eq. (3.127) is checked for every element and for every possible polarization direction to see if switching will occur.

Figure 3.134 shows the fracture initiation loads P_c under different electric fields E_0 obtained from the experiment. Although the test data show large scatter, positive electric field increases the fracture initiation load. Decrease in the fracture initiation load at $E_0 = 1.0$ MV/m is attributed to irreversible damage such as grain breakaway of PZT ceramics. The behavior of the fracture initiation load under negative electric fields $E_0 = -0.4, -0.8 (= -E_c), -1.0$ MV/m is very complicated because of localized polarization switching of the MSP specimens under electromechanical loading. Figure 3.135 shows CLSM images of fracture surfaces for the MSP specimens tested at (a) $E_0 = -1.0$ MV/m and (b) $E_0 = 1.0$ MV/m. The MSP specimen at $E_0 = -1.0$ MV/m has a relatively flat fracture surface. Under $E_0 = 1.0$ MV/m, the fracture surface appears rougher with more intergranular fracture. The samples under $E_0 = 1.0$ MV/m are found to possess conduction channels terminated with craters at both sides of the specimen surfaces (not shown here).

Figure 3.136 shows the measured load–displacement curve for various values of E_0. The FEA results of the load–displacement curve up to average P_c are also shown. Good agreement between the predictions and test data is observed except for low loads. The discrepancy at low

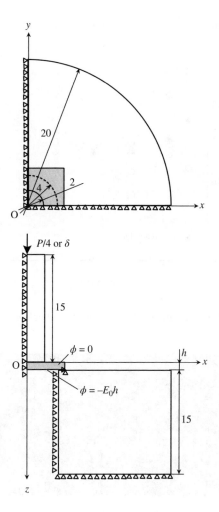

Figure 3.133   Finite element model of the MSP test

load levels may be caused by a nonuniform contact between the punch/specimen holder and specimen during testing. The critical MSP energy $E_{\mathrm{MSP}}^{\mathrm{c}}$ is calculated from the area under the load–displacement curve up to the maximum load (absorbed energy to failure), and Fig. 3.137 presents the $E_{\mathrm{MSP}}^{\mathrm{c}}$ including 180° and 90° switching effects for various electric fields E_0. It is interesting to note that the MSP energies for $E_0 = 0.8$ MV/m and $E_0 = -0.8$ MV/m have very nearly the same values, and a similar phenomenon is observed for the maximum strain energy density (no figure shown). The maximum strain energy density is defined as the maximum of the strain energy absorbed per unit volume and occurs at the observed crack initiation location. Figure 3.138 shows the 180° and 90° switching zones of the specimen under $P = 13.7$ N and $E_0 = -0.5$ MV/m. The localized polarization switching under electromechanical loading affects the fracture behavior of MSP specimen.

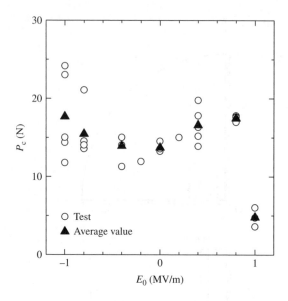

Figure 3.134   Fracture initiation load versus electric field of the MSP specimen

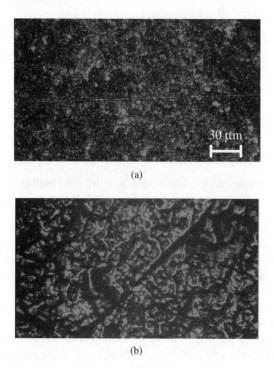

Figure 3.135   Fracture surfaces of the MSP specimen under (a) $E_0 = -1.0$ MV/m and (b) $E_0 = 1.0$ MV/m

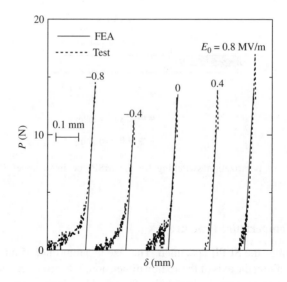

Figure 3.136   Load–displacement curves of the MSP specimen

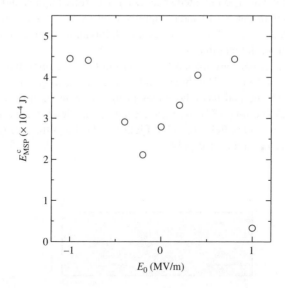

Figure 3.137   Critical MSP energy versus electric field of the MSP specimen

3.10.5   Single-Edge Precracked Beam Test

There has been a significant effort aimed at the understanding of piezoelectric fracture in cracked specimens. One important candidate for standardization of fracture toughness evaluation procedure for ceramics is the SEPB method, which was adopted in 1990 as a Japanese industrial standard (JIS R 1607) [124]. Here, we present the results of SEPB test and corresponding FEA of piezoelectric ceramics under electric field.

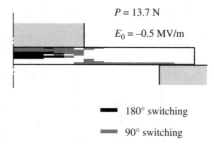

$P = 13.7\,\mathrm{N}$

$E_0 = -0.5\,\mathrm{MV/m}$

■ 180° switching

■ 90° switching

Figure 3.138 Image of polarization switching for the MSP specimen under $P = 13.7\,\mathrm{N}$ and $E_0 = -0.5\,\mathrm{MV/m}$

3.10.5.1   Room-Temperature Fracture

Figure 3.139 illustrates the SEPB specimen. The size is 5 mm thick, 5 mm wide, and 15 mm long. Poling is done along the axis of the 15 mm dimension. We introduce Vickers indents using a commercial microhardness testing machine. At least 11 indents are placed at the midspan on the polished surface of the specimen along a line, with indent diagonals normal to the edges of the specimen bar. The indented specimen is carefully aligned and centered on a steel bridge-anvil as shown in Fig. 3.140 and compressed until a precrack is formed. The specimen is unloaded immediately after "pop-in" to avoid additional slow crack extension. The crack thus produced has an initial length of a.

The precracked specimens are loaded (load P) to failure in a three-point flexure apparatus with a support span of 13 mm. The three-point loaded beam, in general, is more stable than the four-point loaded one and has advantages in the determination of a consistent fracture toughness for brittle materials [132]. We measure the fracture load P_c for each set of specimens under various uniform electric fields E_0 at RT. Figure 3.141 shows the testing setup. To generate the electric fields, we use a power supply.

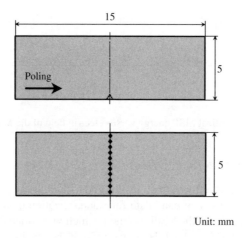

Unit: mm

Figure 3.139   SEPB Specimen

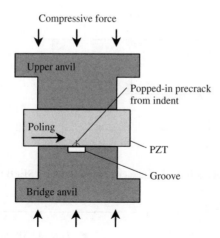

Figure 3.140   Schematic drawing of precracking technique

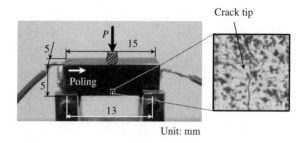

Unit: mm

Figure 3.141   SEPB test setup

We make plane strain calculation for the SEPB specimen under concentrated load to determine the energy release rate. Figure 3.142 shows a schematic representation of the finite element model. Let the coordinate axes $x = x_1$ and $z = x_3$ be chosen such that the $y = x_2$ axis coincides with the thickness direction. The z axis is oriented parallel to the poling direction. The three-point flexure specimen is a beam of width $W = 5$ mm and length $L = 15$ mm containing a crack of length a, and the span length is $S = 13$ mm. A mechanical load is produced by the application of a prescribed force P at $x = 0, z = 0$ along the x-direction. For electrical load, the electric potentials $\phi = -E_0L/2$ and $\phi = E_0L/2$ are applied at the edges $z = L/2, 0 \leq x \leq W$ and $z = -L/2, 0 \leq x \leq W$, respectively. In the case of the permeable crack model, the $z = 0$ plane is equivalent to the ground, that is, $\phi = 0$. Because of symmetry, only the right half of the model is used in the FEA. In the analysis, the energy release rate is computed using the J-integral approach. The switching criterion of Eq. (3.127) is checked for every element and for every possible polarization direction to see if switching will occur.

We briefly discuss the results of the crack length dependence of the fracture load P_c for PZT P-7 [133]. Figure 3.143 shows the fracture load P_c versus crack length to specimen width ratio a/W for P-7 under electric fields $E_0 = -0.1, 0, 0.2$ MV/m obtained from the experiment. Averaged values of two or three data are presented. Decrease in the fracture load with increasing

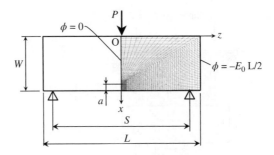

Figure 3.142   Finite element model of the SEPB test

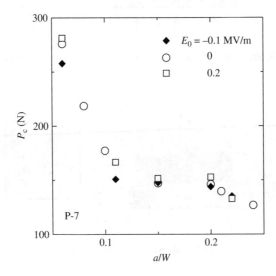

Figure 3.143   Fracture load versus crack length to specimen width ratio of the SEPB specimen

crack length is observed. Crack length dependence of the fracture load becomes small at about $a/W \geq 0.1$. In addition, the electric field can have little effect on the fracture load.

To understand the electric field dependence of the fracture further, we discuss the results for PZT PCM-80 (Panasonic Electronic Devices Co. Ltd. Japan) [134]. The material properties of PZT PCM-80 are

$$c_{11} = 17.0 \times 10^{10} \text{ N/m}^2$$

$$c_{12} = 10.6 \times 10^{10} \text{ N/m}^2$$

$$c_{13} = 11.5 \times 10^{10} \text{ N/m}^2$$

$$c_{33} = 16.5 \times 10^{10} \text{ N/m}^2$$

$$c_{44} = 3.05 \times 10^{10} \text{ N/m}^2$$

$$e_{15} = 13.7 \text{ C/m}^2 \tag{3.405}$$

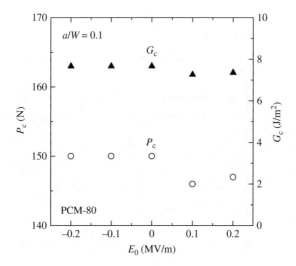

Figure 3.144 Fracture load and critical energy release rate versus electric field of the SEPB specimen

$$e_{31} = -5.99 \text{ C/m}^2$$

$$e_{33} = 15.6 \text{ C/m}^2$$

$$\epsilon_{11} = 95.2 \times 10^{-10} \text{ C/Vm}$$

$$\epsilon_{33} = 68.4 \times 10^{-10} \text{ C/Vm}$$

$$E_c = 2 \text{ MV/m}$$

Figure 3.144 shows the fracture load P_c versus electric field E_0 for PCM-80 with a crack length of $a = 0.5$ mm ($a/W = 0.1$) obtained from the experiment. Also shown is the critical energy release rate G_c. The critical energy release rate is calculated by the FEA using measured fracture load. The fracture load and critical energy release rate are not very much affected by the electric field.

Table 3.6 lists the energy release rate G for the SEPB specimens with permeable, impermeable, open, and discharging cracks of length $a = 0.5$ mm ($a/W = 0.1$) under load $P = 100$

Table 3.6 Comparison of energy release rates for four crack models of the SEPB specimen

Permeable	Impermeable	Open	Discharging
3.27 (3.26)	1.66 (1.60)	2.99 (3.03)	3.27 (3.27)

G (J/m²)

Values in parentheses are the results from the domain integral method

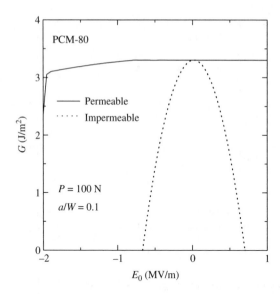

Figure 3.145 Energy release rate versus electric field of the SEPB specimen

N and electric field $E_0 = 0.5$ MV/m. Average values of the four contours are presented, and for the calculation of the energy release rate for the discharging crack model, the standard air breakdown strength $E_d = 3$ MV/m is used. Also shown are the results determined from the domain integral method [135]. Although the data are not shown here, the contribution from the crack interior $(2H^e(x^\Gamma)u_z^+(x^\Gamma)$ in Eq. (3.397)) is small for the open and discharging crack models. Note that the energy release rates predicted by the permeable and discharging crack models are not significantly different. From the table, it is clear that the values from the line and domain integral methods are in good agreement. Figure 3.145 displays the variation of the energy release rate G with the electric field E_0 of the permeable and impermeable crack models under $P = 100$ N for $a/W = 0.1$. For the permeable crack model, a monotonically increasing negative electric field causes polarization switching and decreases the value of the energy release rate. The localized polarization switching leads to a sudden decrease in G after E_0 reaches about -1.95 MV/m (below the coercive electric field $E_c = 2$ MV/m). In the impermeable case, a negative energy release rate is induced with positive and negative electric fields as is expected.

3.10.5.2 Cryogenic Fracture

Here, we deal with the cryogenic fracture behavior of cracked piezoelectric ceramics [136]. We cut PZT C-91 ceramics of 5 mm × 5 mm × 5 mm and produce the SEPB specimen by first poling a PZT and then bonding it between two unpoled PZTs by epoxy adhesive. The size of the specimen is 5 mm thick, 5 mm wide, and 15 mm long as shown in Fig. 3.146. We produce a precrack with an initial length of about $a = 1$ mm using the method previously described. The material properties of C-91 at RT are listed in (3.229).

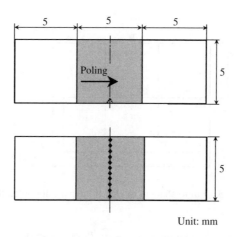

Figure 3.146 Partially poled SEPB specimen

Figure 3.147 SEPB setup at 77 K

We use the three-point bending apparatus with a span of 13 mm and perform the fracture test under uniform electric field E_0 at 77 K. Testing at 77 K is accomplished by submerging the test fixture and specimen in liquid nitrogen. Figure 3.147 shows the cryogenic testing setup. We measure the fracture load P_c, and after testing, we examine the fracture surfaces by scanning electron microscopy (SEM).

We perform plane strain FEA for the partially poled SEPB specimen under concentrated load to calculate the energy release rate. Figure 3.148 shows a schematic representation of the finite element model. Let the coordinate axes $x = x_1$ and $z = x_3$ be chosen such that the $y = x_2$ axis coincides with the thickness direction. The z axis is oriented parallel to the poling direction. The three-point flexure specimen with a span $S = 13$ mm is a beam of width $W = 5$ mm and length $L = 15$ mm containing a crack of length $a = 1$ mm. The length between two electrodes is $L_0 = 5$ mm. A mechanical load is produced by the application of a prescribed force P at $x = 0, z = 0$ along the x-direction. For electrical load, the electric potentials $\phi = -E_0 L_0/2$

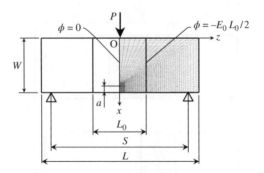

Figure 3.148 Finite element model of the partially poled SEPB test

and $\phi = E_0 L_0/2$ are applied at the interfaces $z = L_0/2, 0 \leq x \leq W$ and $z = -L_0/2, 0 \leq x \leq W$, respectively. In the case of the permeable crack model, the $z = 0$ plane is equivalent to the ground, that is, $\phi = 0$. Because of symmetry, only the right half of the model is used. In the FEA, the energy release rate is computed using the J-integral approach. At 77 K, we employ the temperature-dependent piezoelectric coefficient (Eq. (3.279)). In the elastic compliance s_{33} of C-91 at 77 K, the measured value by compression test is about 8.6×10^{-12} m²/N [137]. Here, the elastic compliance s_{33} is assumed to have a linear relationship with temperature T between RT and 77 K [138], and the other elastic compliances $s_{11}, s_{12}, s_{13}, s_{44}$ are presumed to have the same temperature dependence. For simplicity, the permittivities $\epsilon_{11}^T, \epsilon_{33}^T$ of C-91 are assumed to be independent of temperature.

Table 3.7 lists the measured fracture load P_c for C-91 ceramics under various electric fields E_0 at RT and 77 K. Average values of two to four measurements are shown. It is found that the fracture loads at 77 K are larger than those at RT. This seems to be due to the tetragonal-to-monoclinic phase transformation as shown in Fig. 3.20. Figure 3.149 shows the

Table 3.7 Fracture loads of the SEPB specimen at RT and 77 K

	P_c (N)			
E_0 (MV/m)	-0.4	-0.2	0	0.2
RT	90.2	100	121	125
77 K	203	191	188	198

(a)   (b)

Figure 3.149 Fracture appearances of the SEPB specimen under $E_0 = 0$ V/m at (a) RT and (b) 77 K

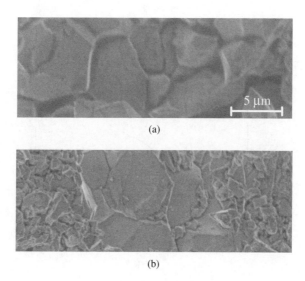

(a)

(b)

Figure 3.150   Fracture surfaces of the SEPB specimen under $E_0 = -0.4$ MV/m at (a) RT and (b) 77 K

fracture appearance of C-91 ceramics under $E_0 = 0$ V/m at RT and 77 K. All specimens fail in a brittle manner. It is interesting to note that C-91 ceramics at 77 K tend to fracture in multiple places (Fig. 3.149(b)). Figure 3.150 shows the SEM images of the fracture surfaces near the crack tip in C-91 ceramics under $E_0 = -0.4$ MV/m at RT and 77 K. At 77 K, formations of precipitates on the surface are clearly visible (Fig. 3.150(b)). This is due to the phase transformation and the transformation increases the fracture load.

Figure 3.151 shows the dependence of the energy release rate G on the temperature T in C-91 ceramics with a crack of length $a = 1$ mm ($a/W = 0.2$) for the permeable crack model under load $P = 100$ N and electric field $E_0 = 0$ V/m. The energy release rate decreases with decreasing temperature, and one of the reasons is a shift in the MPB between the tetragonal and rhombohedral/monoclinic phases, that is, tetragonal-to-monoclinic phase transformation. Hence, the crack propagation is retarded at cryogenic temperatures.

3.10.6   Double Torsion Test

Double torsion (DT) technique for evaluating the fracture mechanics parameters has a number of advantages compared to other available methods [139]. For example, the stress intensity factor is independent of crack length. Here, we present the results of DT test and corresponding FEA of piezoelectric ceramics under electric field [140].

Specimen geometry used is shown in Fig. 3.152. The rectangular Cartesian coordinates $x = x_1, y = x_2$, and $z = x_3$ are used with the z axis coinciding with the poling direction. PZT sample of width $W_p = 5$ mm, thickness 5 mm, and length 30 mm is cut. The specimen is produced by first poling a 5 mm wide PZT beam and then bonding it between two wider brass beams of width 7.5 mm, thickness 5 mm, and length 30 mm with high-strength epoxy. A side groove of depth 2.5 mm and width 1 mm is machined in the PZT beam. Before testing, a thin notch is cut at the end of the PZT beam to a depth of 2.5 mm and a length of $a = 5$ mm. The used PZTs are commercially supplied P-7 and C-91.

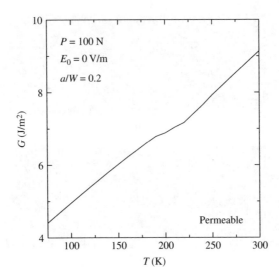

Figure 3.151 Energy release rate versus temperature of the SEPB specimen

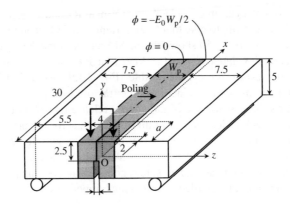

Figure 3.152 DT specimen

We apply concentrated loads $P/2$ to the specimen at $x = 0$ mm, $y = 2.5$ mm, and $z = \pm 2$ mm and measure the load that causes fracture for each set of specimens under various uniform electric fields E_0. To generate the electric field, we use a power supply.

3D finite element calculations are made to determine the fracture mechanics parameters such as energy release rate for the DT specimen. A mechanical load is produced by the application of prescribed forces $P/2$ at $x = 0$ mm, $y = 2.5$ mm, and $z = \pm 2$ mm. For electrical load, the electric potentials $\phi = -E_0 W_p/2$ and $\phi = E_0 W_p/2$ are applied at the interfaces $z = W_p/2$ and $z = -W_p/2$, respectively. In the case of the permeable crack model, the $z = 0$ plane is equivalent to the ground, that is, $\phi = 0$. Because of symmetry, only the right half of the model is used in the FEA.

Table 3.8 lists the measured fracture loads P_c of P-7 and C-91 ceramics under various values of electric field E_0. The fracture loads depend on the material properties. Positive electric field

Table 3.8 Fracture loads of the DT specimens

	P_c (N)						
E_0 (MV/m)	−0.2	−0.1	−0.05	0	0.05	0.1	0.2
P-7	–	188	197	205	222	246	–
C-91	167	–	–	179	–	–	194

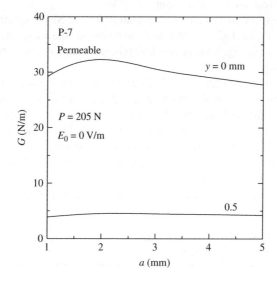

Figure 3.153 Energy release rate versus crack length of the DT specimen

increases the fracture load, and negative one decreases it. Hence, the crack opens less under positive electric field than under negative electric field.

Figure 3.153 shows a plot of the energy release rate G versus crack length a for P-7 ceramics of the permeable crack model at $y = 0$ and 0.5 mm under load $P = 205$ N and electric field $E_0 = 0$ V/m. The magnitude of the energy release rate increases with the crack length reaching a maximum value and then decreases. The energy release rate is highest on the $y = 0$ mm plane. Although the results are not shown here, positive electric field decreases the energy release rate for the permeable crack model, while negative electric field has an opposite effect. The increase in the fracture load with increasing positive electric field is attributed to the decrease of the energy release rate with increasing positive electric field.

3.10.7 Fatigue of SEPB Specimens

One critical limitation on the performance of piezoelectric materials is caused by electric fatigue [141]. In this section, we review the results of the fatigue behavior of piezoelectric ceramics under electromechanical loads using SEPB method.

3.10.7.1   Static Fatigue under DC Electric Field

It is known that the electric field can affect the fatigue life of the piezoelectric devices under sustained mechanical loading. Here, we present results of the static fatigue behavior of piezoelectric ceramics under constant mechanical load and DC electric field [142]. We use the SEPB specimen as shown in Fig. 3.139. Specimen is commercially supplied PZT PCM-80. The precrack length is about 0.5 mm. We perform the static fatigue test using the three-point bending apparatus with a span of 13 mm as shown in Fig. 3.141 and measure the time-to-failure under constant mechanical load $P = P_0$ and DC electric field E_0. After testing, we examine the static fatigue and overload fracture surfaces using the SEM.

We perform plane strain finite element calculation for the SEPB specimen under constant mechanical load $P = P_0$ and DC electric field E_0 to determine the energy release rate. The finite element model is shown in Fig. 3.142. In the analysis, the energy release rate is computed using the J-integral approach.

The measured times-to-failure t_f of PCM-80 ceramics under DC electric fields $E_0 = -0.1, 0, 0.1$ MV/m are listed with standard deviations from two to four experiments in Table 3.9. Although measured values show large scatter, times-to-failure under $E_0 = 0.1$ MV/m and $E_0 = -0.1$ MV/m occur significantly lower and higher than that under $E_0 = 0$ V/m, respectively. The difference between the static fatigue fracture surface and catastrophic quasistatic fracture surface is illustrated in Fig. 3.154 for $E_0 = -0.1$ MV/m. The micrograph reveals a predominantly intergranular fracture path near the precrack tip as shown in Fig. 3.154(a). The PZT ceramics under negative electric field have the greatest

Table 3.9   Times-to-failure of the SEPB specimens under DC electric field

		t_f (s)	
E_0 (MV/m)	-0.1	0	0.1
$P_0 = 135$ N	1573 ± 473	776 ± 482	361 ± 10
140 N	203 ± 120	108 ± 42	43 ± 51

 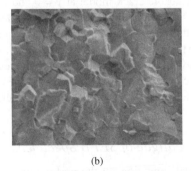

(a)   (b)

Figure 3.154   Fracture surfaces (a) near the precrack tip and (b) away from the precrack tip of the SEPB specimen under $E_0 = -0.1$ MV/m

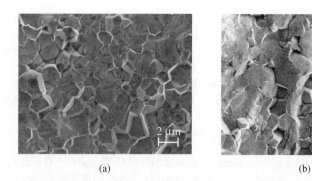

(a)   (b)

Figure 3.155   Fracture surfaces (a) near the precrack tip and (b) away from the precrack tip of the SEPB specimen under $E_0 = 0.1$ MV/m

resistance for long time loads and exhibit a large degree of intergranular cracking. On the other hand, the crack path formed in the catastrophic fracture is predominantly transgranular fracture (Fig. 3.154(b)). Figure 3.155 shows similar micrographs for $E_0 = 0.1$ MV/m. The PZT ceramics under positive electric field show intergranular and transgranular regions. The time-to-failure of PCM-80 ceramics decreases as the mechanical load increases similar to that of other nonpiezoelectric brittle materials. Although the results are not shown here, the time-to-failure also decreases with the increase in the energy release rate.

3.10.7.2   Static Fatigue under AC Electric Field

Here, we present the results of the static fatigue behavior of piezoelectric ceramics under constant mechanical load and AC electric field [143]. We use the SEPB specimen as shown in Fig. 3.146. Specimens are commercially supplied PZTs C-91 and C-203. The material properties of C-91 are listed in (3.229), and the material properties of C-203 are

$$s_{11} = 13.9 \times 10^{-12} \ \text{m}^2/\text{N}$$
$$s_{12} = -4.1 \times 10^{-12} \ \text{m}^2/\text{N}$$
$$s_{13} = -6.4 \times 10^{-12} \ \text{m}^2/\text{N}$$
$$s_{33} = 16.7 \times 10^{-12} \ \text{m}^2/\text{N}$$
$$s_{44} = 42.7 \times 10^{-12} \ \text{m}^2/\text{N}$$
$$d_{15} = 522 \times 10^{-12} \ \text{m/V}$$
$$d_{31} = -144 \times 10^{-12} \ \text{m/V} \qquad\qquad (3.406)$$
$$d_{33} = 325 \times 10^{-12} \ \text{m/V}$$
$$\epsilon_{11}^{T} = 131 \times 10^{-10} \ \text{C/Vm}$$
$$\epsilon_{33}^{T} = 129 \times 10^{-10} \ \text{C/Vm}$$
$$\rho = 7700 \ \text{kg/m}^3$$
$$E_{c} = 2 \ \text{MV/m}$$

The precrack length is about 0.5 mm. We perform the static fatigue test using the three-point bending apparatus with a span of 13 mm, and measure the time-to-failure under constant mechanical load $P = P_0$ and AC electric field $E_0 \exp(i\omega t)$. After testing, we examine the static fatigue and overload fracture surfaces using the SEM.

We perform plane strain finite element calculation for the SEPB specimen under constant mechanical load $P = P_0$ and AC electric field $E_0 \exp(i\omega t)$ to determine the energy release rate. The finite element model is shown in Fig. 3.148, where $\phi = -E_0 L_0 / 2$ is replaced by $\phi = -(E_0 L_0 / 2) \exp(i\omega t)$. In the analysis, the energy release rate with the domain wall motion effect is computed using the J-integral approach.

The average times-to-failure t_f from three measured data of C-91 and C-203 ceramics under $P_0 = 115$ N and various electric fields are summarized in Table 3.10. The values of average time-to-failure of C-91 ceramics decrease by about 30 %, 56 %, and 89 % due to the electric fields of $E_0 = 0.04$ MV/m at frequency $f = 0, 50$, and 400 Hz, respectively. It is also found that the times-to-failure of C-203 ceramics are significantly lower than those of C-91 ceramics. Figure 3.156 shows the dependence of the energy release rate G on the AC electric field

Table 3.10 Times-to-failure of the SEPB specimens under AC electric field

		t_f (s)		
E_0 (MV/m)	0	0.04	0.04	0.04
f (Hz)	–	0	50	400
C-91	3511	2444	1533	381
C-203	68	–	38	–

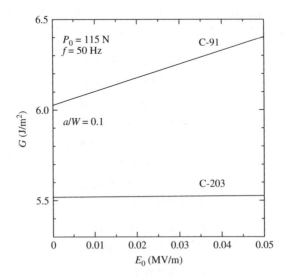

Figure 3.156 Energy release rate versus AC electric field of the SEPB specimen

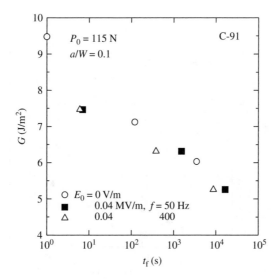

Figure 3.157 Energy release rate versus time-to-failure of the SEPB specimen

amplitude E_0 at $f = 50\,\text{Hz}$ for C-91 and C-203 ceramics with a crack of length $a = 0.5\,\text{mm}$ $(a/W = 0.1)$ for the permeable crack model under $P_0 = 115\,\text{N}$. The energy release rate with the domain wall motion effect increases as the AC electric field amplitude increases. The decrease in the time-to-failure under AC electric field seems to be attributed to the increase of the energy release rate under constant mechanical load. It is also shown that the domain wall motion contribution to the energy release rate for C-203 ceramics is small in comparison with the case for C-91 ceramics. This is because the coercive electric field of C-203 ceramics is larger than that of C-91 ceramics. Figure 3.157 shows the energy release rate G versus average time-to-failure t_f of C-91 ceramics under $P_0 = 115\,\text{N}$ and $E_0 = 0, 0.04\,\text{M/m}$ $(f = 50, 400\,\text{Hz})$ for $a/W = 0.1$. The time-to-failure is observed to decrease with the increase in the energy release rate.

3.10.7.3 Dynamic Fatigue under DC Electric Field

Certain environments may affect formation and extension of cracks over time and at stress levels well below the final failure load. This process is called dynamic fatigue or slow crack growth. Here, we present the results of the dynamic fatigue behavior of piezoelectric ceramics under DC electric field [144]. We use the SEPB specimen as shown in Fig. 3.139 and consider the commercially supplied PZT PCM-80. The precrack length is about 0.5 mm. We conduct the constant load-rate test using the three-point bending apparatus with a span of 13 mm as shown in Fig. 3.141 and measure the fracture loads for various load rates dP/dt under DC electric field E_0.

We perform plane strain finite element calculation for the SEPB specimen under mechanical load P and DC electric field E_0 to determine the energy release rate. The finite element model is shown in Fig. 3.142, and the energy release rate is computed using the J-integral approach.

A summary of the results of constant load-rate testing for PCM-80 ceramics under DC electric field $E_0 = 0, 0.1\,\text{MV/m}$ is shown in Fig. 3.158, where $\log P_c$ (fracture load) is plotted

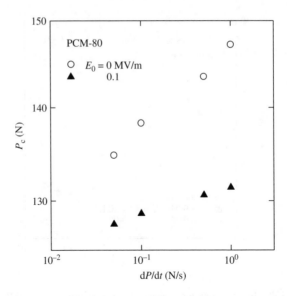

Figure 3.158   Fracture load versus load rate of the SEPB specimen

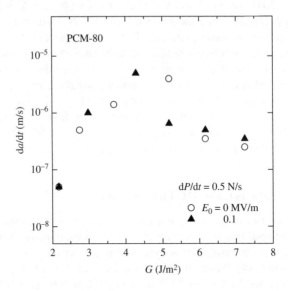

Figure 3.159   Crack propagation velocity versus energy release rate of the SEPB specimen

as a function of log (dP/dt) (load-rate). Average values of two or three data are shown. The PZT ceramics show an increase in the fracture load as the load rate increases. The results also indicate that an overall decrease in the fracture load occurs when testing under $E_0 = 0.1$ MV/m. Figure 3.159 shows a comparison of crack propagation velocities da/dt under $E_0 = 0, 0.1$ MV/m at $dP/dt = 0.5$ N/s, plotted as a function of the energy release rate G. The crack propagation velocity is obtained by the experiment, whereas the energy release rate is calculated by the FEA. The crack propagation velocity increases very quickly with the energy release rate, reaching a peak and then decreases before final failure (the rightmost dots). The decrease in the crack propagation velocity is probably associated with microcrack nucleation and crack bridging. The value of the critical energy release rate G_c for PCM-80 ceramics under $E_0 = 0$ and 0.1 MV/m is about 7.24 J/m^2.

3.10.7.4 Dynamic Fatigue under AC Electric Field

Here, we present results of the dynamic fatigue behavior of piezoelectric ceramics under AC electric field [145]. We use the SEPB specimen as shown in Fig. 3.146 and consider the commercially supplied PZTs C-91 and C-203. The precrack length is about 0.5 mm. We conduct the constant load-rate test using the three-point bending apparatus with a span of 13 mm and measure the fracture loads for various load rates dP/dt under AC electric field $E_0 \exp(i\omega t)$.

We perform plane strain finite element calculation for the SEPB specimen under mechanical load P and AC electric field $E_0 \exp(i\omega t)$ to determine the energy release rate. The finite element model is shown in Fig. 3.148, where $\phi = -E_0 L_0/2$ is replaced by $\phi = -(E_0 L_0/2) \exp(i\omega t)$. In the analysis, the energy release rate with the domain wall motion effect is computed using the J-integral approach.

Table 3.11 lists the experimental results of the fracture load P_c for C-91 and C-203 ceramics at the load rate $dP/dt = 0.05, 3$ N/s under various electric fields. Average values of two or three data are shown. The fracture loads of C-91 and C-203 ceramics under $E_0 = 0$ V/m at $dP/dt = 0.05$ N/s decrease about 11 and 3.5% compared to the case at $dP/dt = 3$ N/s. That is, the effect of load rate on the fracture load for C-203 ceramics is small. It is also shown that the fracture load at $dP/dt = 0.05$ N/s decreases due to the AC electric field, and that the fracture load at 400 Hz is lower than that at 50 Hz. Table 3.12 lists the critical energy release rate G_c calculated by the FEA at $dP/dt = 0.05$ N/s under various electric fields. It seems that the critical energy release rate is not very much affected by the AC electric field.

3.10.7.5 Cyclic Fatigue under DC Electric Field

Literature studies of cyclic fatigue in piezoelectric ceramics under both mechanical and electrical loads are sparse and inconclusive. Here, we present results of the cyclic fatigue behavior

Table 3.11 Fracture loads of the SEPB specimens under AC electric field

	dP/dt (N/s)	P_c (N)		
E_0 (MV/m)		0	0.04	0.04
f (Hz)		–	50	400
C-91	0.05	128	125	119
	3	144	–	–
C-203	0.05	137	133	127
	3	142	–	–

Table 3.12   Critical energy release rates of the SEPB specimens under AC electric field

	dP/dt (N/s)		G_c (N)	
E_0 (MV/m)		0	0.04	0.04
f (Hz)		–	50	400
C-91	0.05	7.0	7.1	6.5

of piezoelectric ceramics under sinusoidal mechanical load and DC electric field [146]. We use the SEPB specimen as shown in Fig. 3.139 and consider the commercially supplied PZT PCM-80. The precrack length is about 0.5 mm. We carry out the cyclic crack growth test using the three-point bending apparatus with a span of 13 mm as shown in Fig. 3.141 and measure the fatigue crack length under DC electric field E_0. Constant amplitude sinusoidal loads are applied to the specimen at a constant load frequency of $f_M = 1$ or 10 Hz. The load ratio, defined as the ratio of minimum load P_{min} to maximum load P_{max} of the fatigue cycle, is $R = 0.2$ or 0.5. Following cyclic experiments, the SEM is used to examine fatigue fracture surfaces.

We perform plane strain finite element calculation for the SEPB specimen under maximum load $P = P_{max}$ and DC electric field E_0 to determine the maximum energy release rate. The finite element model is shown in Fig. 3.142, and the energy release rate is computed using the J-integral approach.

Figure 3.160 shows the plots of fatigue crack growth rate da/dN as a function of the maximum energy release rate G_{max} under DC electric fields $E_0 = -0.1, 0, 0.1$ MV/m at the load frequency $f_M = 1, 10$ Hz with the load ratio $R = 0.5$. At $f_M = 1$ Hz, the crack growth rates exhibit a local minimum resulting in V-shaped growth rate behavior. In the two parts of the

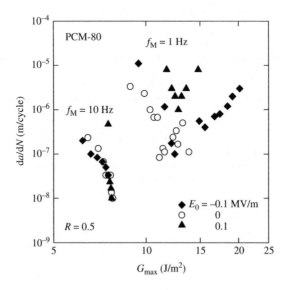

Figure 3.160   Fatigue crack growth rate versus maximum energy release rate of the SEPB specimen

V-shape, that is, before and after the minimum crack growth rate, the data exhibit a linear relation in the log–log plot. Although the micrographs are not shown here, the region of crack initiation exhibits a significant degree of intergranular cracking (before the minimum da/dN), and the surface topography changes to transgranular fatigue crack propagation (after the minimum da/dN) as the crack continues to propagate. The experimental data also suggest that there is a large influence of electric field on the fatigue crack growth rate. At $f_M = 10\,\text{Hz}$, the crack growth rates appear to display a negative dependency of the maximum energy release rate. Although the results are not shown here, the crack growth rate at $R = 0.2$ decreases strongly with increasing maximum energy release rate. That is, the cracks are retarded during cyclic loading in PCM-80 ceramics, and no additional crack growth is observed.

The aforementioned material PCM-80 has high coercive electric field ($E_c = 2\,\text{MV/m}$), which is a characteristic of hard-type PZT. Recently, Shindo et al. [147] considered soft-type PZT C-91 ($E_c = 0.35\,\text{MV/m}$) and discussed the effect of negative electric field on the cyclic fatigue behavior of piezoelectric ceramics under sinusoidal mechanical load and DC electric field. They concluded that the number of cycles to failure under low negative electric field ($E_0 = -0.04\,\text{MV/m}$) and high negative electric field ($E_0 = -0.2\,\text{MV/m}$) are significantly higher and lower than that under no electric field. On the microscopic level, the mode of crack growth under low negative electric field was primarily intergranular. Trace of transgranular patterns was observed under high negative electric field.

3.10.7.6 Cyclic Fatigue under AC Electric Field

Here, we present results of the cyclic fatigue behavior of piezoelectric ceramics under sinusoidal mechanical load and AC electric field [148]. We use the SEPB specimen as shown in Fig. 3.146 and consider the commercially supplied PZT C-91. The precrack length is about 0.5 mm. We conduct the cyclic fatigue test using the three-point bending apparatus with a span of 13 mm and measure the numbers of cycles to failure for various maximum loads $P = P_{\text{max}}$ under AC electric field $E_0 \exp(i\omega t)$. Constant amplitude sinusoidal loads are applied to the specimen at a constant load frequency of $f_M = 50\,\text{Hz}$. The load ratio is $R = 0.5$. The AC electric field frequency is also set to $f = 50\,\text{Hz}$, and the phase between the sinusoidal mechanical load and AC electric field is intended to be 0. Following cyclic experiments, the SEM is used to examine fatigue fracture surfaces.

We perform plane strain finite element calculation for the SEPB specimen under maximum load $P = P_{\text{max}}$ and AC electric field $E_0 \exp(i\omega t)$ to determine the maximum energy release rate. The finite element model is shown in Fig. 3.148, where $\phi = -E_0 L_0/2$ is replaced by $\phi = -(E_0 L_0/2) \exp(i\omega t)$. In the analysis, the energy release rate with the domain wall motion effect is computed using the J-integral approach.

Figure 3.161 shows the maximum energy release rate G_{max} versus number of cycles to failure N_f under AC electric field amplitudes $E_0 = 0, 0.04\,\text{MV/m}$ ($f = 50\,\text{Hz}$). We plot only the average N_f of two measured data. We also show the value under $E_0 = -0.04, -0.2\,\text{MV/m}$ at $f = 0\,\text{Hz}$ [147] for comparison. It is interesting to note that the slope of the maximum energy release rate versus number of cycles to failure curve becomes steeper due to AC electric field. As mentioned earlier, the numbers of cycles to failure under $E_0 = -0.04$ and $-0.2\,\text{MV/m}$ at $f = 0\,\text{Hz}$ occur higher and lower than that determined for $E_0 = 0\,\text{V/m}$. Although the micrographs are not shown here, C-91 ceramics under $E_0 = 0\,\text{V/m}$ for a G_{max} of about 5.2 J/m^2 show intergranular and transgranular fracture regions, and the crack paths in the fracture for C-91

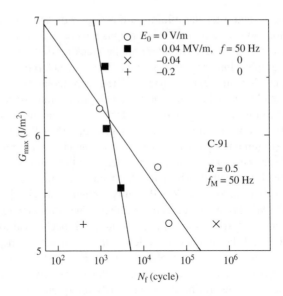

Figure 3.161 Maximum energy release rate versus number of cycles to failure of the SEPB specimen

ceramics under $E_0 = 0.04$ MV/m at $f = 50$ Hz for a $G_{\max}$ of about 5.5 J/m^2 are predominantly transgranular.

3.11   Summary

We have discussed the basic macroscopic response while accounting for microscale phenomena, such as polarization switching and domain wall motion, in piezoelectric material systems and structures. We have also described the cryogenic and high-temperature electromechanical responses of piezoelectric devices. In addition, we have reported the results of the electric field dependence of cracking of piezoelectric materials. We have shown that, in modeling cracks in piezoelectric materials, the impermeable and open crack surface assumptions can lead to significant errors and the permeable crack surface is better assumption.

The combined influence of electromechanical loading requires more research to provide a deeper understanding of piezoelectric response for future demanding micro-/nanodevice applications. There is a great interest in accurate multiscale computational methods and piezoelectric material models for linking atomistic, domain, and macroscale behavior. We believe that future investigation will concentrate on this area.

An area of future emphasis should be on the microstructural analysis of lead-free ferroelectric materials. Although grain size [149, 150] and oxygen activity [151, 152] have been found to play a significant role in piezomechanical properties and microstructures of lead-free barium titanate (BaTiO$_3$) ceramics, it is difficult to find the optimal conditions experimentally. Phase field approach can serve as an efficient tool for the mentioned purpose [153]. We are currently developing the phase field simulation framework for studying the effects of grain size and oxygen vacancy density on the electromechanical response of BaTiO$_3$ polycrystals [154].

References

[1] R. A. Toupin, "The elastic dielectric," *J. Ration. Mech. Anal.* **5**(6), 849 (1956).

[2] R. A. Toupin, "A dynamical theory of elastic dielectrics," *Int. J. Eng. Sci.* **1**(1), 101 (1963).

[3] A. C. Eringen, "On the foundations of electroelastostatics," *Int. J. Eng. Sci.* **1**(1), 127 (1963).

[4] Z. T. Kurlandzka, "Influence of electrostatic field on crack propagation in elastic dielectric," *Bull. Acad. Pol. Sci., Ser. Sci. Tech.* **23**(7), 333 (1975).

[5] Y. E. Pak and G. Herrmann, "Conservation laws and the material momentum tensor for the elastic dielectric," *Int. J. Eng. Sci.* **24**(8), 1365 (1986).

[6] Y. E. Pak and G. Herrmann, "Crack extension force in a dielectric medium," *Int. J. Eng. Sci.* **24**(8), 1375 (1986).

[7] I. N. Sneddon, *Fourier Transforms*, McGraw-Hill, New York, 1951.

[8] Y. Shindo and F. Narita, "The planar crack problem for a dielectric medium," *Arch. Mech.* **56**(6), 425 (2004).

[9] J. D. Eshelby, "The force on an elastic singularity," *Philos. Trans. R. Soc. London* **A244**(877), 87 (1951).

[10] J. R. Rice, "A path independent integral and the approximate analysis of strain concentration by notches and cracks," *ASME J. Appl. Mech.* **35**(2), 379 (1968).

[11] G. C. Sih, "Strain-energy-density factor applied to mixed mode crack problems," *Int. J. Fract.* **10**(3), 305 (1974).

[12] Y. Shindo and F. Narita, "Wave scattering in a cracked dielectric polymer under a uniform electric field," *Boundary Value Prob.* **2009**, 949124 (2009).

[13] Y. Shindo, H. Katsura and W. Yan, "Dynamic stress intensity factor of a cracked dielectric medium in a uniform electric field," *Acta Mech.* **117**(1/4), 1 (1996).

[14] R. D. Mindlin, "On the equations of motion of piezoelectric crystals," *Problems of Continuum Mechanics*, I. E. Block and J. R. M. Radok (eds.), SIAM, Philadelphia, PA, p. 282 (1961).

[15] J. F. Nye, *Physical Properties of Crystals*, The Clarendon Press, Oxford, 1957.

[16] H. F. Tiersten, *Linear Piezoelectric Plate Vibration*, Plenum Press, New York, 1969.

[17] G. A. Mausin, *Continuum Mechanics of Electromagnetic Solid*, Elsevier Science, Amsterdam, 1988.

[18] B. Jaffe, W. R. Cook and H. Jahhe, *Piezoelectric Ceramics*, Academic Press, New York, p. 135, 1971.

[19] B. Noheda, D. E. Cox, G. Shirane, R. Guo, B. Jones and L. E. Cross, "Stability of the monoclinic phase in the ferroelectric perovskite $PbZr_{1-x}Ti_xO_3$," *Phy. Rev. B* **63**(1), 014103 (2000).

[20] P. J. Chen and T. J. Tucker, "One dimensional polar mechanical and dielectric responses of the ferroelectric ceramic PZT 65/35 due to domain switching," *Int. J. Eng. Sci.* **19**(1), 147 (1981).

[21] S. C. Hwang, C. S. Lynch and R. M. McMeeking, "Ferroelectric/ferroelastic interactions and a polarization switching model," *Acta Metall. Mater.* **43**(5), 2073 (1995).

[22] R. E. Loge and Z. Suo, "Nonequilibrium thermodynamics of ferroelectric domain evolution," *Acta Mater.* **44**(8), 3429 (1996).

[23] J. E. Huber and N. A. Fleck, "Multi-axial electrical switching of a ferroelectric: theory versus experiment," *J. Mech. Phys. Solids* **49**(4), 785 (2001).

[24] C. T. Sun and A. Achuthan, "Domain-switching criteria for ferroelectric materials subjected to electrical and mechanical loads," *J. Am. Ceram. Soc.* **87**(3), 395 (2004).

[25] G. Arlt and H. Dederichs, "Complex elastic, dielectric and piezoelectric constants by domain wall damping in ferroelectric ceramics," *Ferroelectrics* **29**, 47 (1980).

[26] S. Li, W. Cao and L. E. Cross, "The extrinsic nature of nonlinear behavior observed in lead zirconate titanate ferroelectric ceramic," *J. Appl. Phys.* **69**(10), 7219 (1991).

[27] C. Kittel, "Domain boundary motion in ferroelectric crystals and the dielectric constant at high frequency," *Phys. Rev.* **83**(2), 458 (1951).

[28] J. Fousek and B. Brezina, "Relaxation of 90° domain walls of $BaTiO_3$ and their equation of motion," *J. Phys. Soc. Jpn.* **19**(6), 830 (1964).

[29] A. G. Luchaninov, A. V. Shil'Nikov, L. A. Shuvalov, and I. J. U. Shipkova, "The domain processes and piezoeffect in polycrystalline ferroelectrics," *Ferroelectrics* **98**, 123 (1989).

[30] S. Li, A. S. Bhalla, R. E. Newnham and L. E. Cross, "Quantitative evaluation of extrinsic contribution to piezoelectric coefficient d_{33} in ferroelectric PZT ceramics," *Mater. Lett.* **17**(1/2), 21 (1993).

[31] F. Narita, Y. Shindo and M. Mikami, "Analytical and experimental study of nonlinear bending response and domain wall motion in piezoelectric laminated actuators under ac electric fields," *Acta Mater.* **53**(17), 4523 (2005).

[32] W. Nowacki, *Dynamic Problems of Thermoelasticity*, Noordhoff, Leiden, 1975.

[33] T. R. Tauchert, "Piezothermoelastic behavior of a laminated plate," *J. Therm. Stresses* **15**(1), 25 (1992).

[34] R. M. Jones, *Mechanics of Composite Materials*, Hemisphere, New York, 1975.

[35] J. G. Smits, S. I. Dalke and T. K. Cooney, "The constituent equations of piezoelectric bimorphs," *Sens. Actuators, A* **28**(1), 41 (1991).

[36] K. Hayashi, Y. Shindo and F. Narita, "Displacement and polarization switching properties of piezoelectric laminated actuators under bending," *J. Appl. Phys.* **94**(7), 4603 (2003).

[37] ANSYS http://www.ansys.com/.

[38] F. Narita, Y. Shindo and K. Hayashi, "Bending and polarization switching of piezoelectric laminated actuators under electromechanical loading," *Comput. Struct.* **83**(15/16), 1164 (2005).

[39] Q. M. Wang, Q. Zhang, B. Xu, R. Liu and L. E. Cross, "Nonlinear piezoelectric behavior of ceramic bending mode actuators under strong electric fields," *J. Appl. Phys.* **86**(6), 3352 (1999).

[40] K. Takagi, J.-F. Li, S. Yokoyama, R. Watanabe, A. Almajid and M. Taya, "Design and fabrication of functionally graded PZT/Pt piezoelectric bimorph actuator," *Sci. Technol. Adv. Mater.* **3**(2), 217 (2002).

[41] J. Qiu, J. Tani, T. Ueno, T. Morita, H. Takahashi and H. Du, "Fabrication and high durability of functionally graded piezoelectric bending actuators," *Smart Mater. Struct.* **12**, 115 (2003).

[42] M. Taya, A. A. Almajid, M. Dunn and H. Takahashi, "Design of bimorph piezo-composite actuators with functionally graded microstructure," *Sens. Actuators, A* **107**(3), 248 (2003).

[43] S. Fang, Y. Shindo, F. Narita and S. Lin, "Three-dimensional electroelastic analysis of functionally graded piezoelectric plate via state vector approach," *ZAMM* **86**(8), 628 (2006).

[44] Y. Shindo, F. Narita, M. Mikami and F. Saito, "Nonlinear dynamic bending and domain wall motion in functionally graded piezoelectric actuators under ac electric fields: simulation and experiment," *JSME Int. J., A* **49**(2), 188 (2006).

[45] Y. Shindo, F. Narita and J. Nakagawa, "Dynamic electromechanical response and self-sensing of functionally graded piezoelectric cantilever transducers," *J. Intell. Mater. Syst. Struct.* **20**(1), 119 (2009).

[46] Y. Shindo, F. Narita and J. Nakagawa, "Nonlinear bending characteristics and sound pressure level of functionally graded piezoelectric actuators by ac electric fields: simulation and experiment," *Smart Mater. Struct.* **17**, 2296 (2007).

[47] Y. Shindo, F. Narita and J. Nakagawa, "Nonlinear dynamic bending and self-sensing of clamped-clamped functionally graded piezoelectric transducers under ac voltage," *Mech. Adv. Mater. Struct.* **16**(7), 536 (2009).

[48] H. S. Tzou and C. I. Tseng, "Distributed piezoelectric sensor/actuator design for dynamic measurement/control of distributed parameter systems: a piezoelectric finite element approach," *J. Sound Vib.* **138**(1), 17 (1990).

[49] A. Furuta and K. Uchino, "Dynamic observation of crack propagation in piezoelectric multilayer actuators," *J. Am. Ceram. Soc.* **76**(6), 1615 (1993).

[50] C. Q. Ru, "Exact solution for finite electrode layers embedded at the interface of two piezoelectric half-planes," *J. Mech. Phys. Solids* **48**(4), 693 (2000).

[51] F. Narita, M. Yoshida and Y. Shindo, "Electroelastic effect induced by electrode embedded at the interface of two piezoelectric half-planes," *Mech. Mater.* **36**(10), 999 (2005).

[52] Y. Shindo, F. Narita and H. Sosa, "Electroelastic analysis of piezoelectric ceramics with surface electrodes," *Int. J. Eng. Sci.* **36**(9), 1001 (1998).

[53] C. L. Hom and N. Shankar, "Numerical analysis of relaxor ferroelectric multilayered actuators and 2-2 composite arrays," *Smart Mater. Struct.* **4**, 305 (1995).

[54] X. Gong and Z. Suo, "Reliability of ceramic multilayer actuators: a nonlinear finite element simulation," *J. Mech. Phys. Solids* **44**(5), 751 (1996).

[55] Y. Shindo, M. Yoshida, F. Narita and K. Horiguchi, "Electroelastic field concentrations ahead of electrodes in multilayer piezoelectric actuators: experiment and finite element simulation," *J. Mech. Phys. Solids* **52**(5), 1109 (2004).

[56] Y. Shindo, F. Narita and M. Hirama, "Electromechanical field concentrations near the electrode tip in partially poled multilayer piezo-film actuators," *Smart Mater. Struct.* **18**, 085020 (2009).

[57] A. C. Dent, C. R. Bowen, R. Stevens, M. G. Cain and M. Stewart, "Effective elastic properties for unpoled barium titanate," *J. Eur. Ceram. Soc.* **27**(13/15), 3739 (2007).

[58] J. L. Jones, A. B. Kounga, E. Aulbach and T. Granzow, "Domain switching during electromechanical poling in lead zirconate titanate ceramics," *J. Am. Ceram. Soc.* **91**, 1586 (2008).

[59] B. A. Kudriavtsev, V. Z. Parton and V. I. Rakitin, "Fracture mechanics of piezoelectric materials. Axisymmetric crack on the boundary with a conductor," *J. Appl. Math. Mech.* **39**(2), 328 (1975).

[60] Y. Shindo, S. Lin and F. Narita, "Electroelastic response of a flat annular crack in a piezoelectric fiber surrounded by an elastic medium," *J. Eng. Math.* **59**(1), 83 (2007).

[61] F. Narita, Y. Shindo and M. Mikami, "Electroelastic field concentrations and polarization switching induced by circular electrode at the interface of piezoelectric disk composites," *Eur. J. Mech. A. Solids* **26**(3), 394 (2007).

[62] M. Yoshida, F. Narita, Y. Shindo, M. Karaiwa and K. Horiguchi, "Electroelastic field concentration by circular electrodes in piezoelectric ceramics," *Smart Mater. Struct.* **12**(6), 394 (2003).

[63] Y. Shindo and R. Togawa, "Multiple scattering of antiplane shear waves in a piezoelectric fibrous composite medium with slip at interfaces," *Wave Motion* **30**(3), 225 (1999).

[64] Y. Shindo, F. Narita and T. Watanabe, "Nonlinear electromechanical fields and localized polarization switching of 1-3 piezoelectric/polymer composites," *Eur. J. Mech. A. Solids* **29**(4), 647 (2010).

[65] F. Narita, Y. Shindo and T. Watanabe, "Dynamic characteristics and electromechanical fields of 1-3 piezoelectric/polymer composites under ac electric fields," *Smart Mater. Struct.* **19**, 075004 (2010).

[66] Y. Shindo, F. Narita, K. Sato and T. Takeda, "Nonlinear electromechanical fields and localized polarization switching of piezoelectric macrofiber composites," *J. Mech. Mater. Struct.* **6**(7/8), 1089 (2011).

[67] A. Deraemaeker, H. Nasser, A. Benjeddou and A. Preumont, "Mixing rules for the piezoelectric properties of macro fiber composites", *J. Intell. Mater. Syst. Struct.* **20**(12), 1475 (2009).

[68] P. Tan and L. Tong, "Micro-electromechanics models for piezoelectric-fiber-reinforced composite materials", *Compos. Sci. Technol.* **61**(5), 759 (2001).

[69] F. Narita, Y. Shindo and K. Sato, "Evaluation of electromechanical properties and field concentrations near electrodes in piezoelectric thick films for MEMS mirrors by simulations and tests," *Comput. Struct.* **89**(11/12), 1077 (2011).

[70] Y. Shindo, F. Narita and K. Sato, "Dynamic electromechanical field concentrations near electrodes in piezoelectric thick films for MEMS mirrors," *ASME J. Mech. Des.* **134**(5), 051005 (2012).

[71] M. S. Senousy, F. X. Li, D. Mumford, M. Gadala and R. K. N. D. Rajapakse, "Thermo-electro-mechanical performance of piezoelectric stack actuators for fuel injector applications," *J. Intell. Mater. Syst. Struct.* **20**(4), 387 (2009).

[72] M. S. Senousy, R. K. N. D. Rajapakse, D. Mumford and M. S. Gadala, "Self-heat generation in piezoelectric stack actuators used in fuel injectors," *Smart Mater. Struct.* **18**, 045008 (2009).

[73] Y. Shindo, F. Narita and T. Sasakura, "Cryogenic electromechanical behavior of multilayer piezo-actuators for fuel injector applications," *J. Appl. Phys.* **110**(8), 084510 (2011).

[74] A. J. Bell and E. Furman, "A two-parameter thermodynamic model for PZT," *Ferroelectrics* **19**, 293 (2003).

[75] D. Pandey, A. K. Singh and S. Baik, "Stability of ferroic phases in the highly piezoelectric $Pb(Zr_xTi_{1-x})O_3$ ceramics," *Acta Crystallogr., Sect. A: Found. Crystallogr.* **64**(1), 192 (2008).

[76] E. Boucher, B. Guiffard, L. Lebrun and D. Guyomar, "Effects of Zr/Ti ratio on structural, dielectric and piezoelectric properties of Mn- and (Mn, F)-doped lead zirconate titanate ceramics," *Ceram. Int.* **32**(5), 479 (2006).

[77] M. W. Hooker, "Properties of PZT-based piezoelectric ceramics between -150 and 250°C" NASA/CR-1998-208708, 1998.

[78] A. Kumar and U. V. Waghmare, "First-principles free energies and Ginzburg-Landau theory of domains and ferroelectric phase transitions in $BaTiO_3$," *Phys. Rev. B* **82**, 054117 (2010).

[79] Y. Shindo, T. Sasakura and F. Narita, "Dynamic electromechanical response of multilayered piezoelectric composites from room to cryogenic temperatures for fuel injector applications," *ASME J. Eng. Mater. Technol.* **134**(3), 031007 (2012).

[80] R. C. Turner, P. A. Fuierer, R. E. Newnham and T. R. Shrout, "Materials for high temperature acoustic and vibration sensors: a review," *Appl. Acoust.* **41**(4), 299 (1994).

[81] F. Narita, R. Hasegawa and Y. Shindo, "Electromechanical response of multilayer piezoelectric actuators for fuel injectors at high temperatures," *J. Appl. Phys.* **115**(18), 184103 (2014).

[82] Q. H. Qin, *Fracture Mechanics of Piezoelectric Materials*, WIT Press, Southampton, 2001.

[83] Y. Shindo, "Fracture and crack mechanics," *Special Topics in the Theory of Piezoelectricity*, J. S. Yang (ed.), Springer-Verlag, Dordrecht, p. 81 (2009).

[84] T.-Y. Zhang, M. H. Zhao and P. Tong, "Fracture of piezoelectric ceramics," *Adv. Appl. Mech.* **38**, 147 (2002).

[85] Y.-H. Chen and N. Hasebe, "Current understanding on fracture behaviors of ferroelectric/piezoelectric," *J. Intell. Mater. Syst. Struct.* **16**(1), 673 (2005).

[86] Y. Shindo, E. Ozawa and J. P. Nowacki, "Singular stress and electric fields of a cracked piezoelectric strip," *Int. J. Appl. Electromagn. Mater.* **1**(1), 77 (1990).

[87] F. Narita, Y. Shindo and K. Horiguchi, "Electroelastic fracture mechanics of piezoelectric ceramics," *Mechanics of Electromagnetic Material Systems and Structures*, Y. Shindo (ed.), WIT Press, Southampton, p. 89 (2003).

[88] H. A. Sosa and N. Khutoryansky, "New development concerning piezoelectric materials with defects," *Int. J. Solids Struct.* **33**(23), 3399 (1996).

[89] T.-H. Hao and Z.-Y. Shen, "A new electric boundary condition of electric fracture mechanics and its applications," *Eng. Fract. Mech.* **47**(6), 793 (1994).

[90] Y. Shindo, K. Watanabe and F. Narita, "Electroelastic analysis of a piezoelectric ceramic strip with a central crack," *Int. J. Eng. Sci.* **38**(1), 1 (2000).

[91] Y. E. Pak, "Crack extension force in a piezoelectric material," *ASME J. Appl. Mech.* **57**(3), 647 (1990).

[92] G. C. Sih, *Mechanics of Fracture Initiation and Propagation*, Kluwer Academic Publishers, The Netherlands, 1991.

[93] J. Z. Zuo and G. C. Sih, "Energy density theory formulation and interpretation of cracking behavior for piezoelectric ceramics," *Theor. Appl. Fract. Mech.* **34**(1), 17 (2000).

[94] S. Lin, F. Narita and Y. Shindo, "Comparison of energy release rate and energy density criteria for a piezoelectric layered composite with a crack normal to interface," *Theor. Appl. Fract. Mech.* **39**(3), 229 (2003).

[95] Y. E. Pak, "Linear electro-elastic fracture mechanics of piezoelectric materials," *Int. J. Fract.*, **54**(1), 79 (1992).

[96] B. L. Wang and Y.-W. Mai, "On the electrical boundary conditions on the crack surfaces in piezoelectric ceramics," *Int. J. Eng. Sci.* **41**(6), 633 (2003).

[97] Y. Shindo, M. Oka and K. Horiguchi, "Analysis and testing of indentation fracture behavior of piezoelectric ceramics under an electric field," *ASME J. Eng. Mater. Technol.* **123**(3), 293 (2001).

[98] F. Narita and Y. Shindo, "Mode I crack growth rate for yield strip model of a narrow piezoelectric ceramic body," *Theor. Appl. Fract. Mech.* **36**(1), 73 (2001).

[99] F. Narita, S. Lin and Y. Shindo, "Electroelastic fracture mechanics analysis of central active piezoelectric transformer," *Eur. J. Mech. A. Solids* **24**(3), 377 (2005).

[100] Y. Shindo and E. Ozawa, "Dynamic analysis of a cracked piezoelectric material," *Proceedings of IUTAM Symposium on Mechanical Modellings of New Electromagnetic Materials*, R. K. T. Hsieh (ed.), Elsevier, Amsterdam, p. 297 (1990).

[101] Y. Shindo, F. Narita and E. Ozawa, "Impact response of a finite crack in an orthotropic piezoelectric ceramic," *Acta Mech.* **137**(1/2), 99 (1999).

[102] S. Lin, F. Narita and Y. Shindo, "Electroelastic analysis of a penny-shaped crack in a piezoelectric ceramic under mode I loading," *Mech. Res. Commun.* **30**(4), 371 (2003).

[103] F. Narita, S. Lin and Y. Shindo, "Penny-shaped crack in a piezoceramic cylinder under mode I loading," *Arch. Mech.* **55**(3), 275 (2003).

[104] S. Lin, F. Narita and Y. Shindo, "Electroelastic analysis of a piezoelectric cylindrical fiber with a penny-shaped crack embedded in a matrix," *Int. J. Solids Struct.* **40**(19), 5157 (2003).

[105] Y. Shindo, F. Narita and K. Tanaka, "Electroelastic intensification near anti-plane shear crack in orthotropic piezoelectric ceramic strip," *Theor. Appl. Fract. Mech.* **25**(1), 65 (1996).

[106] Y. Shindo, K. Tanaka and F. Narita, "Singular stress and electric fields of a piezoelectric ceramic strip with a finite crack under longitudinal shear," *Acta Mech.* **120**(1/4), 31 (1997).

[107] F. Narita and Y. Shindo, "The interface crack problem for bonded piezoelectric and orthotropic layers under antiplane shear loading," *Int. J. Fract.* **98**(1), 87 (1999).

[108] F. Narita and Y. Shindo, "Dynamic anti-plane shear of a cracked piezoelectric ceramic," *Theor. Appl. Fract. Mech.* **29**(3), 169 (1998).

[109] F. Narita and Y. Shindo, "Scattering of antiplane shear waves by a finite crack in piezoelectric laminates," *Acta Mech.* **134**(1/2), 27 (1999).

[110] K. Minamida, F. Narita and Y. Shindo, "Dynamic antiplane shear of a circular piezoelectric fiber embedded in an elastic matrix with curved interface cracks," *Mech. Adv. Mater. Struct.* **11**(2), 133 (2004).

[111] G. C. Sih and E. P. Chen, "Dynamic analysis of cracked plates in bending and extension," *Mechanics of Fracture Vol. 3*, G. C. Sih (ed.), Noordhoff International Publishing, Leyden, p. 231 (1977).

[112] Y. Shindo, W. Domon and F. Narita, "Bending of a symmetric piezothermoelastic laminated plate with a through crack," *Arch. Mech.* **49**(2), 403 (1997).

[113] Y. Shindo, W. Domon and F. Narita, "Dynamic bending of a symmetric piezoelectric laminated plate with a through crack," *Theor. Appl. Fract. Mech.* **28**(3), 175 (1998).

[114] Y. Shindo, F. Narita and F. Saito, "Electroelastic intensification and domain switching near plane strain crack in rectangular piezoelectric material," *J. Mech. Mater. Struct.* **2**(8), 1525 (2007).

[115] C. M. Landis, "Energetically consistent boundary conditions for electromechanical fracture," *Int. J. Solids Struct.* **41**(22/23), 6291 (2004).

[116] D. Weichert and M. D. Schulz, "J-integral concept for multiphase materials," *Comput. Mater. Sci.* **1**(3), 241 (1993).

[117] Y. Shindo, F. Narita and M. Mikami, "Electroelastic fracture mechanics of piezoelectric layered composites," *J. Intell. Mater. Syst. Struct.* **16**(7/8), 573 (2005).

[118] R. M. McMeeking, "Crack tip energy release rate for a piezoelectric compact tension specimen," *Eng. Fract. Mech.* **64**(2), 217 (1999).

[119] R. M. McMeeking, "The energy release rate for a Griffith crack in a piezoelectric material," *Eng. Fract. Mech.* **71**(7/8), 1149 (2004).

[120] Y. Shindo, F. Narita and M. Hirama, "Effect of the electrical boundary condition at the crack face on the mode I energy release rate in piezoelectric ceramics," *Appl. Phys. Lett.* **94**, 081902 (2009).

[121] Y. Shindo, F. Narita and M. Sato, "Effects of electric field and poling on the mode I energy release rate in cracked piezoelectric ceramics at cryogenic temperatures," *Acta Mech.* **224**(11), 2547 (2013).

[122] Y. Shindo, F. Narita and T. Matsuda, "Electric field dependence of the mode I energy release rate in single-edge cracked piezoelectric ceramics: effect due to polarization switching/dielectric breakdown," *Acta Mech.* **219**(1), 129 (2013).

[123] T. Tani, M. Asai, K. Takatori and N. Kamiya, "Evaluation of dielectric strength and breakdown behavior for Sr-, Nb-doped PZT ceramics with various shapes of electrodes," *J. Ceram. Soc. Jpn.* **105**(4), 308 (1997).

[124] JIS R 1607-1990, *Testing Methods for Fracture Toughness of Fine Ceramics*, Japanese Standards Association, Tokyo, Japan, 2010.

[125] G. G. Pisarenko, V. M. Chushko and S. P. Kovalev, "Anisotropy of fracture toughness of piezoelectric ceramics," *J. Am. Ceram. Soc.* **68**(5), 259 (1985).

[126] R. Fu and T. Y. Zhang, "Effect of an applied electric field on the fracture toughness of poled lead zirconate titanate ceramics," *J. Am. Ceram. Soc.* **83**(5), 1215 (2000).

[127] Z. Zhang and R. Raj, "Influence of grain size on ferroelastic toughening and piezoelectric behavior of lead zirconate titanate," *J. Am. Ceram. Soc.* **78**(12), 3363 (1995).

[128] X. Mao, T. Shoji and H. Takahashi, "Characterization of fracture behaviour in small punch test by combined recrystallization-etch method and rigid plastic analysis," *ASTM J. Test. Eval.* **15**(1), 30 (1987).

[129] X. Mao, M. Saito and H. Takahashi, "Small punch test to predict ductile fracture toughness J_{IC} and brittle fracture toughness K_{IC}," *Scr. Metal. Mater.* **25**(11), 2481 (1991).

[130] Y. Shindo, F. Narita, K. Horiguchi, Y. Magara and M. Yoshida, "Electric fracture and polarization switching properties of piezoelectric ceramic PZT studied by the modified small punch test," *Acta Mater.* **51**(15), 4773 (2003).

[131] Y. Magara, F. Narita, Y. Shindo and M. Karaiwa, "Finite element analysis of electric fracture properties in modified small punch testing of piezoceramic plates," *J. Appl. Phys.* **95**(8), 4303 (2004).

[132] F. I. Baratta and W. A. Dunlay, "Crack stability in simply supported four-point and three-point loaded beams of brittle materials," *Mech. Mater.* **10**(1/2), 149 (1999).

[133] Y. Shindo, H. Murakami, K. Horiguchi and F. Narita, "Evaluation of electric fracture properties of piezoelectric ceramics using the finite element and single-edge precracked-beam methods," *J. Am. Ceram. Soc.* **85**(5), 1243 (2002).

[134] F. Narita, Y. Shindo and M. Hirama, "Electric delayed fracture and localized polarization switching of cracked piezoelectric ceramics in three-point bending," *Int. J. Damage Mech.* **19**, 285 (2010).

[135] F. Z. Li, C. F. Shih and A. Needleman, "A comparison of methods for calculating energy release rates," *Eng. Fract. Mech.* **21**(2), 405 (1985).

[136] Y. Shindo and F. Narita, "Cryogenic fracture of cracked piezoelectric ceramics in three-point bending under electric fields," *Acta Mech.* **225**(4/5), 1313 (2014).

[137] Y. Shindo, S. Watanabe, T. Takeda, M. Miura and F. Narita, "Controllability of cryogenic mode I delamination behavior in woven fabric composites using piezoelectric actuators," *Eng. Fract. Mech.* **102**, 171 (2013).

[138] R. G. Sabat, B. K. Mukherjee, W. Ren and G. Yang, "Temperature dependence of the complete material coefficients matrix of soft and hard doped piezoelectric lead zirconate titanate ceramics," *J. Appl. Phys.* **101**(6), 064111 (2007).

[139] A. G. Evans, "A method for evaluating the time-dependent failure characteristics of brittle materials - and its application to polycrystalline alumina," *J. Mater. Sci.* **7**(10), 1137 (1972).

[140] Y. Shindo, F. Narita and M. Mikami, "Double torsion testing and finite element analysis for determining the electric fracture properties of piezoelectric ceramics," *J. Appl. Phys.*, **97**, 114109 (2005).

[141] D. Fang, B. Liu and C. T. Sun, "Fatigue crack growth in ferroelectric ceramics driven by alternating electric fields," *J. Am. Ceram. Soc.* **87**(5), 840 (2004).

[142] Y. Shindo, F. Narita and F. Saito, "Static fatigue behavior of cracked piezoelectric ceramics in three-point bending under electric fields," *J. Eur. Ceram. Soc.* **27**(10), 3135 (2007).

[143] Y. Shindo, F. Narita and Y. Morikawa, "Static fatigue of three-point bending piezoelectric ceramics with a single-edge crack under AC electric fields," *J. Am. Ceram. Soc.* **95**(4), 1326 (2012).

[144] Y. Shindo, F. Narita and M. Hirama, "Dynamic fatigue of cracked piezoelectric ceramics under electromechanical loading: three-point bending test and finite element analysis," *J. Mech. Mater. Struct.* **4**(4), 719 (2009).

[145] F. Narita, Y. Morikawa, Y. Shindo and M. Sato, "Dynamic fatigue behavior of cracked piezoelectric ceramics in three-point bending under AC electric fields," *J. Eur. Ceram. Soc.* **32**(14), 3759 (2012).

[146] F. Narita, Y. Shindo and F. Saito, "Cyclic fatigue crack growth in three-point bending PZT ceramics under electromechanical loading," *J. Am. Ceram. Soc.* **90**(8), 2517 (2007).

[147] Y. Shindo, M. Sato and F. Narita, "Effects of electric field and poling on the fatigue of cracked piezoelectric ceramics in cyclic three-point bending," *J. Electroceram.*, **31**(1/2), 8 (2013).

[148] F. Narita, Y. Shindo and M. Sato, "Fatigue behavior of cracked piezoelectric ceramics in cyclic three-point bending under AC electric fields," *J. Solid Mech. Mater. Eng.* **7**(5), 530 (2013).

[149] P. Zheng, J. L. Zhang, Y. Q. Tan and C. L. Wang, "Grain-size effects on dielectric and piezoelectric properties of poled $BaTiO_3$ ceramics," *Acta Mater.* **60**(13/14), 5022 (2012).

[150] Y. Huan, X. Wang, J. Fang and L. Li, "Grain size effect on piezoelectric and ferroelectric properties of $BaTiO_3$ ceramics," *J. Eur. Ceram. Soc.* **34**(15), 1445 (2014).

[151] Q. Feng, C. J. McConville, D. D. Edwards, D. E. McCauley and M. Chu, "Effect of oxygen partial pressure on the dielectric properties and microstructures of cofired base-metal-electrode multilayer ceramic capacitors," *J. Am. Ceram. Soc.* **89**(3), 894 (2006).

[152] H. T. Langhammer, D. Makovec, Y. Pu, H.-P. Abicht and M. Drofenik, "Grain boundary reoxidation of donor-doped barium titanate ceramics," *J. Eur. Ceram. Soc.* **26**(14), 2899 (2006).

[153] Y. Su and G. J Weng, "Phase field approach and micromechanics in ferroelectric crystals," *Handbook of Micromechanics and Nanomechanics*, S. Li and X.-L. Gao (Eds.), Pan Stanford Publishing, Singapore, p. 73 (2013).

[154] Y. Shindo, F. Narita and T. Kobayashi, "Phase field simulation on the electromechanical response of poled barium titanate polycrystals with oxygen vacancies," *J. Appl. Phys.*, in press.

4

Ferromagnetic Material Systems and Structures

If ferromagnetic materials are used in magnetic field, the effect of induced magnetization should be considered. In Chapter 4, Part 4.1, first, magnetoelastic instability and magnetoelastic vibrations and waves of soft ferromagnetic and magnetically saturated materials are discussed analytically and experimentally. Next, the linear magnetoelastic problems for cracked materials are analyzed. Cracked specimens are also used, and bending and tensile tests are conducted in the bore of a superconducting magnet at room temperature to obtain the magnetic moment and stress intensity factors. Theoretical predictions are compared with experimental data, and good agreement is observed. Furthermore, a combined analytical and experimental study on the fracture and fatigue of cracked materials is reported, and the effect of magnetic field on the fracture toughness and the crack growth rate is discussed.

Magnetostrictive material systems are recognized for their potential utility in a wide variety of recent sensor and actuator applications. In Part 4.2, the nonlinear magneto-mechanical response of magnetostrictive/metal and magnetostrictive/piezoelectric laminates under electromagneto-mechanical loading is discussed. The model for the response of magnetostrictive and piezoelectric particle-reinforced composites under electromagneto-mechanical loading is also examined.

Part 4.1   Ferromagnetics

When ferromagnetic materials are placed under mechanical load and magnetic field, they will deform as a result of the interaction between the magnetic field and induced magnetization as shown in Fig. 4.1. The force of a magnetized material in magnetic field can be expressed in a number of ways including the pole, dipole, Maxwell stress, and Amperian-current models and has been discussed by Brown [1].

Electromagneto-Mechanics of Material Systems and Structures, First Edition. Yasuhide Shindo.
© 2015 John Wiley & Sons Singapore Pte Ltd. Published 2015 by John Wiley & Sons Singapore Pte Ltd.

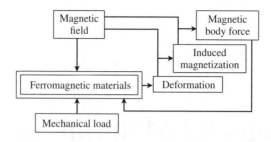

Figure 4.1   Magnetoelastic interactions of ferromagnetic materials

4.1   Basic Equations of Magnetoelasticity

4.1.1   Soft Ferromagnetic Materials

Consider soft ferromagnetic materials, and denote the rectangular Cartesian coordinates $x_i(O\text{-}x_1, x_2, x_3)$. We will consider small perturbations characterized by the displacement vector **u** produced in the material. All magnetic quantities are divided into two parts, those in the rigid body state and those in the perturbation state as follows:

$$\mathbf{B} = \mathbf{B}_0 + \mathbf{b}, \quad \mathbf{M} = \mathbf{M}_0 + \mathbf{m}, \quad \mathbf{H} = \mathbf{H}_0 + \mathbf{h} \tag{4.1}$$

where **B**, **M**, and **H** are the magnetic induction, magnetization, and magnetic field intensity vectors, respectively. The first parts, which are indicated by the subscript 0, are magnetic quantities in the undeformed material. The second parts, which are represented by lowercase letters, are corrections to account for the additional changes in magnetic quantities due to deformations.

We assume the dipole model for the magnetization [2] and neglect the magnetostrictive effect. The linearized field equations can be expressed in the following forms for the case of $|M_{0j}u_{i,j}| \ll |m_i|$

$$t_{ji,j} + \kappa_0(M_{0j}H_{0i,j} + M_{0j}h_{i,j} + m_jH_{0i,j}) = \rho u_{i,tt} \tag{4.2}$$

$$\varepsilon_{ijk}H_{0k,j} = 0, \quad B_{0i,i} = 0 \tag{4.3}$$

$$\varepsilon_{ijk}h_{k,j} = 0, \quad b_{i,i} = 0 \tag{4.4}$$

where u_i is the displacement vector component, t_{ij} is the magnetoelastic stress tensor component, $B_{0i}, M_{0i}, H_{0i}, b_i, m_i, h_i$ are the components of $\mathbf{B}_0, \mathbf{M}_0, \mathbf{H}_0, \mathbf{b}, \mathbf{m}, \mathbf{h}$, respectively, ρ is the mass density, ε_{ijk} is the permutation symbol, a comma followed by an index denotes partial differentiation with respect to the space coordinate x_i or the time t, and the summation convention over repeated indices is used.

We have the following linearized constitutive equations:

$$\sigma_{ij} = \lambda u_{k,k}\delta_{ij} + \mu(u_{i,j} + u_{j,i}) \tag{4.5}$$

$$t_{ij} = \sigma_{ij} + \frac{\kappa_0}{\chi}M_{0i}M_{0j} + \frac{\kappa_0}{\chi}(M_{0i}m_j + M_{0j}m_i) \tag{4.6}$$

$$\sigma_{ij}^M = \kappa_0(1+\chi)H_{0i}H_{0j} - \frac{1}{2}\kappa_0 H_{0k}H_{0k}\delta_{ij}$$

$$+ \kappa_0(1+\chi)(H_{0i}h_j + h_i H_{0j}) - \kappa_0 H_{0k}h_k\delta_{ij} \tag{4.7}$$

$$H_{0i} = \frac{1}{\chi}M_{0i}, \quad h_i = \frac{1}{\chi}m_i, \tag{4.8}$$

$$B_{0i} = \kappa_0(H_{0i} + M_{0i}) = \kappa_0\kappa_r H_{0i}, \quad b_i = \kappa_0(h_i + m_i) = \kappa_0\kappa_r h_i \tag{4.9}$$

where σ_{ij}^M is the Maxwell stress tensor component, σ_{ij} is the stress tensor component, $\lambda = 2G\nu/(1-2\nu)$ and $\mu = G$ are the Lamé constants, $G = E/2(1+\nu)$ is the shear modulus, E and ν are the Young's modulus and Poisson's ratio, respectively, $\kappa_0 = 1.26 \times 10^{-6}$ H/m is the magnetic permeability of free space, $\kappa_r = 1 + \chi$ is the specific magnetic permeability, χ is the magnetic susceptibility, and δ_{ij} is the Kronecker delta.

The linearized boundary conditions are

$$[\![t_{ji} + \sigma_{ji}^M]\!]n_j = 0 \tag{4.10}$$

$$\varepsilon_{ijk}n_j[\![H_{0k}]\!] = 0$$
$$[\![B_{0i}]\!]n_i = 0 \tag{4.11}$$

$$\varepsilon_{ijk}(n_j[\![h_k]\!] - n_l u_{l,j}[\![H_{0k}]\!]) = 0$$
$$[\![b_i]\!]n_i - [\![B_{0i}]\!]u_{i,j}n_j = 0 \tag{4.12}$$

where n_i is the component of the outer unit vector $\mathbf{n}$ normal to an undeformed material as shown in Fig. 2.2, and $[\![f_i]\!]$ means the jump in any field quantity f_i across the boundary; that is, $[\![f_i]\!] = f_i^e - f_i$. The superscript e denotes the quantity outside the material. Note that although $|H_{0j}u_{i,j}| \ll |h_i|$ within one material, the jump of both quantities is still kept in the first of Eqs. (4.12) because $|H_{0j}u_{i,j}|$ on one side of the boundary may be of the same order of magnitude as $|h_i|$ on the other side.

With regard to the aforementioned theory, some literature about thermodynamic treatments of ferromagnetic materials can be found. See, for instance, Parkus [3] and Alblas [4].

4.1.2 Magnetically Saturated Materials

Here, we treat magnetically saturated materials. Let us now consider the rectangular Cartesian coordinates $x_i(\text{O-}x_1, x_2, x_3)$. All magnetic quantities are divided into two parts, as given in Eq. (4.1).

The rigid body is supposed to be magnetically saturated in the x_3-direction by a biasing field $\mathbf{B}_0$ in this direction. The solution of the problem of magnetostatics of the rigid body state is

$$B_{0i} = \delta_{i3}B_{0s}, \quad H_{0i} = \delta_{i3}H_{0s}, \quad M_{0i} = \delta_{i3}M_s \tag{4.13}$$

where $B_{0s} = \kappa_0(H_{0s} + M_s)$ and H_{0s} are the saturation values of $\mathbf{B}_0$ and $\mathbf{H}_0$, respectively, and M_s is the saturation magnetization. Assuming the Maxwell–Minkowski model for the description of magnetoelastic interactions [5], the linearized field equations in the perturbation state can be expressed as

$$t_{ji,j} + \kappa_0 M_s(h_{3,i} - H_{0s}u_{3,3i}) = \rho u_{i,tt} \tag{4.14}$$

$$\varepsilon_{ijk}h_{k,j} = 0, \quad b_{i,i} = 0 \tag{4.15}$$

The linearized constitutive equations can be written as

$$\sigma_{ij} = \lambda u_{k,k}\delta_{ij} + \mu(u_{i,j} + u_{j,i}) \tag{4.16}$$

$$t_{ij} = \sigma_{ij} + T_{0jk}u_{i,k} + 2\kappa_0 M_s\{\bar{b}_1\delta_{ij}m_3 + \bar{b}_2(\delta_{i3}m_j + \delta_{j3}m_i)\} \tag{4.17}$$

$$b_i = \kappa_0 m_i + \kappa_0 h_i - \kappa_0 H_{0s}(\delta_{i3}u_{k,k} - u_{i,3} - u_{3,i}) \tag{4.18}$$

$$m_i = \frac{M_s}{H_{0s}}h_i - M_s(1 + 2\beta_r^2\bar{b}_2)(u_{i,3} + u_{3,i}) \quad (i = 1, 2), \quad m_3 = -M_s u_{3,3} \tag{4.19}$$

where T_{0ij} is the first Piola–Kirchhoff stress tensor component, $\bar{b}_1, \bar{b}_2$ are the magnetostrictive constants, and $\beta_r^2 = B_{0s}/\kappa_0 H_{0s}$. The prestress T_{0ij} is generated by the magnetic stress acting on the surface according to

$$T_{0ij}n_j = \frac{\kappa_0}{2}(M_s\delta_{j3}n_j)^2 n_i \tag{4.20}$$

We may assert that

$$|T_{0ij}| = O(M_s^2) \tag{4.21}$$

The linearized boundary conditions are

$$t_{ji}n_j = T_{0ji}n_j(2u_{l,k}n_in_k - u_{l,l}) - T_{0kj}n_ku_{j,i} + \kappa_0 M_s m_j n_j n_i n_3 \tag{4.22}$$

$$\varepsilon_{ijk}(h_k^e - h_k)n_j = -\varepsilon_{ijk}(H_{0l}^e u_{l,k})n_j \tag{4.23}$$

$$(b_i^e - b_i)n_i = -(B_{0i}^e u_{j,j} - B_{0j}^e u_{i,j})n_i \tag{4.24}$$

4.1.3 Electromagnetic Materials

Here, we present basic equations for magnetizable, electrically and thermally conducting materials in accordance with the Ampère formulation of electromagnetism [6]. For simplicity, we neglect the effect of polarization. Let us now consider the rectangular Cartesian coordinates $x_i(O\text{-}x_1, x_2, x_3)$. All magnetic quantities are divided into two parts:

$$\mathbf{B} = \mathbf{B}_0 + \mathbf{b}, \quad \mathbf{M} = \mathbf{M}_0 + \mathbf{m} \tag{4.25}$$

We now define the potentials a_i and φ as follows:

$$b_i = \varepsilon_{ijk}a_{k,j}, \quad e_i = -a_{i,t} - \varphi_{,i} \tag{4.26}$$

where e_i is the component of electric field intensity vector $\mathbf{e}$. The linearized field equations can be expressed in the following forms for the case of $|B_{0j}u_{i,j}| \ll |b_i|$ and $|E_{0j}u_{i,j}| \ll |e_i|$

$$t_{ji,j} + \rho f_i^M = \rho u_{i,tt} \tag{4.27}$$

$$\frac{1}{\kappa_0}\varepsilon_{ijk}B_{0k,j} = \varepsilon_{ijk}M_{0k,j}, \quad B_{0i,i} = 0 \tag{4.28}$$

$$\frac{1}{c^2}a_{i,tt} - a_{i,jj} = \kappa_0 j_i + \kappa_0\varepsilon_{ijk}\{m_{k,j} + (u_{j,t}M_{0k})_{,t}\},$$

$$\frac{1}{c^2}\varphi_{,tt} - \varphi_{,jj} = \frac{\rho_e}{\epsilon_0} - \kappa_0\varepsilon_{ijk}(u_{j,t}M_{0k})_{,i}, \tag{4.29}$$

$$\frac{1}{c^2}\varphi_{,t} + a_{i,i} = 0$$

where E_{0i}, f_i^M, j_i are the components of electric field intensity vector $\mathbf{E}_0$, electromagnetic body force vector $\mathbf{f}^M$, current density vector $\mathbf{j}$, respectively, ρ_e is the free electric charge density, $\epsilon_0 = 8.85 \times 10^{-12}$ C/Vm is the permittivity of free space, and $c = (\kappa_0\epsilon_0)^{-1/2}$ is the speed of light in free space. The body force is given by

$$\rho f_i^M = M_{0j}B_{0i,j} + \varepsilon_{ijk}\varepsilon_{klm}M_{0j}B_{0m,l} + \varepsilon_{ijk}j_jB_{0k} + m_jB_{0i,j} + M_{0j}b_{i,j}$$

$$+ \varepsilon_{ijk}\varepsilon_{klm}(m_jB_{0m,l} + M_{0j}b_{m,l}) + \frac{1}{c^2}B_{0j}(M_{0i}u_{j,t} - M_{0j}u_{i,t})_{,t}$$

$$+ \varepsilon_{ijk}\varepsilon_{jnm}(M_{0n}u_{m,t}B_{0k})_{,t} \tag{4.30}$$

The linearized constitutive equations are

$$t_{ij} = -\frac{1}{\kappa_a}B_{0i}B_{0j} + v_{klij}B_{0k}B_{0l} + c_{ijkl}u_{k,l}$$

$$+ \left\{2v_{klij}B_{0l} - \frac{1}{\kappa_a}(\delta_{ik}B_{0j} + \delta_{jk}B_{0i})\right\}b_k \tag{4.31}$$

$$\sigma_{ij}^M = \frac{1}{\kappa_0}B_{0i}B_{0j} - \frac{1}{2\kappa_0}B_{0k}B_{0k}\delta_{ij} - B_{0i}M_{0j} + B_{0k}M_{0k}\delta_{ij}$$

$$+ \frac{1}{\kappa_0}(B_{0i}b_j + b_iB_{0j}) - \frac{1}{\kappa_0}B_{0k}b_k\delta_{ij} - (B_{0k}m_k + M_{0k}b_k)\delta_{ij} \tag{4.32}$$

$$M_{0i} = \frac{1}{\kappa_a}B_{0i}, \quad m_i = \frac{1}{\kappa_a}b_i \tag{4.33}$$

$$j_i = \sigma(e_i + \varepsilon_{ijk}u_{j,t}B_{0k}) \tag{4.34}$$

where κ_a is the magnetic constant, v_{ijkl} is the magnetostrictive constant, c_{ijkl} is the elastic stiffness, and σ is electric conductivity.

The linearized boundary conditions are

$$[\![t_{ji} + \sigma_{ji}^M]\!]n_j = 0 \tag{4.35}$$

$$\varepsilon_{ijk}n_j[\![e_k - u_{j,t}B_{0i}]\!] = 0$$

$$\varepsilon_{ijk}n_j\left[\!\!\left[\left(\frac{B_{0l}}{\kappa_0} - M_{0l}\right)(-k_lk_m\varepsilon_{km} + u_{l,k}) + \frac{b_k}{\kappa_0} - m_k\right]\!\!\right] = 0$$

$$[\![b_i + B_{0j}n_jn_k\varepsilon_{ik}]\!]n_i = 0 \tag{4.36}$$

$$\left[\!\!\left[\epsilon_0 e_i + \frac{1}{c^2}\varepsilon_{ijk}u_{j,t}M_{0k}\right]\!\!\right]n_i = 0$$

where the strain tensor component ε_{ij} is related to the displacement vector component by the relation

$$\varepsilon_{ij} = \frac{1}{2}(u_{j,i} + u_{i,j}) \tag{4.37}$$

and k_i is the component of the outer unit tangent vector $\mathbf{k}$ normal to $\mathbf{t}$ and $\mathbf{n}$, which form a right-handed orthogonal triad.

On the basis of the aforementioned theory, Hutter [7, 8] considered the magnetoelastic wave propagation for electromagnetic materials and presented some numerical results for the wave speed and attenuation as a function of wavelength, electrical conductivity, magnetic permeability, and intensity and direction of magnetic induction. On the other hand, Ersoy and Kiral [9] developed a unified dynamic theory for polarizable and magnetizable electromagneto-thermoviscoelastic anisotropic materials with thermal and electrical conductions. Ersoy [10] then investigated the propagation of plane waves in electrically and thermally conductive magneto-thermoviscoelastic materials.

4.2 Magnetoelastic Instability

When a ferromagnetic thin plate is placed in a magnetic field, buckling may occur at a certain critical value of the magnetic field. Moon and Pao [11] performed the analysis of this buckling phenomenon. In this section, the buckling of ferromagnetic materials is discussed. Consider a ferromagnetic plate with thickness $2h$ in a rectangular Cartesian coordinate system (x, y, z) as shown in Fig. 4.2. The coordinate axes x and y are in the middle plane of the plate, and

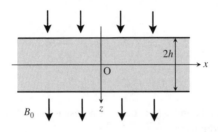

Figure 4.2 A ferromagnetic plate

the z-axis is normal to this plane. A uniform magnetic field of magnetic induction B_0 in the z-direction ($B_{0z} = B_0, B_{0x} = B_{0y} = 0$) is applied. Consequently, we may set $H_{0x} = H_{0y} = 0$ and $M_{0x} = M_{0y} = 0$. Also, we assume plane strain normal to the y-axis.

4.2.1 Buckling of Soft Ferromagnetic Material

Here, we consider a soft ferromagnetic material. The solutions in a rectangular Cartesian coordinate system (x, y, z) for the rigid body state satisfying the field equation, Eqs. (4.3), and boundary condition, Eqs. (4.11), can, with the aid of the first of Eqs. (4.8) and the first of Eqs. (4.9), be written as

$$
\left.
\begin{aligned}
B_{0z}^e = B_0, \quad H_{0z}^e = \frac{B_0}{\kappa_0}, \quad M_{0z}^e = 0 \qquad (|z| > h) \\[2mm]
B_{0z} = B_0, \quad H_{0z} = \frac{B_0}{\kappa_0 \kappa_r}, \quad M_{0z} = \frac{\chi B_0}{\kappa_0 \kappa_r} \quad (|z| \le h)
\end{aligned}
\right\}
\tag{4.38}
$$

The mechanical constitutive equations are taken to be the usual Hooke's law

$$
\begin{aligned}
\sigma_{xx} &= \lambda(u_{x,x} + u_{z,z}) + 2\mu u_{x,x} \\
\sigma_{zz} &= \lambda(u_{x,x} + u_{z,z}) + 2\mu u_{z,z} \\
\sigma_{zx} &= \sigma_{xz} = \mu(u_{z,x} + u_{x,z})
\end{aligned}
\tag{4.39}
$$

The magnetoelastic and Maxwell stresses are

$$
\begin{aligned}
t_{xx} &= \sigma_{xx} \\[1mm]
t_{zz} &= \sigma_{zz} + \frac{\chi B_0^2}{\kappa_0 \kappa_r^2} + \frac{2B_0}{\kappa_r} m_z \\[1mm]
t_{zx} &= t_{xz} = \sigma_{zx} + \frac{B_0}{\kappa_r} m_x
\end{aligned}
\tag{4.40}
$$

$$
\begin{aligned}
\sigma_{xx}^M &= -\frac{B_0^2}{2\kappa_0 \kappa_r^2} - \frac{B_0}{\kappa_r} h_z \\[1mm]
\sigma_{zz}^M &= \frac{(1+2\chi)B_0^2}{2\kappa_0 \kappa_r^2} + \frac{(1+2\chi)B_0}{\kappa_r} h_z \\[1mm]
\sigma_{zx}^M &= \sigma_{xz}^M = B_0 h_x
\end{aligned}
\tag{4.41}
$$

By substituting from Eqs. (4.40) into Eq. (4.2) and considering the second of Eqs. (4.8) and the fifth and sixth of Eqs. (4.38), the governing equations without inertia are given by

$$
\begin{aligned}
\sigma_{xx,x} + \sigma_{zx,z} + \frac{2\chi B_0}{\kappa_r} h_{x,z} = 0 \\[2mm]
\sigma_{xz,x} + \sigma_{zz,z} + \frac{2\chi B_0}{\kappa_r} h_{z,z} = 0
\end{aligned}
\tag{4.42}
$$

Using the second of Eqs. (4.9), the Maxwell's equations, Eqs. (4.4), are

$$h^e_{x,z} - h^e_{z,x} = 0, \quad h_{x,z} - h_{z,x} = 0 \tag{4.43}$$

$$h^e_{x,x} + h^e_{z,z} = 0, \quad h_{x,x} + h_{z,z} = 0 \tag{4.44}$$

The magnetic field equations, Eqs. (4.43) and (4.44), are satisfied by introducing a magnetic potential ϕ_m such that

$$\begin{aligned}
h^e_x &= \phi^e_{m,x}, \quad h^e_z = \phi^e_{m,z}, \\
\phi^e_{m,xx} &+ \phi^e_{m,zz} = 0 \\
h_x &= \phi_{m,x}, \quad h_z = \phi_{m,z}, \\
\phi_{m,xx} &+ \phi_{m,zz} = 0
\end{aligned} \tag{4.45}$$

From Eqs. (4.10) and (4.12), we obtain the linearized boundary conditions

$$\sigma_{zx}(x, \pm h) = -\frac{\chi B_0}{\kappa_r} \phi_{m,x}(x, \pm h)$$

$$\sigma_{zz}(x, \pm h) = \frac{\chi(\chi - 2)}{\kappa_r} \left\{ \frac{B_0^2}{2\kappa_0 \kappa_r} + B_0 \phi_{m,z}(x, \pm h) \right\} \tag{4.46}$$

$$\phi^e_{m,x}(x, \pm h) - \phi_{m,x}(x, \pm h) = -\frac{\chi B_0}{\kappa_0 \kappa_r} u_{z,x}(x, \pm h)$$

$$\phi^e_{m,z}(x, \pm h) - \kappa_r \phi_{m,z}(x, \pm h) = 0 \tag{4.47}$$

Here, we consider an infinite thin plate in the x–y plane and apply the common approximations of the classical theory of plate. Assume that the z-component of the displacement vector u_z is a function of only x. The displacement components, representing the bending of the beam plate, can be expressed as follows:

$$u_x = -zw_{,x}(x), \quad u_z = w(x) \tag{4.48}$$

where $w(x)$ is the deflection of the middle plane of the plate.

The bending moment per unit length M_{xx} and the vertical shear force per unit length Q_x can be expressed as

$$M_{xx} = \int_{-h}^{h} \sigma_{xx} z \, dz \tag{4.49}$$

$$Q_x = \int_{-h}^{h} \sigma_{zx} \, dz \tag{4.50}$$

Multiplying the first of Eqs. (4.42) by $z \, dz$ and integrating from $-h$ to h, with the fourth of Eqs. (4.45), we find

$$M_{xx,x} - Q_x = -m^D_{xx} \tag{4.51}$$

where

$$m^D_{xx} = [z\sigma_{zx}]^h_{-h} + \frac{2\chi B_0}{\kappa_r} \int_{-h}^{h} \phi_{m,xz} z \, dz \tag{4.52}$$

Multiplying the second of Eqs. (4.42) by dz and integrating from $-h$ to h, with the fifth of Eqs. (4.45), we also find

$$Q_{x,x} = -q^D \tag{4.53}$$

where

$$q^D = [\sigma_{zz}]^h_{-h} + \frac{2\chi B_0}{\kappa_r} \int_{-h}^{h} \phi_{m,zz} \, dz \tag{4.54}$$

From Eq. (4.39), with Eq. (4.48), the stress component σ_{xx} becomes

$$\sigma_{xx} = -\frac{\mu}{1-v} zw_{,xx} \tag{4.55}$$

Substituting Eq. (4.55) into Eq. (4.49) gives

$$M_{xx} = -Dw_{,xx} \tag{4.56}$$

where $D = 4\mu h^3 / 3(1 - v)$ is the flexural rigidity of the plate. Eliminating Q_x from Eqs. (4.51) and (4.53) and taking into account Eq. (4.56), we have

$$Dw_{,xxxx} = m^D_{xx,x} + q^D \tag{4.57}$$

We assume the solutions of Eq. (4.57) and the third and sixth of Eqs. (4.45) in the following form:

$$w = w_0 \sin(kx) \tag{4.58}$$

$$\phi_m = A_1 \cosh(kz) \sin(kx) \tag{4.59}$$

$$\left. \begin{array}{l} \phi^e_m = A_2 \exp(-kz) \sin(kx) \quad (z \geq h) \\ A_3 \exp(kz) \sin(kx) \quad (z \leq -h) \end{array} \right\} \tag{4.60}$$

where w_0 is a constant with dimension of deflection, and A_1, A_2, A_3, and k are the unknowns to be solved. From the boundary conditions, Eqs. (4.47), we obtain

$$A_1 = \frac{\chi B_0}{\kappa_0 \kappa_r \{\kappa_r \sinh(kh) + \cosh(kh)\}} w_0$$

$$A_2 = A_3 = -\frac{\chi B_0}{\kappa_0 \{\kappa_r \sinh(kh) + \cosh(kh)\}} \exp(kh) \sinh(kh) w_0 \tag{4.61}$$

Substituting from Eqs. (4.52) and (4.54) into Eq. (4.57), considering Eqs. (4.46), and using Eqs. (4.58)–(4.61), we obtain

$$Dk^2 + \frac{2\chi^2 B_0^2}{\kappa_0 \kappa_r^2 \{\kappa_r \sinh(kh) + \cosh(kh)\}} \left[h \cosh(kh) - \frac{1 + \kappa_r}{k} \sinh(kh) \right] = 0 \tag{4.62}$$

If $kh \ll 1$, the above-mentioned solution is simplified to

$$Dk^2 - \frac{2\chi^2 B_0^2 h}{\kappa_0 \kappa_r (\kappa_r kh + 1)} = 0 \tag{4.63}$$

For a particular wavelength $2\pi/k$, the equilibrium is maintained by a magnetic induction $B_0 = B_{cr}$ as given by Eq. (4.63). Thus, B_{cr} is the critical magnetic induction at which the plate may buckle, and we have

$$\frac{B_{cr}^2}{\kappa_0 \mu} = \frac{2\kappa_r (kh)^2}{3(1-v)\chi^2}(\kappa_r kh + 1) \tag{4.64}$$

One can find Eq. (4.64) in Refs [2, 11]. If χ is so large that $\kappa_r = 1 + \chi \approx \chi$ and $\kappa_r kh \gg 1$, the critical magnetic induction becomes

$$\frac{B_{cr}^2}{\kappa_0 \mu} = \frac{2(kh)^3}{3(1-v)} \tag{4.65}$$

Pao and Yeh [2] indicated that the buckling magnetic induction for the antisymmetric deformation (displacements being antisymmetric about the middle plane of the plate) is much lower than that for the symmetric deformation (displacements being symmetric about the middle plane of the plate). This pointed out that the antisymmetric buckling of the plate in a magnetic field is very easy to observe because buckling initiates at the lowest critical value of the magnetic induction.

There appeared some articles [12, 13] dealing with the buckling behavior of soft ferromagnetic thin plates by the use of Amperian-current model. Some works [14–16] theoretically and experimentally investigated the magnetoelastic buckling of soft ferromagnetic thin plates. van de Ven [17] considered the magnetoelastic buckling of a soft ferromagnetic circular plate and proved that two theories based on Maxwell stress and Amperian-current models yield identical results. Furthermore, van de Ven et al. [18] discussed the interactive effect of two nearby ferromagnetic rods on the magnetoelastic buckling. The buckling analysis was based on a perturbation theory, and the theoretically predicted results were confirmed experimentally.

4.2.2 Buckling of Magnetically Saturated Material

Here, we consider a magnetically saturated material. The perturbed system is simplified when performing the following steps [5]: (1) We neglect all the terms produced by β_r^{-2} because the parameter β_r^2 is very large ($\beta_r^2 \gg 1$) for the ferromagnetic materials. (2) We notice that for the out-of-deformations of the plates, the symmetric part of the displacement gradient is very small as compared to the antisymmetric part, and therefore, we neglect terms, such as $u_{x,x}, u_{y,y}, u_{z,z}, (u_{x,y} + u_{y,x})/2, (u_{y,z} + u_{z,y})/2, (u_{z,x} + u_{x,z})/2$, produced by the symmetric part. Within an approximation holding for the plates, the prestresses no longer occur. On account of Eqs. (4.13), we find the solutions in a rectangular Cartesian coordinate system (x, y, z) for the rigid body state

$$\left.\begin{array}{l} B_{0z}^e = B_{0s}, \quad H_{0z}^e = \dfrac{B_{0s}}{\kappa_0}, \quad M_{0z}^e = 0 \quad (|z| > h) \\[2mm] B_{0z} = B_{0s}, \quad H_{0z} = H_{0s}, \quad M_{0z} = M_s \quad (|z| \le h) \end{array}\right\} \tag{4.66}$$

The mechanical constitutive equations are given by Eqs. (4.39). The magnetoelastic stresses are

$$t_{xx} = \sigma_{xx} + 2\kappa_0 M_s \bar{b}_1 m_z$$
$$t_{zz} = \sigma_{zz} + 2\kappa_0 M_s (\bar{b}_1 + 2\bar{b}_2) m_z \qquad (4.67)$$
$$t_{zx} = \sigma_{zx} + 2\kappa_0 M_s \bar{b}_2 m_x$$

By substituting from Eqs. (4.67) into Eq. (4.14) and considering Eqs. (4.19), the governing equations without inertia are given by

$$\sigma_{xx,x} + \sigma_{zx,z} + \kappa_0 M_s (h_{z,x} + 2\beta_r^2 \bar{b}_2 h_{x,z}) = 0$$
$$\sigma_{xz,x} + \sigma_{zz,z} + \kappa_0 M_s (h_{z,z} + 2\beta_r^2 \bar{b}_2 h_{x,x}) = 0 \qquad (4.68)$$

Using Eq. (4.18), the Maxwell's equations, Eqs. (4.15), are

$$h_{x,z}^e - h_{z,x}^e = 0, \quad h_{x,z} - h_{z,x} = 0 \qquad (4.69)$$

$$h_{x,x}^e + h_{z,z}^e = 0, \quad \beta_r^2 h_{x,x} + h_{z,z} = 0 \qquad (4.70)$$

The magnetic field equations, Eqs. (4.69) and (4.70), are satisfied by using the magnetic potential ϕ_m such that

$$h_x^e = \phi_{m,x}^e, \quad h_z^e = \phi_{m,z}^e,$$
$$\phi_{m,xx}^e + \phi_{m,zz}^e = 0$$
$$h_x = \phi_{m,x}, \quad h_z = \phi_{m,z}, \qquad (4.71)$$
$$\beta_r^2 \phi_{m,xx} + \phi_{m,zz} = 0$$

From Eqs. (4.22)–(4.24), we obtain the linearized boundary conditions

$$\sigma_{zx}(x, \pm h) = -2\kappa_r M_s \beta_r^2 \bar{b}_2 \phi_{m,x}(x, \pm h)$$
$$\sigma_{zz}(x, \pm h) = 0 \qquad (4.72)$$
$$\phi_{m,x}^e(x, \pm h) - \phi_{m,x}(x, \pm h) = -\frac{B_{0s}}{\kappa_0} u_{z,x}(x, \pm h) \qquad (4.73)$$
$$\phi_{m,z}^e(x, \pm h) - \phi_{m,z}(x, \pm h) = 0$$

With the beam-plate theory, the displacement components can be expressed as Eq. (4.48). Multiplying the first of Eqs. (4.68) by $z\,dz$ and integrating from $-h$ to h, with the fourth and fifth of Eqs. (4.71), we find

$$M_{xx,x} - Q_x = -m_{xx}^S \qquad (4.74)$$

where

$$m_{xx}^S = [z\sigma_{zx}]_{-h}^h + \kappa_0 M_s \left(\int_{-h}^h \phi_{m,zx} z\,dz + 2\beta_r^2 \bar{b}_2 \int_{-h}^h \phi_{m,xz} z\,dz \right) \qquad (4.75)$$

Multiplying the second of Eqs. (4.68) by dz and integrating from $-h$ to h, with the fourth and fifth of Eqs. (4.71), we also find

$$Q_{x,x} = -q^S \qquad (4.76)$$

where

$$q^S = [\sigma_{zz}]^h_{-h} + \kappa_0 M_s \left(2\beta_r^2 \bar{b}_2 \int_{-h}^h \phi_{m,xx} \, dz + \int_{-h}^h \phi_{m,zz} \, dz \right) \tag{4.77}$$

Eliminating Q_x from Eqs. (4.74) and (4.76) and taking into account Eq. (4.56), we have

$$Dw_{,xxxx} = m^S_{xx,x} + q^S \tag{4.78}$$

We assume the solutions of Eq. (4.78) and the third and sixth of Eqs. (4.71) in the following form:

$$w = w_0 \sin(kx) \tag{4.79}$$

$$\phi_m = A_1 \cosh(\beta_r kz) \sin(kx) \tag{4.80}$$

$$\left. \begin{array}{l} \phi^e_m = A_2 \exp(-kz) \sin(kx) \quad (z \geq h) \\ \quad\quad A_3 \exp(kz) \sin(kx) \quad (z \leq -h) \end{array} \right\} \tag{4.81}$$

where w_0 is a constant with dimension of deflection, and A_1, A_2, A_3, and k are the unknowns to be solved. From the boundary conditions (4.73), we obtain

$$A_1 = \frac{B_{0s}}{\kappa_0\{\beta_r \sinh(\beta_r kh) + \cosh(\beta_r kh)\}} w_0$$

$$A_2 = A_3 = -\frac{\beta_r B_{0s}}{\kappa_0\{\beta_r \sinh(\beta_r kh) + \cosh(\beta_r kh)\}} \exp(kh) \sinh(\beta_r kh) w_0 \tag{4.82}$$

Substituting from Eqs. (4.75) and (4.77) into Eq. (4.78), considering Eqs. (4.72), and using Eqs. (4.79)–(4.82), we obtain

$$Dk^2 - \frac{2M_s\beta_r \sinh(\beta_r kh)B_{0s}}{k\{\beta_r \sinh(\beta_r kh) + \cosh(\beta_r kh)\}} = 0 \tag{4.83}$$

With $\kappa_0 M_s = B_{0s} - \kappa_0 H_{0s} = B_{0s}(1 - \beta_r^{-2}) \approx B_{0s}$, the above-mentioned solution is approximated by

$$Dk^2 - \frac{2\beta_r \sinh(\beta_r kh)B_{0s}^2}{\kappa_0 k\{\beta_r \sinh(\beta_r kh) + \cosh(\beta_r kh)\}} = 0 \tag{4.84}$$

The critical kh value at saturation is

$$(kh)^3_{cr} = \frac{3(1-v)}{2} \left\{ 1 + \frac{\cosh(\beta_r kh)}{\beta_r \sinh(\beta_r kh)} \right\}^{-1} \frac{B_{0s}^2}{\kappa_0 \mu} \tag{4.85}$$

One can find Eq. (4.85) in Ref. [5]. If β_r is so large that $\cosh(\beta_r kh)/\{\beta_r \sinh(\beta_r kh)\} \ll 1$, Eq. (4.85) is

$$(kh)^3_{cr} = \frac{3(1-v)}{2} \frac{B_{0s}^2}{\kappa_0 \mu} \tag{4.86}$$

4.2.3 Bending of Soft Ferromagnetic Material

Here, we consider a cantilever soft ferromagnetic material, obtain the critical buckling magnetic induction based on the beam-plate theory, and then discuss the bending behavior theoretically and experimentally [19]. Figure 4.3 shows a cantilever soft ferromagnetic plate with length a and thickness $2h$ in a rectangular Cartesian coordinate system (x, y, z). The width is b. The coordinate axes x and y are in the middle plane of the plate, and the z-axis is normal to this plane. A uniform magnetic field of magnetic induction B_0 in the z-direction $(B_{0z} = B_0, B_{0x} = B_{0y} = 0)$ is applied.

Let a normal line load P be applied to the end $(x = a)$. We now consider the problem of $b \to \infty$. A similar procedure to that in Section 4.2.1 can be applied. The differential equation (4.57) can be written by using a force per unit length P/b in a convenient form

$$Dw_{,xxxx} = m^D_{xx,x} + q^D + (P/b)\delta(x - a) \tag{4.87}$$

where $\delta()$ is the Dirac-delta function. The boundary conditions are

$$w(0) = 0, \quad w_{,x}(0) = 0$$
$$M_{xx}(a) = -Dw_{,xx}(a) = 0, \quad Q_x(a) = -Dw_{,xxx}(a) + m^D_{xx}(a) = 0 \tag{4.88}$$

Fourier series approach is used, and the deflection is obtained as

$$w = w_0 \sum_{n=1,3,5,...}^{\infty} \frac{96\cos(n\pi)\{\cos(k_n x) - 1\}}{(n\pi)^2\{(n\pi)^2 - 4(a^2 h b_n/D)\}} \tag{4.89}$$

where $w_0 = Pa^3/3D$ is the deflection at the loaded end $(x = a)$ for $B_0 = 0$ and

$$k_n = \frac{n\pi}{2a} \tag{4.90}$$

$$b_n = \frac{2\chi^2 B_0^2}{\kappa_0 \kappa_r\{\kappa_r \sinh(k_n h) + \cosh(k_n h)\}} \tag{4.91}$$

Equation (4.89) shows that $w = \infty$ when

$$b_n = b_{ncr} = \frac{(n\pi)^2 D}{4a^2 h} \tag{4.92}$$

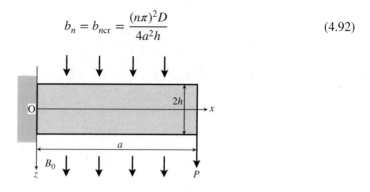

Figure 4.3 A cantilever ferromagnetic plate

which yields the critical magnetic induction B_{cr} in Eq. (4.91). Using Eq. (4.91), we find

$$B_{cr} = \frac{n\pi}{2\chi a} \left[\frac{D\kappa_0 \kappa_r \{\kappa_r \sinh(k_n h) + \cosh(k_n h)\}}{2h} \right]^{1/2} \tag{4.93}$$

In order to validate the prediction, we conduct experiments on stainless steel SUS430 for various values of a, b, and $2h$. The material properties of SUS430 are as follows:

$$\begin{aligned} E &= 162.3 \times 10^9 \text{ N/m}^2 \\ v &= 0.294 \\ \kappa_r &= 122.9 \end{aligned} \tag{4.94}$$

We use a superconducting magnet with a 220-mm-diameter bore to create a magnetic field of magnetic induction B_0 normal to the wide face of the specimen. The specimen is clamped at one end and bent by a concentrated load $P = 9.8$ N applied at the other. The deflection of the loaded end of the specimen is measured under magnetic field at room temperature using a laser displacement meter.

Table 4.1 lists the dimensions of the tested specimens and calculated critical magnetic induction B_{cr}. The critical magnetic induction decreases as the specimen length to thickness ratio $a/2h$ increases. Figure 4.4 gives a plot of the deflection with the magnetic field B_0 showing

Table 4.1 Critical buckling magnetic induction

a (mm)	b (mm)	$2h$ (mm)	$a/2h$	B_{cr} (T)
60	20	3	20.0	2.141
100	20	3	33.3	0.995
100	40	1	100.0	0.191
100	40	2	50.0	0.542
100	40	3	33.3	0.995

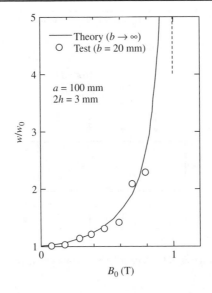

Figure 4.4 Deflection versus magnetic field

both the theoretical solution of $b \rightarrow \infty$ (solid line) and test data of $b = 20\,\text{mm}$ (open circle) for $a = 100\,\text{mm}$ and $2h = 3\,\text{mm}$. In this figure, the deflection w is normalized with respect to w_0, which is the corresponding value for $B_0 = 0$. The dashed line represents the critical magnetic induction. Close agreement between the calculations and test data gives the theory its perspective. The existence of the magnetic field produces larger values of the deflection. This effect becomes more pronounced as the specimen thickness decreases or the specimen length increases (no figure shown). On the other hand, no dependence of the measured deflection with the magnetic field on the width is observed.

4.3 Magnetoelastic Vibrations and Waves

Moon and Pao [20] investigated the vibration and dynamic instability of a ferromagnetic beam plate in a magnetic field. Moon [21] extended this theory to the two-dimensional bending of ferromagnetic thin plates and studied the vibration and dynamic instability in detail. In this section, the magnetoelastic vibrations and waves of ferromagnetic materials are discussed.

4.3.1 Vibrations and Waves of Soft Ferromagnetic Material

Here, consider a soft ferromagnetic plate with thickness $2h$ in a rectangular Cartesian coordinate system (x, y, z). The coordinate axes x and y are in the middle plane of the plate, and the z-axis is normal to this plane. A uniform magnetic field of magnetic induction B_0 in the z-direction $(B_{0z} = B_0, B_{0x} = B_{0y} = 0)$ is applied. Consequently, we may set $H_{0x} = H_{0y} = 0$ and $M_{0x} = M_{0y} = 0$.

The solutions for the rigid body state can be written as Eqs. (4.38). The mechanical constitutive equations are taken to be the usual Hooke's law

$$
\begin{aligned}
\sigma_{xx} &= \lambda(u_{x,x} + u_{y,y} + u_{z,z}) + 2\mu u_{x,x} \\
\sigma_{yy} &= \lambda(u_{x,x} + u_{y,y} + u_{z,z}) + 2\mu u_{y,y} \\
\sigma_{zz} &= \lambda(u_{x,x} + u_{y,y} + u_{z,z}) + 2\mu u_{z,z} \\
\sigma_{xy} &= \sigma_{yx} = \mu(u_{x,y} + u_{y,x}) \\
\sigma_{yz} &= \sigma_{zy} = \mu(u_{y,z} + u_{z,y}) \\
\sigma_{xz} &= \sigma_{zx} = \mu(u_{z,x} + u_{x,z})
\end{aligned}
\tag{4.95}
$$

The magnetoelastic and Maxwell stresses are

$$
\begin{aligned}
t_{xx} &= \sigma_{xx} \\
t_{yy} &= \sigma_{yy} \\
t_{zz} &= \sigma_{zz} + \frac{\chi B_0^2}{\kappa_0 \kappa_r^2} + \frac{2B_0}{\kappa_r} m_z \\
t_{xy} &= t_{yx} \\
t_{yz} &= t_{zy} = \sigma_{yz} + \frac{B_0}{\kappa_r} m_y \\
t_{zx} &= t_{xz} = \sigma_{zx} + \frac{B_0}{\kappa_r} m_x
\end{aligned}
\tag{4.96}
$$

$$\sigma_{xx}^M = -\frac{B_0^2}{2\kappa_0\kappa_r^2} - \frac{B_0}{\kappa_r}h_z$$

$$\sigma_{yy}^M = -\frac{B_0^2}{2\kappa_0\kappa_r^2} - \frac{B_0}{\kappa_r}h_z$$

$$\sigma_{zz}^M = \frac{(1+2\chi)B_0^2}{2\kappa_0\kappa_r^2} + \frac{(1+2\chi)B_0}{\kappa_r}h_z \qquad (4.97)$$

$$\sigma_{xy}^M = \sigma_{yx}^M = 0$$

$$\sigma_{yz}^M = \sigma_{zy}^M = B_0 h_y$$

$$\sigma_{zx}^M = \sigma_{xz}^M = B_0 h_x$$

By substituting from Eqs. (4.96) into Eq. (4.2) and considering the second of Eqs. (4.8) and the fifth and sixth of Eqs. (4.38), the stress equations of motion are given by

$$\sigma_{xx,x} + \sigma_{yx,y} + \sigma_{zx,z} + \frac{2\chi B_0}{\kappa_r}h_{x,z} = \rho u_{x,tt}$$

$$\sigma_{xy,x} + \sigma_{yy,y} + \sigma_{zy,z} + \frac{2\chi B_0}{\kappa_r}h_{y,z} = \rho u_{y,tt} \qquad (4.98)$$

$$\sigma_{xz,x} + \sigma_{yz,y} + \sigma_{zz,z} + \frac{2\chi B_0}{\kappa_r}h_{z,z} = \rho u_{z,tt}$$

Using the second of Eqs. (4.9), the Maxwell's equations, Eqs. (4.4), are

$$h_{z,y}^e - h_{y,z}^e = 0, \quad h_{x,z}^e - h_{z,x}^e = 0, \quad h_{y,x}^e - h_{x,y}^e = 0$$

$$h_{z,y} - h_{y,z} = 0, \quad h_{x,z} - h_{z,x} = 0, \quad h_{y,x} - h_{x,y} = 0 \qquad (4.99)$$

$$h_{x,x}^e + h_{y,y}^e + h_{z,z}^e = 0$$

$$h_{x,x} + h_{y,y} + h_{z,z} = 0 \qquad (4.100)$$

The magnetic field equations (4.99) and (4.100), are satisfied by using the magnetic potential ϕ_m such that

$$h_x^e = \phi_{m,x}^e, \quad h_y^e = \phi_{m,y}^e, \quad h_z^e = \phi_{m,z}^e,$$

$$\phi_{m,xx}^e + \phi_{m,yy}^e + \phi_{m,zz}^e = 0$$

$$h_x = \phi_{m,x}, \quad h_y = \phi_{m,y}, \quad h_z = \phi_{m,z}, \qquad (4.101)$$

$$\phi_{m,xx} + \phi_{m,yy} + \phi_{m,zz} = 0$$

From Eqs. (4.10) and (4.12), we obtain the linearized boundary conditions

$$\sigma_{zx}(x, y, \pm h, t) = -\frac{\chi B_0}{\kappa_r}\phi_{m,x}(x, y, \pm h, t)$$

$$\sigma_{zy}(x, y, \pm h, t) = -\frac{\chi B_0}{\kappa_r}\phi_{m,y}(x, y, \pm h, t) \qquad (4.102)$$

$$\sigma_{zz}(x, y, \pm h, t) = \frac{\chi(\chi - 2)}{\kappa_r}\left\{\frac{B_0^2}{2\kappa_0\kappa_r} + B_0\phi_{m,z}(x, \pm h, t)\right\}$$

$$\phi^e_{m,x}(x, y, \pm h, t) - \phi_{m,x}(x, y, \pm h, t) = -\frac{\chi B_0}{\kappa_0 \kappa_r} u_{z,x}(x, y, \pm h, t)$$

$$\phi^e_{m,y}(x, y, \pm h, t) - \phi_{m,y}(x, y, \pm h, t) = -\frac{\chi B_0}{\kappa_0 \kappa_r} u_{z,y}(x, y, \pm h, t) \qquad (4.103)$$

$$\phi^e_{m,z}(x, y, \pm h, t) - \kappa_r \phi_{m,z}(x, y, \pm h, t) = 0$$

4.3.1.1 Soft Ferromagnetic Thin Plate Bending

Here, consider a soft ferromagnetic thin plate with thickness $2h$ in a rectangular Cartesian coordinate system (x, y, z) as shown in Fig. 4.5. The coordinate axes x and y are in the middle plane of the plate, and the z-axis is normal to this plane. Let magnetic flexural waves be traveled in the y-direction. The plate is permeated by a uniform magnetic field of magnetic induction $B_{0z} = B_0$.

Classical plate bending theory [22] is applied. The rectangular displacement components can be expressed as follows:

$$u_x = -zw_{,x}, \quad u_y = -zw_{,y}, \quad u_z = w(x, y, t) \qquad (4.104)$$

where $w(x, y, t)$ represents the deflection of the middle plane of the plate.

The bending and twisting moments per unit length $(M_{xx}, M_{yy}, M_{xy} = M_{yx})$ and the vertical shear forces per unit length (Q_x, Q_y) can be expressed in terms of w as

$$M_{xx} = \int_{-h}^{h} \sigma_{xx} z \, dz = -D(w_{,xx} + vw_{,yy})$$

$$M_{yy} = \int_{-h}^{h} \sigma_{yy} z \, dz = -D(w_{,yy} + vw_{,xx}) \qquad (4.105)$$

$$M_{xy} = M_{yx} = \int_{-h}^{h} \sigma_{xy} z \, dz = -D(1 - v)w_{,xy}$$

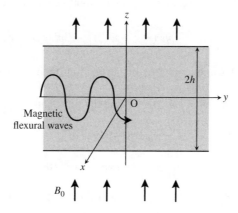

Figure 4.5 A ferromagnetic plate and magnetic flexural waves

$$Q_x = \int_{-h}^{h} \sigma_{zx}\, dz = -D(w_{,xx} + w_{,yy})_{,x}$$

$$Q_y = \int_{-h}^{h} \sigma_{zy}\, dz = -D(w_{,xx} + w_{,yy})_{,y}$$

(4.106)

Now, if we multiply the first and second of Eqs. (4.98) by $z\, dz$ and integrate from $-h$ to h, taking into account the first and second of Eqs. (4.102), we obtain the results

$$M_{xx,x} + M_{yx,y} - Q_x = -\frac{2}{3}\rho h^3 w_{,xtt} - m_{xx}^D$$

$$M_{xy,x} + M_{yy,y} - Q_y = -\frac{2}{3}\rho h^3 w_{,ytt} - m_{yy}^D$$

(4.107)

The moments m_{xx}^D and m_{yy}^D are

$$m_{xx}^D = -\frac{\chi B_0 h}{\kappa_r}\{\phi_{m,x}(x,y,h,t) + \phi_{m,x}(x,y,-h,t)\} + \frac{2\chi B_0}{\kappa_r}\int_{-h}^{h} \phi_{m,xz} z\, dz$$

$$m_{yy}^D = -\frac{\chi B_0 h}{\kappa_r}\{\phi_{m,y}(x,y,h,t) + \phi_{m,y}(x,y,-h,t)\} + \frac{2\chi B_0}{\kappa_r}\int_{-h}^{h} \phi_{m,yz} z\, dz$$

(4.108)

If the third of Eqs. (4.98) is multiplied by dz and integrated from $-h$ to h, taking into account the third of Eqs. (4.102), we obtain

$$Q_{x,x} + Q_{y,y} = 2h\rho w_{,tt} - q^D$$

(4.109)

The load q^D applied to the plate is

$$q^D = -\frac{2\chi(\chi - 2)B_0}{\kappa_r}\{\phi_{m,z}(x,y,h,t) - \phi_{m,z}(x,y,-h,t)\} + \frac{2\chi B_0}{\kappa_r}\int_{-h}^{h} \phi_{m,zz}\, dz$$

(4.110)

Eliminating Q_x, Q_y from Eqs. (4.107) and Eq. (4.109), and taking into account Eq. (4.105), we have the equation of motion for a thin plate under the influence of magnetic field

$$D(w_{,xxxx} + 2w_{,xxyy} + w_{,yyyy}) - \frac{2}{3}\rho h^3(w_{,xx} + w_{,yy})_{,tt} + 2\rho h w_{,tt} - m_{xx,x}^D - m_{yy,y}^D - q^D = 0$$

(4.111)

Equation (4.111) is the basic equation of linear bending theory for soft ferromagnetic thin plates.

We can assume the solutions of Eq. (4.111) and the fourth and eighth of Eqs. (4.101) in the form

$$w = w_0 \exp\{-i(ky + \omega t)\}$$

(4.112)

$$\phi_m = A_1 \cosh(kz)\exp\{-i(ky + \omega t)\}$$

(4.113)

$$\phi_m^e = A_2 \exp\{-kz - i(ky + \omega t)\} \quad (z \geq h) \\ \left. = A_3 \exp\{kz - i(ky + \omega t)\} \quad (z \leq -h) \right\}$$

(4.114)

where w_0 is the amplitude of time harmonic waves, k is the wave number, ω is the angular frequency, and A_1, A_2, A_3 are the unknowns to be solved. From the boundary conditions (4.103), we obtain

$$A_1 = \frac{\chi B_0}{\kappa_0 \kappa_r \{\kappa_r \sinh(kh) + \cosh(kh)\}} w_0$$

$$A_2 = A_3 = -\frac{\chi B_0}{\kappa_0 \{\kappa_r \sinh(kh) + \cosh(kh)\}} \exp(kh) \sinh(kh) w_0$$

(4.115)

Substituting from Eqs. (4.108) and (4.110) into Eq. (4.111) and using Eqs. (4.112)–(4.115), we obtain the dispersion relation

$$\left(\frac{\omega}{kc_2}\right)^2 - \frac{1}{3 + (kh)^2} \left[\frac{2(kh)^2}{1 - \nu} - \frac{3\chi^2 \sinh(kh) b_c^2}{\kappa_r kh \{\kappa_r \sinh(kh) + \cosh(kh)\}}\right] = 0 \qquad (4.116)$$

where $c_2 = (\mu/\rho)^{1/2}$ is the shear wave velocity and

$$b_c^2 = \frac{B_0^2}{\kappa_0 \mu} \qquad (4.117)$$

4.3.1.2 Soft Ferromagnetic Mindlin Plate Bending

Here, consider a soft ferromagnetic Mindlin plate with thickness $2h$ in a rectangular Cartesian coordinate system (x, y, z) as shown in Fig. 4.5. The coordinate axes x and y are in the middle plane of the plate, and the z-axis is normal to this plane. Let magnetic flexural waves be traveled in the y-direction. The plate is permeated by a uniform magnetic field of magnetic induction $B_{0z} = B_0$.

Mindlin's theory of plate bending [23], which accounts for the rotatory inertia and shear effects, is applied. The rectangular displacement components may assume the forms

$$u_x = z\Psi_x(x, y, t), \quad u_y = z\Psi_y(x, y, t), \quad u_z = \Psi_z(x, y, t) \qquad (4.118)$$

where Ψ_z represents the normal displacement of the plate, and Ψ_x and Ψ_y denote the rotations of the normals about the x- and y-axes.

The bending and twisting moments per unit length $(M_{xx}, M_{yy}, M_{xy} = M_{yx})$ and the vertical shear forces per unit length (Q_x, Q_y) can be expressed in terms of Ψ_x, Ψ_y, and Ψ_z as

$$M_{xx} = \int_{-h}^{h} \sigma_{xx} z \, dz = D(\Psi_{x,x} + \nu\Psi_{y,y})$$

$$M_{yy} = \int_{-h}^{h} \sigma_{yy} z \, dz = D(\Psi_{y,y} + \nu\Psi_{x,x}) \qquad (4.119)$$

$$M_{xy} = M_{yx} = \int_{-h}^{h} \sigma_{xy} z \, dz = \frac{1 - \nu}{2} D(\Psi_{y,x} + \Psi_{x,y})$$

$$Q_x = \int_{-h}^{h} \sigma_{zx}\,dz = \frac{\pi^2}{6}\mu h(\Psi_{z,x} + \Psi_x)$$

$$Q_y = \int_{-h}^{h} \sigma_{zy}\,dz = \frac{\pi^2}{6}\mu h(\Psi_{z,y} + \Psi_y)$$

$$(4.120)$$

Now, multiplying the first and second of Eqs. (4.98) by $z\,dz$, integrating from $-h$ to h, and taking into account the first and second of Eqs. (4.102), we find

$$M_{xx,x} + M_{yx,y} - Q_x = \frac{2}{3}\rho h^3 \Psi_{x,tt} - m_{xx}^D$$

$$M_{xy,x} + M_{yy,y} - Q_y = \frac{2}{3}\rho h^3 \Psi_{y,tt} - m_{yy}^D$$

$$(4.121)$$

Multiplying the third of Eqs. (4.98) by dz, integrating from $-h$ to h, and taking into account the third of Eqs. (4.102), we obtain

$$Q_{x,x} + Q_{y,y} = 2h\rho\Psi_{z,tt} - q^D \tag{4.122}$$

Substituting Eqs. (4.119) and (4.120) into Eqs. (4.121) and Eq. (4.122), we have

$$\frac{S}{2}\{(1-v)(\Psi_{x,xx} + \Psi_{x,yy}) + (1+v)\Phi_{,x}\} - \Psi_x - \Psi_{z,x} = \frac{4h^2\rho}{\pi^2\mu}\Psi_{x,tt} - \frac{6}{\pi^2\mu h}m_{xx}^D$$

$$\frac{S}{2}\{(1-v)(\Psi_{y,xx} + \Psi_{y,yy}) + (1+v)\Phi_{,y}\} - \Psi_y - \Psi_{z,y} = \frac{4h^2\rho}{\pi^2\mu}\Psi_{y,tt} - \frac{6}{\pi^2\mu h}m_{yy}^D \quad (4.123)$$

$$\Psi_{z,xx} + \Psi_{z,yy} + \Phi = \frac{4h^2\rho}{\pi^2\mu}\frac{1}{R}\Psi_{z,tt} - \frac{6}{\pi^2\mu h}q^D$$

where

$$\Phi = \Psi_{x,x} + \Psi_{y,y} \tag{4.124}$$

$$R = \frac{h^2}{3}, \quad S = \frac{6D}{\pi^2\mu h} \tag{4.125}$$

Equations (4.123) are the basic equations of linear bending theory for soft ferromagnetic Mindlin plates.

We can assume the solutions of Eq. (4.123) and the fourth and eighth of Eqs. (4.101) in the form

$$\Psi_x = 0$$
$$\Psi_y = \Psi_{y0}\exp\{-i(ky + \omega t)\}$$
$$\Psi_z = \Psi_{z0}\exp\{-i(ky + \omega t)\}$$

$$(4.126)$$

$$\phi_m = A_1\cosh(kz)\exp\{-i(ky + \omega t)\} \tag{4.127}$$

$$\phi_m^e = A_2\exp\{-kz - i(ky + \omega t)\} \quad (z \geq h) \atop = A_3\exp\{kz - i(ky + \omega t)\} \quad (z \leq -h) \Bigg\} \tag{4.128}$$

where Ψ_{y0} and Ψ_{z0} are the amplitudes of the time harmonic waves, and A_1, A_2, A_3 are the unknowns to be solved. From the boundary conditions in Eqs. (4.103), we obtain

$$A_1 = \frac{\chi B_0}{\kappa_0 \kappa_r \{\kappa_r \sinh(kh) + \cosh(kh)\}} \Psi_{z0}$$

$$A_2 = A_3 = -\frac{\chi B_0}{\kappa_0 \{\kappa_r \sinh(kh) + \cosh(kh)\}} \exp(kh) \sinh(kh) \Psi_{z0}$$

(4.129)

Substituting from Eqs. (4.108) and (4.110) into Eq. (4.123) and using Eqs. (4.126)–(4.129), we obtain the dispersion relation and the relation between Ψ_{y0} and Ψ_{z0}

$$(kh)^3 \left(\frac{\omega}{kc_2}\right)^4 + \left\{-\frac{(kh)^3}{2}\left(\frac{4}{1-v} + \frac{\pi^2}{6}\right) + \frac{\chi^3 (kh)^2 \sinh(kh) b_c^2}{\kappa_r^2 \{\kappa_r \sinh(kh) + \cosh(kh)\}}\right.$$

$$\left. -\frac{\pi^2 (kh)}{4}\right\} \left(\frac{\omega}{kc_2}\right)^2$$

$$+ \frac{\pi^2 (kh)^3}{6(1-v)} - \frac{2\chi^3 (kh)^2 \sinh(kh) b_c^2}{(1-v)\kappa_r^2 \{\kappa_r \sinh(kh) + \cosh(kh)\}}$$

$$- \frac{\pi^2 \chi^2 \{(\chi+2)\sinh(kh) - (kh)\cosh(kh)\} b_c^2}{4\kappa_r^2 \{\kappa_r \sinh(kh) + \cosh(kh)\}} = 0$$

(4.130)

$$i\frac{\pi^2 h}{6} \Psi_{y0} = \left\{2(kh)\left(\frac{\omega}{kc_2}\right)^2 - \frac{\pi^2 (kh)}{6} + \frac{2\chi^3 \sinh(kh) b_c^2}{\kappa_r^2 \{\kappa_r \sinh(kh) + \cosh(kh)\}}\right\} \Psi_{z0}$$

(4.131)

4.3.1.3 Soft Ferromagnetic Plane Strain Plate Bending

Here, consider a soft ferromagnetic plane strain plate with thickness $2h$ in a rectangular Cartesian coordinate system (x, y, z) as shown in Fig. 4.5. The coordinate axes x and y are in the middle plane of the plate, and the z-axis is normal to this plane. Let magnetic flexural waves be traveled in the y-direction. The plate is permeated by a uniform magnetic field of magnetic induction $B_{0z} = B_0$.

We assume plane strain normal to the x-axis. The relevant components of the stress tensor follow from the Hooke's law as

$$\sigma_{yy} = \lambda(u_{y,y} + u_{z,z}) + 2\mu u_{y,y}$$
$$\sigma_{zz} = \lambda(u_{y,y} + u_{z,z}) + 2\mu u_{z,z}$$
$$\sigma_{yz} = \mu(u_{y,z} + u_{z,y})$$

(4.132)

The stress equations of motion (4.98) become

$$\sigma_{yy,y} + \sigma_{zy,z} + \frac{2\chi B_0}{\kappa_r} h_{y,z} = \rho u_{y,tt}$$

$$\sigma_{yz,y} + \sigma_{zz,z} + \frac{2\chi B_0}{\kappa_r} h_{z,z} = \rho u_{z,tt}$$

(4.133)

The magnetic field equations (4.101) are

$$h_y^e = \phi_{m,y}^e, \quad h_z^e = \phi_{m,z}^e,$$
$$\phi_{m,yy}^e + \phi_{m,zz}^e = 0$$
$$h_y = \phi_{m,y}, \quad h_z = \phi_{m,z},$$
$$\phi_{m,yy} + \phi_{m,zz} = 0$$

(4.134)

When the constitutive equations (4.132) and the fourth and fifth of Eqs. (4.134) are substituted in the stress equations of motion (4.133), we obtain the displacement equations of motion

$$\{2\mu u_{y,y} + \lambda(u_{y,y} + u_{z,z})\}_{,y} + \mu(u_{y,z} + u_{z,y})_{,z} + \frac{2\chi B_0}{\kappa_r}\phi_{m,yz} = \rho u_{y,tt}$$

$$\mu(u_{y,z} + u_{z,y})_{,y} + \{2\mu u_{z,z} + \lambda(u_{y,y} + u_{z,z})\}_{,z} + \frac{2\chi B_0}{\kappa_r}\phi_{m,zz} = \rho u_{z,tt}$$

(4.135)

From Eqs. (4.102) and (4.103), we obtain the linearized boundary conditions

$$\mu\{u_{y,z}(y, \pm h, t) + u_{z,y}(y, \pm h, t)\} = -\frac{\chi B_0}{\kappa_r}\phi_{m,y}(y, \pm h, t)$$

$$2\mu u_{z,z}(y, \pm h, t) + \lambda\{u_{y,y}(y, \pm h, t) + u_{z,z}(y, \pm h, t)\} =$$

$$\frac{\chi(\chi - 2)}{\kappa_r}\left\{\frac{B_0^2}{2\kappa_0\kappa_r} + B_0\phi_{m,z}(y, \pm h, t)\right\}$$

(4.136)

$$\phi_{m,y}^e(y, \pm h, t) - \phi_{m,y}(y, \pm h, t) = -\frac{\chi B_0}{\kappa_0\kappa_r}u_{z,y}(y, \pm h, t)$$

$$\phi_{m,z}^e(y, \pm h, t) - \kappa_r\phi_{m,z}(y, \pm h, t) = 0$$

(4.137)

To investigate wave motion in the soft ferromagnetic plane strain plate, the solutions for u_y, u_z, ϕ_m, and ϕ_m^e are assumed to be of the form

$$u_y = u_{y0}\exp\{pz - i(ky + \omega t)\}$$
$$u_z = u_{z0}\exp\{pz - i(ky + \omega t)\}$$
$$\phi_m = \phi_{m0}\exp\{pz - i(ky + \omega t)\}$$
$$\phi_m^e = \phi_{m0}^e\exp\{pz - i(ky + \omega t)\}$$

(4.138)

where $u_{y0}, u_{z0}, \phi_{m0}, \phi_{m0}^e$ are the amplitudes of the time harmonic waves and p is the modification factor of the wave amplitude with respect to the thickness. Substitution of the aforementioned solutions for u_y, u_z, ϕ_m, and ϕ_m^e into Eqs. (4.135) and the third and sixth of Eqs. (4.134) yields the following algebraic equations:

$$\begin{bmatrix} a_{11} & a_{12} & a_{13} \\ a_{21} & a_{22} & a_{23} \\ 0 & 0 & a_{33} \end{bmatrix}\begin{bmatrix} u_y \\ u_z \\ \phi_m \end{bmatrix} = 0$$

(4.139)

$$a_{44}\phi_m^e = 0$$

(4.140)

where

$$a_{11} = (\lambda + 2\mu)k^2 - \mu p^2 - \rho\omega^2$$

$$a_{12} = i(\lambda + \mu)kp$$

$$a_{13} = ikp\frac{2\chi B_0}{\mu\kappa_r}$$

$$a_{21} = i(\lambda + \mu)kp$$

$$a_{22} = \mu k^2 - (\lambda + 2\mu)p^2 - \rho\omega^2 \qquad (4.141)$$

$$a_{23} = -p^2\frac{2\chi B_0}{\mu\kappa_r}$$

$$a_{33} = p^2 - k^2$$

$$a_{44} = p^2 - k^2$$

A nontrivial solution of Eq. (4.139) will exist when p is related to k and ω such that the determinant of the coefficient matrix is zero, that is, when

$$\left\{\left(\frac{p}{k}\right)^2 - 1\right\}\left[\frac{c_1^2}{c_2^2}\left(\frac{p}{k}\right)^4 + \left\{\left(1 + \frac{c_1^2}{c_2^2}\right)\left(\frac{\omega}{kc_2}\right)^2 - 2\frac{c_1^2}{c_2^2}\right\}\left(\frac{p}{k}\right)^2\right.$$

$$\left. + \left(\frac{\omega}{kc_2}\right)^4 - \left(1 + \frac{c_1^2}{c_2^2}\right)\left(\frac{\omega}{kc_2}\right)^2 + \frac{c_1^2}{c_2^2}\right] = 0 \qquad (4.142)$$

where $c_1 = \{(\lambda + 2\mu)/\rho\}^{1/2}$ is the longitudinal wave velocity. The solutions take the form

$$\begin{bmatrix} u_y \\ u_z \\ \phi_m \end{bmatrix} = \begin{bmatrix} b_{11} & b_{12} & -b_{13} & -b_{11} & -b_{12} & b_{13} \\ 1 & 1 & 1 & 1 & 1 & 1 \\ 0 & 0 & b_{33} & 0 & 0 & b_{33} \end{bmatrix}$$

$$\times \begin{bmatrix} A_{11}\exp(\lambda_{11}kz)\exp\{-i(ky + \omega t)\} \\ A_{12}\exp(\lambda_{12}kz)\exp\{-i(ky + \omega t)\} \\ A_{13}\exp(kz)\exp\{-i(ky + \omega t)\} \\ B_{11}\exp(-\lambda_{11}kz)\exp\{-i(ky + \omega t)\} \\ B_{12}\exp(-\lambda_{12}kz)\exp\{-i(ky + \omega t)\} \\ B_{13}\exp(-kz)\exp\{-i(ky + \omega t)\} \end{bmatrix} \qquad (4.143)$$

and

$$\begin{aligned} \phi_m^e &= B_{14}\exp(-kz)\exp\{-i(ky + \omega t)\} \quad (z \geq h) \\ &= B_{15}\exp(kz)\exp\{-i(ky + \omega t)\} \quad (z \leq -h) \end{aligned} \right\} \qquad (4.144)$$

where $A_{11}, A_{12}, A_{13}, B_{11}, B_{12}, B_{13}, B_{14}$, and B_{15} are the unknowns, $p/k = \pm 1, \pm\lambda_{11}, \pm\lambda_{12}$ are the roots of Eq. (4.142), and

$$b_{1j} = i\frac{1 - (c_1^2/c_2^2)\lambda_{1j} - (\omega/kc_2)^2}{\lambda_{1j}(c_1^2/c_2^2 - 1)} \quad (j = 1, 2), \qquad b_{13} = i$$

$$\qquad (4.145)$$

$$b_{33} = -\left\{\left(\frac{\omega}{kc_2}\right)^2 + 2\left(\frac{c_1^2}{c_2^2} - 1\right)\right\}\frac{\mu^2\kappa_r}{2\chi B_0}$$

For convenience, in Eqs. (4.143) we set

$$A_{1j} = \frac{1}{2}(B_j + C_j), \quad B_{1j} = \frac{1}{2}(B_j - C_j) \quad (j = 1, 2, 3) \tag{4.146}$$

This leads to

$$
\begin{aligned}
u_y &= \{B_1 b_{11} \sinh(\lambda_{11} kz) + B_2 b_{12} \sinh(\lambda_{12} kz) - B_3 b_{13} \sinh(kz) + C_1 b_{11} \cosh(\lambda_{11} kz) \\
&\quad + C_2 b_{12} \cosh(\lambda_{12} kz) - C_3 b_{13} \cosh(kz)\} \exp\{-i(ky + \omega t)\} \\[6pt]
u_z &= \{B_1 \cosh(\lambda_{11} kz) + B_2 \cosh(\lambda_{12} kz) + B_3 \cosh(kz) + C_1 \sinh(\lambda_{11} kz) \\
&\quad + C_2 \sinh(\lambda_{12} kz) + C_3 \sinh(kz)\} \exp\{-i(ky + \omega t)\} \\[6pt]
\phi_m &= \{B_3 b_{33} \cosh(kz) + C_3 b_{33} \sinh(kz)\} \exp\{-i(ky + \omega t)\}
\end{aligned}
\tag{4.147}
$$

Substituting from Eqs. (4.147) and (4.144) into Eqs. (4.136) and (4.137), it is found for the antisymmetric modes that

$$
\begin{bmatrix}
\alpha_{11} & \alpha_{12} & \alpha_{13} & 0 \\
\alpha_{21} & \alpha_{22} & \alpha_{23} & 0 \\
\alpha_{31} & \alpha_{32} & \alpha_{33} & \alpha_{34} \\
0 & 0 & \alpha_{43} & \alpha_{44}
\end{bmatrix}
\begin{bmatrix}
B_1 \cosh(\lambda_{11} kh) \\
B_2 \cosh(\lambda_{12} kh) \\
B_3 \cosh(kh) \\
B_{14} \cosh(kh)
\end{bmatrix} = 0
\tag{4.148}
$$

and

$$B_{14} = B_{15} \tag{4.149}$$

where

$$
\begin{aligned}
&\alpha_{1j} = \mu \lambda_{1j} b_{1j} - i\mu \ (j = 1, 2), \quad \alpha_{13} = -\mu b_{13} - i\mu - i\frac{\chi B_0}{\kappa_r} b_{33}, \\
&\alpha_{2j} = \{(\lambda + 2\mu)\lambda_{1j} - i\lambda b_{1j}\} \tanh(\lambda_{1j} kh) \ (j = 1, 2), \\
&\alpha_{23} = \left\{ \lambda + 2\mu + i\lambda b_{13} - \frac{\chi(\chi - 2)B_0}{\kappa_r} b_{33} \right\} \tanh(kh), \\
&\alpha_{31} = \alpha_{32} = \frac{\chi B_0}{\kappa_0 \kappa_r}, \quad \alpha_{33} = \frac{\chi B_0}{\kappa_0 \kappa_r} - b_{33}, \quad \alpha_{34} = 1, \\
&\alpha_{43} = \kappa_r b_{33}, \quad \alpha_{44} = -1
\end{aligned}
\tag{4.150}
$$

The frequency equation is obtained by setting the determinant of the coefficients of Eq. (4.148) equal to zero, and frequencies $(f = \omega/2\pi)$ of the antisymmetric waves are solutions of

$$(\alpha_{11}\alpha_{22} - \alpha_{12}\alpha_{21})(\alpha_{33} - \alpha_{43}) + (\alpha_{12}\alpha_{23} - \alpha_{13}\alpha_{22})\alpha_{31}$$

$$+(\alpha_{13}\alpha_{21} - \alpha_{11}\alpha_{23})\alpha_{32} = 0 \tag{4.151}$$

4.3.1.4   Soft Ferromagnetic Plate Bending Solutions

To examine the effect of magnetic field on the flexural waves, we consider a stainless steel HT-9. The material properties of HT-9 are as follows:

$$E = 196 \times 10^9 \text{ N/m}^2$$
$$v = 0.3 \tag{4.152}$$
$$\kappa_r = 70$$

Figure 4.6 shows the variation of the phase velocity ω/kc_2 with the wave number kh of the classical plate under the magnetic inductions of $B_0 = 1.0, 1.6$ T. The dashed line represents the result of $B_0 = 0$ T. The effect of the magnetic field on the phase velocity is observed at low wave number. Figure 4.7 shows the variation of the phase velocity ω/kc_2 with the wave number kh of the classical, Mindlin, and plane strain plates under $B_0 = 1.6$ T. The curves obtained for the cases of classical and Mindlin plates coincide with the case of the plane strain plate.

4.3.2   Vibrations and Waves of Magnetically Saturated Material

Here, consider a magnetically saturated plate with thickness $2h$ in a rectangular Cartesian coordinate system (x, y, z). The coordinate axes x and y are in the middle plane of the plate, and the z-axis is normal to this plane. The plate is permeated by a uniform magnetic field normal to the plate surface and is magnetically saturated in the z-direction ($B_{0z} = B_{0s}, B_{0x} = B_{0y} = 0$).

Similarly to Section 4.2.2, we neglect all terms produced by β_r^{-2} and the symmetric part of the displacement gradient. The solutions for the rigid body state can be written as

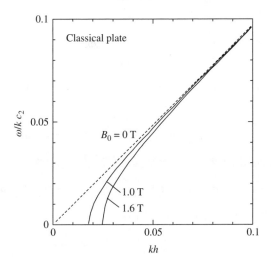

Figure 4.6   Phase velocity versus wave number of the classical plate

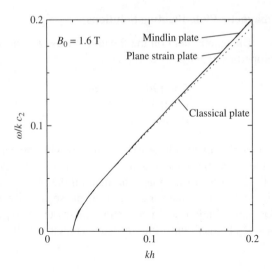

Figure 4.7 Phase velocity versus wave number of the classical, Mindlin, and plane strain plates

Eqs. (4.66). The mechanical constitutive equations are given by Eqs. (4.95), and the magnetoelastic stresses are

$$t_{xx} = \sigma_{xx} + 2\kappa_0 M_s \bar{b}_1 m_z$$
$$t_{yy} = \sigma_{yy} + 2\kappa_0 M_s \bar{b}_1 m_z$$
$$t_{zz} = \sigma_{zz} + 2\kappa_0 M_s (\bar{b}_1 + 2\bar{b}_2) m_z \qquad (4.153)$$
$$t_{xy} = t_{yx} = \sigma_{xy}$$
$$t_{yz} = t_{zy} = \sigma_{yz} + 2\kappa_0 M_s \bar{b}_2 m_y$$
$$t_{zx} = t_{xz} = \sigma_{zx} + 2\kappa_0 M_s \bar{b}_2 m_x$$

By substituting from Eqs. (4.153) into Eq. (4.14) and considering Eqs. (4.19), the stress equations of motion are given by

$$\sigma_{xx,x} + \sigma_{yx,y} + \sigma_{zx,z} + \kappa_0 M_s (h_{z,x} + 2\beta_r^2 \bar{b}_2 h_{x,z}) = \rho u_{x,tt}$$
$$\sigma_{xy,x} + \sigma_{yy,y} + \sigma_{zy,z} + \kappa_0 M_s (h_{z,y} + 2\beta_r^2 \bar{b}_2 h_{y,z}) = \rho u_{y,tt} \qquad (4.154)$$
$$\sigma_{xz,x} + \sigma_{yz,y} + \sigma_{zz,z} + \kappa_0 M_s \{h_{z,z} + 2\beta_r^2 \bar{b}_2 (h_{x,x} + h_{y,y})\} = \rho u_{z,tt}$$

Using Eq. (4.18), the Maxwell's equations (4.15) are

$$h_{z,y}^e - h_{y,z}^e = 0, \quad h_{x,z}^e - h_{z,x}^e = 0, \quad h_{y,x}^e - h_{x,y}^e = 0$$
$$h_{z,y} - h_{y,z} = 0, \quad h_{x,z} - h_{z,x} = 0, \quad h_{y,x} - h_{x,y} = 0 \qquad (4.155)$$

$$h_{x,x}^e + h_{y,y}^e + h_{z,z}^e = 0$$
$$\beta_r^2 (h_{x,x} + h_{y,y}) + h_{z,z} = 0 \qquad (4.156)$$

The magnetic field equations (4.155) and (4.156) are satisfied by using the magnetic potential ϕ_m such that

$$h_x^e = \phi_{m,x}^e, \quad h_y^e = \phi_{m,y}^e, \quad h_z^e = \phi_{m,z}^e,$$

$$\phi_{m,xx}^e + \phi_{m,yy}^e + \phi_{m,zz}^e = 0$$

$$h_x = \phi_{m,x}, \quad h_y = \phi_{m,y}, \quad h_z = \phi_{m,z},$$

$$\beta_r^2(\phi_{m,xx} + \phi_{m,yy}) + \phi_{m,zz} = 0$$

(4.157)

From Eqs. (4.22)–(4.24), we obtain the linearized boundary conditions

$$\sigma_{zx}(x, y, \pm h, t) = -2\kappa_r M_s \beta_r^2 \bar{b}_2 \phi_{m,x}(x, y, \pm h, t)$$

$$\sigma_{zy}(x, y, \pm h, t) = -2\kappa_r M_s \beta_r^2 \bar{b}_2 \phi_{m,y}(x, y, \pm h, t)$$

(4.158)

$$\sigma_{zz}(x, y, \pm h, t) = 0$$

$$\phi_{m,x}^e(x, y, \pm h, t) - \phi_{m,x}(x, y, \pm h, t) = -\frac{B_{0s}}{\kappa_0} u_{z,x}(x, y, \pm h, t)$$

$$\phi_{m,y}^e(x, y, \pm h, t) - \phi_{m,y}(x, y, \pm h, t) = -\frac{B_{0s}}{\kappa_0} u_{z,y}(x, y, \pm h, t)$$

(4.159)

$$\phi_{m,z}^e(x, y, \pm h, t) - \kappa_r \phi_{m,z}(x, y, \pm h, t) = 0$$

4.3.2.1   Magnetically Saturated Thin Plate Bending

Here, consider a magnetically saturated thin plate with thickness $2h$ in a rectangular Cartesian coordinate system (x, y, z) as shown in Fig. 4.5. The coordinate axes x and y are in the middle plane of the plate, and the z-axis is normal to this plane. Let magnetic flexural waves be traveled in the y-direction. The plate is permeated by a uniform magnetic field of magnetic induction $B_{0z} = B_{0s}$.

By using the classical plate bending theory [22], the rectangular displacement components can be given by Eq. (4.104). If we multiply the first and second of Eqs. (4.154) by $z\,dz$ and integrate from $-h$ to h, taking into account the first and second of Eqs. (4.158), we obtain the results

$$M_{xx,x} + M_{yx,y} - Q_x = -\frac{2}{3}\rho h^3 w_{,xtt} - m_{xx}^S$$

(4.160)

$$M_{xy,x} + M_{yy,y} - Q_y = -\frac{2}{3}\rho h^3 w_{,ytt} - m_{yy}^S$$

where

$$m_{xx}^S = -2\kappa_0 M_s \beta_r^2 \bar{b}_2 h\{\phi_{m,x}(x, y, h, t) + \phi_{m,x}(x, y, -h, t)\}$$

$$+ \kappa_0 M_s(1 + 2\beta_r^2 \bar{b}_2) \int_{-h}^{h} \phi_{m,xz} z\,dz$$

(4.161)

$$m_{yy}^S = -2\kappa_0 M_s \beta_r^2 \bar{b}_2 h\{\phi_{m,y}(x, y, h, t) + \phi_{m,y}(x, y, -h, t)\}$$

$$+ \kappa_0 M_s(1 + 2\beta_r^2 \bar{b}_2) \int_{-h}^{h} \phi_{m,yz} z\,dz$$

If the third of Eqs. (4.154) is multiplied by dz and integrated from $-h$ to h, taking into account the third of Eqs. (4.158), we obtain

$$Q_{x,x} + Q_{y,y} = 2h\rho w_{,tt} - q^S \tag{4.162}$$

where

$$q^S = \kappa_0 M_s \beta_r^2 (1 - 2\bar{b}_2) \int_{-h}^{h} (\phi_{m,xx} + \phi_{m,yy})dz \tag{4.163}$$

Eliminating Q_x, Q_y from Eqs.(4.160) and Eq.(4.162), and taking into account Eq.(4.105), we have the equation of motion for a thin plate under the influence of magnetic field

$$D(w_{,xxxx} + 2w_{,xxyy} + w_{,yyyy}) - \frac{2}{3}\rho h^3(w_{,xx} + w_{,yy})_{,tt} + 2\rho h w_{,tt} - m_{xx,x}^S - m_{yy,y}^S - q^S = 0 \tag{4.164}$$

Equation (4.164) is the basic equation of linear bending theory for magnetically saturated thin plates.

If we assume the solutions of Eq. (4.164) and the fourth and eighth of Eqs. (4.157) to be Eqs. (4.112)–(4.114), we obtain the unknowns from the boundary conditions (4.159) in the form

$$A_1 = \frac{B_{0s}}{\kappa_0\{\beta_r \sinh(\beta_r kh) + \cosh(\beta_r kh)\}} w_0$$

$$A_2 = A_3 = -\frac{\beta_r B_{0s}}{\kappa_0\{\beta_r \sinh(\beta_r kh) + \cosh(\beta_r kh)\}} \exp(kh)\sinh(\beta_r kh)w_0 \tag{4.165}$$

Substituting from Eqs. (4.161) and (4.163) into Eq. (4.164) and using Eqs. (4.112)–(4.114) with Eqs. (4.165), we obtain the dispersion relation

$$\left(\frac{\omega}{kc_2}\right)^2 - \frac{1}{3 + (kh)^2}\left[\frac{2(kh)^2}{1 - v} - \frac{3\beta_r \sinh(\beta_r kh)b_c^2}{kh\{\beta_r \sinh(\beta_r kh) + \cosh(\beta_r kh)\}}\right] = 0 \tag{4.166}$$

4.3.2.2   Magnetically Saturated Mindlin Plate Bending

Here, consider a magnetically saturated Mindlin plate with thickness $2h$ in a rectangular Cartesian coordinate system (x, y, z) as shown in Fig. 4.5. The coordinate axes x and y are in the middle plane of the plate, and the z-axis is normal to this plane. Let magnetic flexural waves be traveled in the y-direction. The plate is permeated by a uniform magnetic field of magnetic induction $B_{0z} = B_{0s}$.

Mindlin's theory of plate bending [23] is applied, and the rectangular displacement components can be given by Eq. (4.118). If we multiply the first and second of Eqs. (4.154) by $z\,dz$ and integrate from $-h$ to h, taking into account the first and second of Eqs. (4.158), we obtain

$$M_{xx,x} + M_{yx,y} - Q_x = \frac{2}{3}\rho h^3 \Psi_{x,tt} - m_{xx}^S$$

$$M_{xy,x} + M_{yy,y} - Q_y = \frac{2}{3}\rho h^3 \Psi_{y,tt} - m_{yy}^S \tag{4.167}$$

If the third of Eqs. (4.154) is multiplied by dz and integrated from $-h$ to h, taking into account the third of Eqs. (4.158), we find

$$Q_{x,x} + Q_{y,y} = 2h\rho\Psi_{z,tt} - q^S \tag{4.168}$$

Substituting Eqs. (4.119) and (4.120) into Eqs. (4.167) and Eq. (4.168), we have

$$\frac{S}{2}\{(1-v)(\Psi_{x,xx} + \Psi_{x,yy}) + (1+v)\Phi_{,x}\} - \Psi_x - \Psi_{z,x} = \frac{4h^2\rho}{\pi^2\mu}\Psi_{x,tt} - \frac{6}{\pi^2\mu h}m^S_{xx}$$

$$\frac{S}{2}\{(1-v)(\Psi_{y,xx} + \Psi_{y,yy}) + (1+v)\Phi_{,y}\} - \Psi_y - \Psi_{z,y} = \frac{4h^2\rho}{\pi^2\mu}\Psi_{y,tt} - \frac{6}{\pi^2\mu h}m^S_{yy} \tag{4.169}$$

$$\Psi_{z,xx} + \Psi_{z,yy} + \Phi = \frac{4h^2\rho}{\pi^2\mu}\frac{1}{R}\Psi_{z,tt} - \frac{6}{\pi^2\mu h}q^S$$

If we assume the solutions of Eq. (4.169) and the fourth and eighth of Eqs. (4.157) to be Eqs. (4.126)–(4.128), we obtain the unknowns from the boundary conditions (4.159) in the form

$$A_1 = \frac{B_{0s}}{\kappa_0\{\beta_r\sinh(\beta_r kh) + \cosh(\beta_r kh)\}}\Psi_{z0}$$

$$A_2 = A_3 = -\frac{\beta_r B_{0s}}{\kappa_0\{\beta_r\sinh(\beta_r kh) + \cosh(\beta_r kh)\}}\exp(kh)\sinh(\beta_r kh)\Psi_{z0} \tag{4.170}$$

Substituting from Eqs. (4.161) and (4.163) into Eq. (4.169) and using Eqs. (4.126)–(4.128) with Eqs. (4.170), we obtain

$$(kh)^3\left(\frac{\omega}{kc_2}\right)^4 + \left\{-\frac{(kh)^3}{2}\left(\frac{4}{1-v} + \frac{\pi^2}{6}\right) + \frac{\beta_r(1-2\bar{b}_2)(kh)^2\sinh(\beta_r kh)b_c^2}{\{\beta_r\sinh(\beta_r kh) + \cosh(\beta_r kh)\}}\right.$$

$$\left.-\frac{\pi^2(kh)}{4}\right\}\left(\frac{\omega}{kc_2}\right)^2$$

$$+\frac{\pi^2(kh)^3}{6(1-v)} - \frac{2\beta_r(1-2\bar{b}_2)(kh)^2\sinh(\beta_r kh)b_c^2}{(1-v)\{\beta_r\sinh(\beta_r kh) + \cosh(\beta_r kh)\}}$$

$$-\frac{\pi^2\{(\beta_r^2+1)\sinh(\beta_r kh) - \beta_r(kh)\cosh(\beta_r kh)\}b_c^2}{4\beta_r\{\beta_r\sinh(\beta_r kh) + \cosh(\beta_r kh)\}} = 0 \tag{4.171}$$

$$i\frac{\pi^2 h}{6}\Psi_{y0} = \left\{2(kh)\left(\frac{\omega}{kc_2}\right)^2 - \frac{\pi^2(kh)}{6} + \frac{2\beta_r(1-2\bar{b}_2)\sinh(\beta_r kh)b_c^2}{\{\beta_r\sinh(\beta_r kh) + \cosh(\beta_r kh)\}}\right\}\Psi_{z0}$$

$$\tag{4.172}$$

4.3.2.3 Magnetically Saturated Plane Strain Plate Bending

Here, consider a magnetically saturated plane strain plate with thickness $2h$ in a rectangular Cartesian coordinate system (x, y, z) as shown in Fig. 4.5. The coordinate axes x and y are in the middle plane of the plate, and the z-axis is normal to this plane. Let magnetic flexural waves be traveled in the y-direction. The plate is permeated by a uniform magnetic field of magnetic induction $B_{0z} = B_{0s}$.

We assume plane strain normal to the x-axis. The relevant components of the stress tensor follow from Hooke's law as Eq. (4.132). The stress equations of motion, Eqs. (4.154), become

$$\sigma_{yy,y} + \sigma_{zy,z} + \kappa_0 M_s (h_{z,y} + 2\beta_r^2 \bar{b}_2 h_{y,z}) = \rho u_{y,tt}$$

$$\sigma_{yz,y} + \sigma_{zz,z} + \kappa_0 M_s \{h_{z,z} + 2\beta_r^2 \bar{b}_2 h_{y,y}\} = \rho u_{z,tt} \qquad (4.173)$$

The magnetic field equations, Eqs. (4.157), are

$$h_y^e = \phi_{m,y}^e, \quad h_z^e = \phi_{m,z}^e,$$

$$\phi_{m,yy}^e + \phi_{m,zz}^e = 0,$$

$$h_y = \phi_{m,y}, \quad h_z = \phi_{m,z}, \qquad (4.174)$$

$$\beta_r^2 \phi_{m,yy} + \phi_{m,zz} = 0$$

When the constitutive equations (4.132) and the fourth and fifth of Eqs. (4.174) are substituted in the stress equations of motion, Eqs. (4.173), we obtain the displacement equations of motion

$$\{2\mu u_{y,y} + \lambda(u_{y,y} + u_{z,z})\}_{,y} + \mu(u_{y,z} + u_{z,y})_{,z} + B_{0s}(1 + 2\beta_r^2 \bar{b}_2)\phi_{m,yz} = \rho u_{y,tt}$$

$$\mu(u_{y,z} + u_{z,y})_{,y} + \{2\mu u_{z,z} + \lambda(u_{y,y} + u_{z,z})\}_{,z} + B_{0s}(1 + 2\beta_r^2 \bar{b}_2)\phi_{m,zz} = \rho u_{z,tt} \qquad (4.175)$$

From Eqs. (4.158) and (4.159), we obtain the linearized boundary conditions

$$\mu\{u_{y,z}(y, \pm h, t) + u_{z,y}(y, \pm h, t)\} = -2B_{0s}\beta_r^2 \bar{b}_2 \phi_{m,y}(y, \pm h, t)$$

$$2\mu u_{z,z}(y, \pm h, t) + \lambda\{u_{y,y}(y, \pm h, t) + u_{z,z}(y, \pm h, t)\} = 0 \qquad (4.176)$$

$$\phi_{m,y}^e(y, \pm h, t) - \phi_{m,y}(y, \pm h, t) = -\frac{B_{0s}}{\kappa_0} u_{z,y}(y, \pm h, t)$$

$$\phi_{m,z}^e(y, \pm h, t) - \kappa_r \phi_{m,z}(y, \pm h, t) = 0 \qquad (4.177)$$

To investigate wave motion in the magnetically saturated plane strain plate, the solutions for u_y, u_z, ϕ_m, and ϕ_m^e are assumed to be Eqs. (4.138). Substitution of these solutions into Eqs. (4.175) and the third and sixth of Eqs. (4.174) yields

$$\begin{bmatrix} f_{11} & f_{12} & f_{13} \\ f_{21} & f_{22} & f_{23} \\ 0 & 0 & f_{33} \end{bmatrix} \begin{bmatrix} u_y \\ u_z \\ \phi_m \end{bmatrix} = 0 \qquad (4.178)$$

$$f_{44}\phi_m^e = 0 \qquad (4.179)$$

where

$$
\begin{aligned}
f_{11} &= (\lambda + 2\mu)k^2 - \mu p^2 - \rho\omega^2 \\
f_{12} &= i(\lambda + \mu)kp \\
f_{13} &= ikpB_{0s}(1 + 2\beta_r^2\bar{b}_2) \\
f_{21} &= i(\lambda + \mu)kp \\
f_{22} &= \mu k^2 - (\lambda + 2\mu)p^2 - \rho\omega^2 \\
f_{23} &= -p^2 B_{0s}(1 + 2\beta_r^2\bar{b}_2) \\
f_{33} &= p^2 - (\beta_r k)^2 \\
f_{44} &= p^2 - (\beta_r k)^2
\end{aligned}
\tag{4.180}
$$

A nontrivial solution of Eq. (4.178) will exist when p is related to k and ω such that the determinant of the coefficient matrix is zero, that is, when

$$
\left\{ \left(\frac{p}{k}\right)^2 - \beta_r^2 \right\} \left[\frac{c_1^2}{c_2^2} \left(\frac{p}{k}\right)^4 + \left\{ \left(1 + \frac{c_1^2}{c_2^2}\right) \left(\frac{\omega}{kc_2}\right)^2 - 2\frac{c_1^2}{c_2^2} \right\} \left(\frac{p}{k}\right)^2 \right.
$$

$$
\left. + \left(\frac{\omega}{kc_2}\right)^4 - \left(1 + \frac{c_1^2}{c_2^2}\right) \left(\frac{\omega}{kc_2}\right)^2 + \frac{c_1^2}{c_2^2} \right] = 0
\tag{4.181}
$$

The solutions can be compressed into the form

$$
\begin{bmatrix} u_y \\ u_z \\ \phi_m \end{bmatrix} =
\begin{bmatrix}
g_{11} & g_{12} & -g_{13} & -g_{11} & -g_{12} & g_{13} \\
1 & 1 & 1 & 1 & 1 & 1 \\
0 & 0 & g_{33} & 0 & 0 & g_{33}
\end{bmatrix}
$$

$$
\times
\begin{bmatrix}
A_{11} \exp(\lambda_{21}kz) \exp\{-i(ky + \omega t)\} \\
A_{12} \exp(\lambda_{22}kz) \exp\{-i(ky + \omega t)\} \\
A_{13} \exp(\beta_r kz) \exp\{-i(ky + \omega t)\} \\
B_{11} \exp(-\lambda_{21}kz) \exp\{-i(ky + \omega t)\} \\
B_{12} \exp(-\lambda_{22}kz) \exp\{-i(ky + \omega t)\} \\
B_{13} \exp(-\beta_r kz) \exp\{-i(ky + \omega t)\}
\end{bmatrix}
\tag{4.182}
$$

and

$$
\begin{aligned}
\phi_m^e &= B_{14} \exp(-\beta_r kz) \exp\{-i(ky + \omega t)\} \quad (z \geq h) \\
&= B_{15} \exp(\beta_r kz) \exp\{-i(ky + \omega t)\} \quad (z \leq -h)
\end{aligned}
\tag{4.183}
$$

where $A_{11}, A_{12}, A_{13}, B_{11}, B_{12}, B_{13}, B_{14}$, and B_{15} are the unknowns, $p/k = \pm 1, \pm\lambda_{21}, \pm\lambda_{22}$ are the roots of Eq. (4.181), and

$$
g_{1j} = \frac{i}{\lambda_{2j}} \ (j = 1, 2), \quad g_{13} = \frac{i}{\beta_r}
$$

$$
g_{33} = -\left\{ \left(\frac{\omega}{kc_2}\right)^2 + \beta_r^2 \left(2 - \frac{c_1^2}{c_2^2}\right) - \frac{c_1^2}{c_2^2} \right\} \frac{\mu}{\beta_r^2(1 + 2\beta_r^2\bar{b}_2)B_{0s}}
\tag{4.184}
$$

For convenience, in Eqs. (4.182) we set

$$A_{1j} = \frac{1}{2}(F_j + G_j), \quad B_{1j} = \frac{1}{2}(F_j - G_j) \ (j = 1, 2, 3) \tag{4.185}$$

Hence

$$u_y = \{F_1 g_{11} \sinh(\lambda_{21}kz) + F_2 g_{12} \sinh(\lambda_{22}kz) - F_3 g_{13} \sinh(\beta_r kz) + G_1 g_{11} \cosh(\lambda_{21}kz)$$
$$+ G_2 g_{12} \cosh(\lambda_{22}kz) - G_3 g_{13} \cosh(\beta_r kz)\} \exp\{-i(ky + \omega t)\}$$

$$u_z = \{F_1 \cosh(\lambda_{21}kz) + F_2 \cosh(\lambda_{22}kz) + F_3 \cosh(\beta_r kz) + G_1 \sinh(\lambda_{21}kz)$$
$$+ G_2 \sinh(\lambda_{22}kz) + G_3 \sinh(\beta_r kz)\} \exp\{-i(ky + \omega t)\}$$

$$\phi_m = \{F_3 g_{33} \cosh(\beta_r kz) + G_3 g_{33} \sinh(\beta_r kz)\} \exp\{-i(ky + \omega t)\} \tag{4.186}$$

Substituting from Eqs. (4.186) and (4.183) into Eqs. (4.176) and (4.177), it is found for the antisymmetric modes that

$$\begin{bmatrix} \beta_{11} & \beta_{12} & \beta_{13} & 0 \\ \beta_{21} & \beta_{22} & \beta_{23} & 0 \\ \beta_{31} & \beta_{32} & \beta_{33} & \beta_{34} \\ 0 & 0 & \beta_{43} & \beta_{44} \end{bmatrix} \begin{bmatrix} F_1 \cosh(\lambda_{21}kh) \\ F_2 \cosh(\lambda_{22}kh) \\ F_3 \cosh(\beta_r kh) \\ B_{14} \cosh(\beta_r kh) \end{bmatrix} = 0 \tag{4.187}$$

and

$$B_{14} = B_{15} \tag{4.188}$$

where

$$\beta_{1j} = \mu \lambda_{2j} g_{1j} - i\mu \ (j = 1, 2), \quad \beta_{13} = -\mu g_{13} - i\mu - 2i\beta_r^2 \bar{b}_2 B_{0s} g_{33},$$
$$\beta_{2j} = \{(\lambda + 2\mu)\lambda_{2j} - i\lambda g_{1j}\} \tanh(\lambda_{2j}kh) \ (j = 1, 2),$$
$$\beta_{23} = \{\beta_r(\lambda + 2\mu) + i\lambda g_{13}\} \tanh(\beta_r kh), \tag{4.189}$$
$$\beta_{31} = \beta_{32} = \frac{B_{0s}}{\kappa_0}, \quad \beta_{33} = \frac{B_{0s}}{\kappa_0} - g_{33}, \quad \beta_{34} = 1,$$
$$\beta_{43} = \kappa_r g_{33}, \quad \beta_{44} = -1$$

The frequency equation is obtained by setting the determinant of the coefficients of Eq. (4.187) equal to zero, and frequencies $(f = \omega/2\pi)$ of the antisymmetric waves are solutions of

$$(\beta_{11}\beta_{22} - \beta_{12}\beta_{21})(\beta_{33} - \beta_{43}) + (\beta_{12}\beta_{23} - \beta_{13}\beta_{22})\beta_{31}$$
$$+(\beta_{13}\beta_{21} - \beta_{11}\beta_{23})\beta_{32} = 0 \tag{4.190}$$

4.4 Magnetic Moment Intensity Factor

When cracked ferromagnetic plates are subjected to bending and placed in magnetic field, the existence of the magnetic field may produce higher singular moments near the crack tip. In this section, the magnetoelastic problems in ferromagnetic plates with a through crack under bending are dealt with, and the static and dynamic moment intensity factors are discussed.

4.4.1 Simply Supported Plate Under Static Bending

Here, we consider a soft ferromagnetic thin plate of length l and thickness $2h$ with a through crack of length $2a$ in a rectangular Cartesian coordinate system (x, y, z) as shown in Fig. 4.8. The coordinate axes x and y are in the middle plane of the plate, and the z-axis is normal to this plane. The plate is permeated by a uniform magnetic field of magnetic induction $B_{0z} = B_0$ and is deformed by bending moments per unit length M_0 at the simply supported ends $y = \pm l/2$.

We can find the basic equations in Section 4.3.1 by neglecting the inertia terms. The boundary conditions can be written as

$$V_y(x, 0) = 0 \qquad (0 \le |x| < \infty) \tag{4.191}$$

$$\begin{cases} M_{yy}(x, 0) = 0 & (0 \le |x| < a) \\ u_y(x, 0) = 0 & (a \le |x| < \infty) \end{cases} \tag{4.192}$$

$$\begin{aligned} w(x, \pm l/2) &= 0 & (0 \le |x| < \infty) \\ M_{yy}(x, \pm l/2) &= M_0 & (0 \le |x| < \infty) \end{aligned} \tag{4.193}$$

where $V_y = Q_y + M_{xy,x}$ is the equivalent shear force [24].

The deflection of the middle plane of the uncracked plate is given by Shindo et al. [25]

$$w^0 = \frac{M_0}{Dk^2} \left\{ \left[\cos\left\{ k\left(y + \frac{l}{2}\right) \right\} + \frac{1 - \cos(kl)}{\sin(kl)} \sin\left\{ k\left(y + \frac{l}{2}\right) \right\} \right] - 1 \right\} \tag{4.194}$$

where the superscript 0 refers to the uncracked plate, and the constant k is obtained as

$$(kh)^3 - \frac{3\chi^2(1 - v)b_c^2\{(\chi + 2)\sinh(kh) - kh\cosh(kh)\}}{2\kappa_r^2\{\kappa_r \sinh(kh) + \cosh(kh)\}} = 0 \tag{4.195}$$

For the soft ferromagnetic material, where $\chi \gg 1$, we obtain

$$(kh)^3 - \frac{3\chi^3(1 - v)b_c^2 \sinh(kh)}{2\kappa_r^2\{\kappa_r \sinh(kh) + \cosh(kh)\}} = 0 \tag{4.196}$$

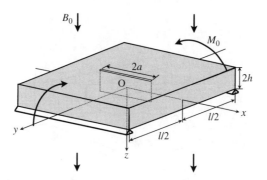

Figure 4.8   A simply supported soft ferromagnetic plate with a through crack

By substituting from Eq. (4.194) into the second of Eq. (4.105), the bending moment of the uncracked plate is obtained as

$$M_{yy}^0 = M_0 \left[\cos \left\{ k \left(y + \frac{l}{2} \right) \right\} + \frac{1 - \cos(kl)}{\sin(kl)} \sin \left\{ k \left(y + \frac{l}{2} \right) \right\} \right] \qquad (4.197)$$

For the cracked plate, by using Fourier transforms [26], we reduce the problem to solving a pair of dual integral equations. The solution of the dual integral equations is then expressed in terms of a Fredholm integral equation of the second kind. The moment intensity factor is defined by

$$K_{\mathrm{I}} = \lim_{x \to a^+} \{2\pi(x - a)\}^{1/2} M_{yy}(x, 0) \qquad (4.198)$$

Inspections of the numerical results indicate that the moment intensity factor increases with increasing magnetic field, and the effect of the magnetic field on the moment intensity factor is more pronounced with increasing crack length to plate thickness ratio a/h, plate length to thickness ratio $l/2h$, and magnetic susceptibility χ (not shown here). Similar results can be found for the magnetically saturated thin plate with a through crack [27].

4.4.2   Fixed-End Plate Under Static Bending

Here, we consider a soft ferromagnetic thin plate of length l, width b, and thickness $2h$ with a through crack of length $2a$ in a rectangular Cartesian coordinate system (x, y, z) as shown in Fig. 4.9. The coordinate axes x and y are in the middle plane of the plate, and the z-axis is normal to this plane. The plate is permeated by a uniform magnetic field of magnetic induction $B_{0z} = B_0$ and is bent by a normal line load P applied at the middle of the span.

We can find the basic equations in Section 4.3.1 by neglecting the inertia terms. The differential equation (4.111) can be written in a convenient form

$$D(w_{,xxxx} + 2w_{,xxyy} + w_{,yyyy}) = m_{xx,x}^D + m_{yy,y}^D + q^D + (P/b)\delta(y) \qquad (4.199)$$

Let us now consider $b \to \infty$. The boundary conditions can be expressed as

$$V_y(x, 0) = 0 \qquad (0 \le |x| < \infty) \qquad (4.200)$$

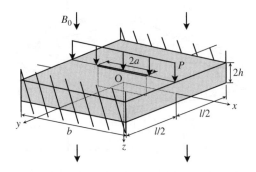

Figure 4.9   A fixed-end soft ferromagnetic plate with a through crack

$$\begin{cases} M_{yy}(x,0) = 0 & (0 \le |x| < a) \\ u_y(x,0) = 0 & (a \le |x| < \infty) \end{cases} \tag{4.201}$$

$$w(x, \pm l/2) = 0 \quad (0 \le |x| < \infty)$$
$$w_{,y}(x, \pm l/2) = 0 \quad (0 \le |x| < \infty) \tag{4.202}$$

The force per unit length P/b is applied at the center $y = 0$.

The deflection of the middle plane of the uncracked plate is given by

$$w^0 = \frac{2Pl^3}{Db} \sum_{n=1,3,5,\ldots}^{\infty} \frac{\cos(n\pi)[\cos\{k_n(y+l/2)\} - 1]}{(2n\pi)^4 + (l/h)^3(2n\pi)\{(\chi + 2)\sinh(k_n h) - k_n h \cosh(k_n h)\}B_n} \tag{4.203}$$

where

$$B_n = -\frac{3(1-v)\chi^2 b_c^2}{2\kappa_r^2\{\kappa_r \sinh(k_n h) + \cosh(k_n h)\}} \tag{4.204}$$

$$k_n = \frac{2n\pi}{l} \tag{4.205}$$

By substituting from Eq. (4.203) into the second of Eq. (4.105), the bending moment of the uncracked plate is obtained as

$$M_{yy}^0 = \frac{2Pl}{b} \sum_{n=1,3,5,\ldots}^{\infty} \frac{(2n\pi)\cos(n\pi)^2 \cos\{k_n(y+l/2)\}}{(2n\pi)^3 + (l/h)^3\{(\chi + 2)\sinh(k_n h) - k_n h \cosh(k_n h)\}B_n} \tag{4.206}$$

Fourier transform method is used to solve the crack problem, and the solution is expressed in terms of a pair of dual integral equations. The dual integral equations are further reduced to a Fredholm integral equation of the second kind [28]. The moment intensity factor is defined by Eq. (4.198).

In order to validate the prediction, we evaluate the moment intensity factor of stainless steel SUS430 for various values of a and $2h$. The length and width of the specimen are $l = 100$ mm and $b = 40$ mm, respectively. A five-element strip gage is installed along the $90°$ line as shown in Fig. 4.10, and the center point of the element closest to the crack tip is

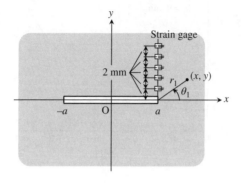

Figure 4.10 Position of strain gage and coordinate system used to express crack tip strain

$x = a, y = 2\,\text{mm}$. The values of K_I may be evaluated conveniently by measuring the local strain at selected positions [29]. We use a superconducting magnet with a 220-mm-diameter bore to create the magnetic field of magnetic induction B_0 normal to the wide face of the specimen. The fixed-end cracked specimen is bent by a normal line load P, which consists of weights. This line load is applied at the center of the plate. The experiments are conducted for various values of the magnetic field, and the strain is measured.

The radial strain ε_{rr} near the crack tip is given by

$$E\varepsilon_{rr} = -\frac{(3/2h^2)(1+v)zK_I}{4(3+v)h}\left\{(7+v)\cos\left(\frac{3\theta_1}{2}\right)\right.$$

$$\left. + 3(v-1)\cos\left(\frac{\theta_1}{2}\right)\right\}\frac{1}{(2r_1)^{1/2}}$$

$$-4z\{(1-v)+(1+v)\cos(2\theta_1)\}A + \cdots \tag{4.207}$$

where A is a constant. Setting $\theta_1 = \pi/2$ and $z = h$ gives

$$\frac{8(3+v)Eh^2}{3(5-v)(1+v)}\varepsilon_{rr}r_1^{1/2} = K_I + \frac{64(3+v)vh^3}{3(5-v)(1+v)}Ar_1^{1/2} + \cdots \tag{4.208}$$

From Eq. (4.208), a plot of $\{8(3+v)Eh^2/3(5-v)(1+v)\}\varepsilon_{rr}r_1^{1/2}$ versus $r_1^{1/2}$ is linear for the small values of r_1, and the intercept at $r_1 = 0$, at the crack tip, gives the moment intensity factor K_I.

Figure 4.11 shows a plot of the moment intensity factor K_I with the magnetic field B_0 for the specimen of $2h = 1\,\text{mm}$ and $2a = 9\,\text{mm}$. Solid line and open circle show the theoretical solution ($b \to \infty$) and experimental data ($b = 40\,\text{mm}$, $P = 49\,\text{N}$), respectively. The moment intensity factor is normalized with respect to K_{I0}, which is the corresponding value for $B_0 = 0$.

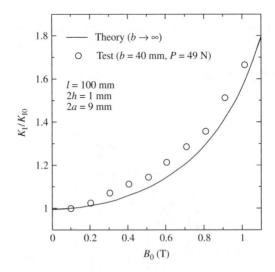

Figure 4.11   Moment intensity factor versus magnetic field

The analysis and experiment show the increase in the moment intensity factor with increasing magnetic field. The theoretical result agrees very well with the experimental data, and the result obtained from the classical plate bending theory for the soft ferromagnetic materials is quite acceptable.

4.4.3 Infinite Plate Under Dynamic Bending

Shindo and Horiguchi [30] applied the classical plate bending theory to solve the dynamic magnetoelastic problem of a soft ferromagnetic plate with a through crack. Here, we consider a soft ferromagnetic Mindlin plate of thickness $2h$ with a through crack of length $2a$ in a rectangular Cartesian coordinate system (x, y, z) as shown in Fig. 4.12. The coordinate axes x and y are in the middle plane of the plate, and the z-axis is normal to this plane. The plate is permeated by a uniform magnetic field of magnetic induction $B_{0z} = B_0$. Incident waves give rise to moments symmetric about the crack plane $y = 0$.

Let incident flexural waves be direct at an angle $\pi/2$ with the x-axis so that

$$\Psi_x^i = 0$$

$$\Psi_y^i = \frac{1}{2}\Psi_{y0}[\exp\{-i(ky + \omega t)\} - \exp\{i(ky - \omega t)\}] \tag{4.209}$$

$$\Psi_z^i = \frac{1}{2}\Psi_{z0}[\exp\{-i(ky + \omega t)\} + \exp\{i(ky - \omega t)\}]$$

$$\phi_m^i = \frac{A_1}{2}\cosh(kz)[\exp\{-i(ky + \omega t)\} + \exp\{i(ky - \omega t)\}] \tag{4.210}$$

$$\left.\begin{aligned}
\phi_m^{ei} &= \frac{A_2}{2}\exp(-kz)[\exp\{-i(ky + \omega t)\} + \exp\{i(ky - \omega t)\}] \quad (z \geq h) \\
&= \frac{A_3}{2}\exp(kz)[\exp\{-i(ky + \omega t)\} + \exp\{i(ky - \omega t)\}] \quad (z \leq -h)
\end{aligned}\right\} \tag{4.211}$$

where Ψ_{y0}, Ψ_{z0} are the amplitudes of the incident waves, A_1, A_2, A_3 are the unknowns to be solved, and the superscript i stands for the incident component. From the boundary conditions

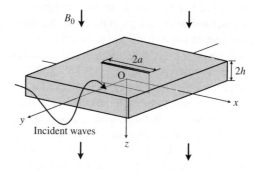

Figure 4.12 A soft ferromagnetic plate with a through crack and incident waves

(4.103), we obtain Eqs. (4.129). The dispersion relation is given by Eq. (4.130), and the relation between Ψ_{y0} and Ψ_{z0} is obtained as Eq. (4.131).

The complete solution of the waves as diffracted by the through crack is obtained by adding the incident and scattered waves. Likewise, the plate displacements, moments, and shears can also be found by superposing the incident and scattered parts. For a traction-free crack, the quantities M_{yy}, M_{yx}, Q_y must each vanish for $|x| < a, y = 0$. The conditions to be specified on the crack for the scattered field become

$$M_{yx}^s(x, 0, t) = 0 \qquad (0 \leq |x| < \infty) \tag{4.212}$$

$$Q_y^s(x, 0, t) = 0 \qquad (0 \leq |x| < \infty) \tag{4.213}$$

$$\begin{cases} M_{yy}^s(x, 0, t) = -M_{yy}^i & (0 \leq |x| < a) \\ \Psi_y^s(x, 0, t) = 0 & (a \leq |x| < \infty) \end{cases} \tag{4.214}$$

where the superscript s stands for the scattered component, and

$$M_{yy}^i = -ikD\Psi_{y0}\exp(-i\omega t) \tag{4.215}$$

The field quantities will all contain the time factor $\exp(-i\omega t)$.

Fourier transforms are applied reducing the mixed boundary value problem to a Fredholm integral equation that can be solved numerically [31]. The moment intensity factor is defined by

$$K_I = \lim_{x \to a^+} \{2\pi(x - a)\}^{1/2} M_{yy}(x, 0, t) \tag{4.216}$$

The numerical results indicate that the significant increase in the dynamic moment intensity factor due to magnetic field occurs at low frequency and the magnetic field effect dies out gradually as the frequency is increased. Similar results can be found for the magnetically saturated Mindlin plate with a through crack [32].

4.5   Tensile Fracture and Fatigue

If ferromagnetic materials are used in magnetic field, the combination of mechanical and magnetic forces could produce elevated stresses and strains, and the materials may be degraded in such a stress level. The strength of the ferromagnetic materials is also weakened by the presence of defects such as voids and cracks. It is therefore important to understand the degradation phenomena of the ferromagnetic materials. Shindo [33–36] applied a linear theory of magnetoelasticity to various crack geometries to study the influence of the magnetic field on the stress intensity factor of infinite soft ferromagnetic materials under tensile loading. Shindo [37] also considered a soft ferromagnetic strip with two coplanar cracks under the magnetic field normal to the edges of the strip. Furthermore, Shindo [38–40] analyzed the dynamic problems of cracked soft ferromagnetic materials. In this section, we overview the theoretical and experimental observations in the tensile fracture and fatigue behavior of soft ferromagnetic materials under the magnetic field.

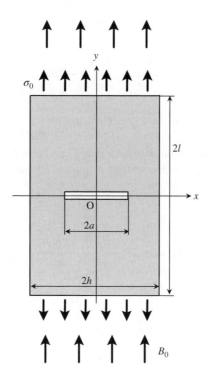

Figure 4.13   A rectangular soft ferromagnetic material with a central crack

4.5.1   Cracked Rectangular Soft Ferromagnetic Material

Consider a rectangular soft ferromagnetic material of width $2h$ and length $2l$, which contains a central crack of length $2a$ aligned with its plane normal to the free edges as shown in Fig. 4.13. The thickness is t. A rectangular Cartesian coordinate system (x, y, z) is attached to be the center of the crack for reference purposes. We consider a uniform normal stress, $\sigma_{yy} = \sigma_0$, applied with a uniform magnetic field of magnetic induction B_0 in the y-direction ($B_{0y} = B_0$, $B_{0x} = B_{0z} = 0$). Consequently, we may set $H_{0x} = H_{0z} = 0$ and $M_{0x} = M_{0z} = 0$.

Let us now consider $l \to \infty$. The solutions in a rectangular Cartesian coordinate system (x, y, z) for the rigid body state satisfying the field equation (4.3) and boundary condition (4.11) can, with the aid of the first of Eqs. (4.8) and the first of Eqs. (4.9), be written as

$$
\left.
\begin{aligned}
B^e_{0y} &= \frac{B_0}{\kappa_r}, \quad H^e_{0y} = \frac{B_0}{\kappa_0 \kappa_r}, \quad M^e_{0y} = 0 \qquad (|x| > h) \\[2mm]
B_{0y} &= B_0, \quad H_{0y} = \frac{B_0}{\kappa_0 \kappa_r}, \quad M_{0y} = \frac{\chi B_0}{\kappa_0 \kappa_r} \quad (|x| \le h)
\end{aligned}
\right\} \tag{4.217}
$$

$$
B^e_{0y} = B_0, \quad H^e_{0y} = \frac{B_0}{\kappa_0}, \quad M^e_{0y} = 0 \qquad (0 \le |x| < a, y = 0)
$$

The problem is reducible to two dimensions. The relevant components of the stress tensor follow from Hooke's law as

$$\sigma_{xx} = \kappa(u_{x,x} + u_{y,y}) + 2\mu u_{x,x}$$
$$\sigma_{yy} = \kappa(u_{x,x} + u_{y,y}) + 2\mu u_{y,y} \qquad (4.218)$$
$$\sigma_{xy} = \sigma_{yx} = \mu(u_{x,y} + u_{y,x})$$

where $\kappa = \lambda$ for plane strain (t is much greater than h) and $\kappa = 2\lambda\mu/(\lambda + 2\mu)$ for plane stress (t is much smaller than h). The components of magnetoelastic and Maxwell stress tensors are given by

$$t_{xx} = \sigma_{xx}$$
$$t_{yy} = \sigma_{yy} + \frac{\chi B_0^2}{\kappa_0 \kappa_r^2} + \frac{2\chi B_0 h_y}{\kappa_r} \qquad (4.219)$$
$$t_{xy} = t_{yx} = \sigma_{xy} + \frac{\chi B_0 h_x}{\kappa_r}$$

$$\sigma_{xx}^M = -\frac{B_0 h_y}{\kappa_r} - \frac{B_0^2}{2\kappa_0 \kappa_r^2}$$
$$\sigma_{yy}^M = \frac{(1 + 2\chi)B_0 h_y}{\kappa_r} + \frac{(1 + 2\chi)B_0^2}{2\kappa_0 \kappa_r^2} \qquad (4.220)$$

$$\sigma_{xy}^M = \sigma_{yx}^M = B_0 h_x$$

From Eqs. (4.98), the stress equations without inertia become

$$\sigma_{xx,x} + \sigma_{yx,y} + \frac{2\chi B_0}{\kappa_r} h_{x,y} = 0$$
$$\qquad (4.221)$$
$$\sigma_{xy,x} + \sigma_{yy,y} + \frac{2\chi B_0}{\kappa_r} h_{y,y} = 0$$

The magnetic field equations (4.101) are

$$h_x^e = \phi_{m,x}^e, \quad h_y^e = \phi_{m,y}^e,$$
$$\phi_{m,xx}^e + \phi_{m,yy}^e = 0$$
$$\qquad (4.222)$$
$$h_x = \phi_{m,x}, \quad h_y = \phi_{m,y},$$
$$\phi_{m,xx} + \phi_{m,yy} = 0$$

When the constitutive equations (4.218) and the fourth and fifth of Eqs. (4.222) are substituted in the stress equations (4.221), we obtain the displacement equations

$$u_{x,xx} + u_{x,yy} + \left(\frac{\kappa}{\mu} + 1\right)(u_{x,x} + u_{y,y})_{,x} + \frac{2\chi B_0}{\mu\mu_r}\phi_{m,xy} = 0$$

$$u_{y,xx} + u_{y,yy} + \left(\frac{\kappa}{\mu} + 1\right)(u_{x,x} + u_{y,y})_{,y} + \frac{2\chi B_0}{\mu\mu_r}\phi_{m,yy} = 0$$

(4.223)

The mixed boundary conditions in the perturbation state may be expressed as follows:

$$\sigma_{yx}(x,0) = -\frac{\chi B_0}{\kappa_r}h_x(x,0) \quad (0 \le |x| \le h)$$

(4.224)

$$\sigma_{yy}(x,0) = \frac{\chi(\chi-2)B_0}{\kappa_r}\left\{h_y(x,0) + \frac{B_0}{\kappa_0\kappa_r}\right\} \quad (0 \le |x| < a)$$

$$u_y(x,0) = 0 \qquad\qquad\qquad (a \le |x| \le h)$$

(4.225)

$$h_x^{ec}(x,0) - h_x(x,0) = -\frac{\chi B_0}{\kappa_0\kappa_r}u_{y,x}(x,0) \quad (0 \le |x| < a)$$

$$\phi_m(x,0) = 0 \qquad\qquad\qquad (a \le |x| \le h)$$

(4.226)

$$b_y^e(x,0) - b_y(x,0) = 0 \quad (0 \le |x| < a)$$

$$\phi_m^e(x,0) = 0 \qquad\qquad (0 \le |x| < a)$$

(4.227)

$$\sigma_{xx}(\pm h, y) = 0$$

$$\sigma_{xy}(\pm h, y) = -\frac{\chi B_0}{\kappa_r}h_x(\pm h, y)$$

$$h_x^e(\pm h, y) - \kappa_r h_x(\pm h, y) = -\frac{\chi B_0}{\kappa_0\kappa_r}u_{x,y}(\pm h, y)$$

$$h_y^e(\pm h, y) - h_y(\pm h, y) = 0$$

(4.228)

The far-field loading condition is

$$\sigma_{yy}(x,y) = \frac{\chi(\chi-2)B_0^2}{\kappa_0\kappa_r^2} + \sigma_0$$

(4.229)

Only the first quadrant with appropriate boundary conditions needs to be analyzed owing to symmetry.

The problem is formulated by means of integral transforms and reduced to the solution of a Fredholm integral equation of the second kind [41]. The magnetic stress intensity factor is defined by

$$K_I = \lim_{x \to a^+} \{2\pi(x-a)\}^{1/2} \{t_{yy}(x,0) + \sigma_{yy}^M(x,0)\} \tag{4.230}$$

When $h \to \infty$, the magnetic stress intensity factor is obtained exactly [33], and

$$K_I = \sigma_0(\pi a)^{1/2} \frac{2(\kappa+\mu) + \chi \{(\kappa+2\mu) + \chi(2\kappa+5\mu)\} \left(b_c/\kappa_r\right)^2}{2(\kappa+\mu) - (\kappa+2\mu)\{\chi - \mu/(\kappa+2\mu)\} \left(\chi b_c/\kappa_r\right)^2} \tag{4.231}$$

In order to validate the prediction, we evaluate the magnetic stress intensity factor of nickel-irons TMC-V and TMH-B developed by NEC/Tokin Co. Ltd. The material properties of TMC-V are given by

$$E = 182 \times 10^9 \text{ N/m}^2$$
$$v = 0.146 \tag{4.232}$$
$$\kappa_r = 27900$$

The material properties of TMH-B are

$$E = 203 \times 10^9 \text{ N/m}^2$$
$$v = 0.279 \tag{4.233}$$
$$\kappa_r = 10690$$

The length and thickness of the specimen are $l = 140$ mm and $t = 1$ mm, respectively. The crack length is varied ($2a = 10, 15, 20$ mm) while keeping the specimen width fixed at $2h = 40$ mm. Initial through-thickness notch is machined by using the electro-discharge machining. The specimen is then fatigue precracked. A five-element strip gage is installed along the 90° line as shown in Fig. 4.10. We use a superconducting magnet with a 220-mm-diameter bore to create the magnetic field of magnetic induction B_0 normal to the crack surface. The specimen is loaded by $P = \sigma_0 ht = 29.4$ N load that consists of weights. The strains are recorded as a function of the magnetic field. The magnetic stress intensity factor K_I is then obtained using the strain ε_{yy} near the crack tip for the plane stress case by a similar method to that in Section 4.4.2.

Figure 4.14 gives a comparison of the theoretical results ($l \to \infty$, plane stress) of the normalized stress intensity factor $K_I/\sigma_0(\pi a)^{1/2}$ versus strip width to crack length ratio h/a with experimental data ($l = 140$ mm, $t = 1$ mm) for TMC-V ($2h = 40$ mm) under the magnetic field $B_0 = 0, 1.0, 1.5$ T. The dotted curve obtained for $B_0 = 0$ T coincides with that of the purely elastic plane stress case. The stress intensity factor decreases as the h/a ratio increases. Applying the magnetic field increases the stress intensity factor. The theoretical results agree very well with the experimental data. The calculated $K_I/\sigma_0(\pi a)^{1/2}$ values ($l \to \infty$, plane stress) of TMC-V and TMH-B for $h/a = 2$ ($2h = 40$ mm, $2a = 20$ mm) under various values of B_0 are compared to the experimental data ($l = 140$ mm, $t = 1$ mm) in Fig. 4.15. A large value of B_0 tends to increase the stress intensity factor depending on the material. Similar results can be found for the single-edge cracked [42] and double cantilever beam [43] specimens.

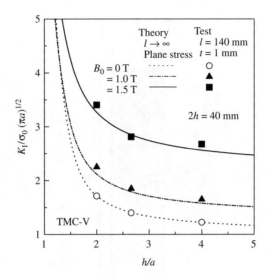

Figure 4.14   Stress intensity factor versus strip width to crack length ratio

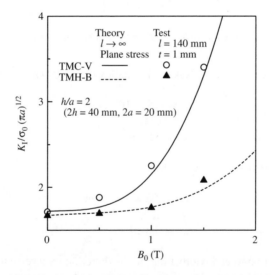

Figure 4.15   Stress intensity factor versus magnetic field

4.5.2   Fracture Test

Here, we present results of the tensile fracture behavior of soft ferromagnetic materials under magnetic field [44]. We use the single-edge cracked specimen of length 110 mm, width 30 mm, and thickness 1 mm as shown in Fig. 4.16. Specimens are 45% nickel permalloy PB and 78%

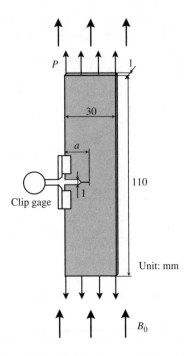

Figure 4.16   Single-edge cracked specimen

nickel permalloy PC developed by Nakano Permalloy Co., Ltd. The material properties of PB
are given by

$$E = 169 \times 10^9 \ \text{N/m}^2$$
$$v = 0.243 \quad\quad\quad\quad\quad\quad (4.234)$$
$$\kappa_r = 35799$$

The material properties of PC are given by

$$E = 202 \times 10^9 \ \text{N/m}^2$$
$$v = 0.254 \quad\quad\quad\quad\quad\quad (4.235)$$
$$\kappa_r = 35014$$

A saw cut of length 13 mm and width 1 mm in the direction transverse to the long axis of the
specimen is introduced at one edge of the specimen, and the specimen is precracked to the
length of 2 mm by applying tension-to-tension cyclic loads. So the initial crack is about $a =$
15 mm length. We use a cryocooler-cooled superconducting magnet with a 100-mm diameter working bore, as shown in Fig. 4.17, to create the magnetic field of magnetic induction
B_0 normal to the crack surface. The specimen is loaded to failure on a 30-kN axial loading
capacity servo-hydraulic testing machine, and the load P and crack opening displacement δ
are automatically logged by computer from the load cell and clip-gage, respectively.

   We perform the plane stress magnetoelastic analysis of a soft ferromagnetic strip with a
single-edge crack [42] by a similar method to that in Section 4.5.1 and calculate the magnetic
stress intensity factor K_I. Then, we determine the fracture toughness K_Q by using the critical
load P_Q from the measured load (P) versus crack opening displacement (δ) curve [45].

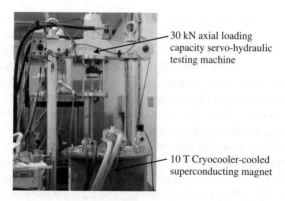

30 kN axial loading
capacity servo-hydraulic
testing machine

10 T Cryocooler-cooled
superconducting magnet

Figure 4.17   Fracture toughness test setup

Table 4.2   Fracture loads of the single-edge cracked specimens under the magnetic field

	P_Q (kN)			
B_0 (T)	0	0.5	1.0	1.5
PB	3.56	–	2.32	1.51
PC	4.52	3.92	–	–

Table 4.3   Fracture toughnesses of the single-edge cracked specimens under the magnetic field

	K_Q (N)			
B_0 (T)	0	0.5	1.0	1.5
PB	20.3	–	21.2	19.2
PC	25.8	24.2	–	–

The values of P_Q for PB and PC are summarized in Table 4.2. The magnetic field significantly decreases the critical load. Table 4.3 lists the obtained values of fracture toughness K_Q for PB and PC. There is little difference in the fracture toughness for the different magnetic fields. The decrease in the critical load with increasing magnetic field may be related to the increase in the stress intensity factor with increasing magnetic field [42].

4.5.3   Fatigue Crack Growth Test

Shindo et al. [46] obtained a crack growth rate equation for a finite crack in the soft ferromagnetic material under cyclic tensile load and uniform magnetic field, by using the Dugdale's model regarding the plastic zone in metals, and discussed the effect of magnetic field on the crack growth rate. Here, we present results on the tensile fatigue crack growth behavior of soft ferromagnetic materials under magnetic field [47]. We use the single-edge cracked specimen as shown in Fig. 4.16 and consider the PB, PC, and TMC-V alloys. A saw cut of length 10 mm and width 1 mm in the direction transverse to the long axis of the specimen is introduced at one edge of the specimen, and the specimen is precracked to the length of 1 mm by applying tension-to-tension cyclic loads. So the initial crack length is about $a = 11$ mm. We use a cryocooler-cooled superconducting magnet with a 100-mm diameter working bore, as shown in Fig. 4.17, to create the magnetic field of magnetic induction B_0 normal to the crack surface. The constant force amplitude test is conducted by using a 30-kN axial loading capacity servo-hydraulic testing machine under sinusoidal load P control at a frequency of $f_M = 10$ Hz with a load ratio $R = P_{min}/P_{max} = 0.1$, where P_{min} and P_{max} are the minimum and maximum loads applied over the fatigue cycle, respectively. The crack opening displacement δ is monitored via the clip-gage, and the instantaneous crack length a as a function of load cycles N is inferred from changes in the load-displacement slope. The crack growth rate could be related to magnetic stress intensity factor range $\Delta K_I = K_I(P_{max}) - K_I(P_{min})$. The $K_I(P)$ is derived by the plane stress magnetoelastic analysis of a soft ferromagnetic strip with a single-edge crack [42].

The effect of magnetic field B_0 on the fatigue crack growth rate da/dN for PB is shown in Fig. 4.18. The values of B_0 chosen for the experiments on PB are 0 and 1 T. The da/dN is

$$da/dN = \begin{cases} 1.09 \times 10^6 \Delta K_I^{1.59} & B_0 = 0 \text{ T} \\ 2.53 \times 10^6 \Delta K_I^{1.59} & B_0 = 1 \text{ T} \end{cases} \tag{4.236}$$

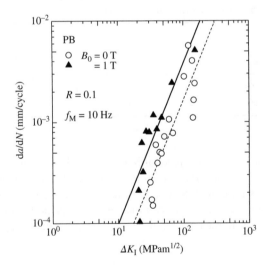

Figure 4.18   Fatigue crack growth rate versus stress intensity factor range

Although experimental data show large scatter, the crack growth rate of PB in the magnetic field is larger than that in no magnetic field. Similar results for PC and TMC-V are obtained (not shown here).

4.6 Summary

The mechanical behavior of uncracked and cracked ferromagnetic materials under the magnetic field has been studied theoretically. Also, some experiments have been conducted on ferromagnetic specimens, and the effect of magnetic field on the mechanical behavior has been summarized in a drawing. Close agreement between the theoretical predictions and experimental data gives the theory its perspective.

Part 4.2   Magnetostriction

Magnetostriction is a change in the shape of materials under the influence of the external magnetic field [48]. The magnetostrictive effect, as shown in Fig. 4.19, was first described in the 19th century by James Joule. He observed that a sample of ferromagnetic material, that is, iron, changes its length in the presence of the magnetic field. This change in length is the result of the rotation and reorientation of small magnetic domains. This rotation and reorientation causes internal strains in the material, and the strains lead to the stretching of the material in the direction of the magnetic field [49].

4.7   Basic Equations of Magnetostriction

Let us now take the rectangular Cartesian coordinates $x_i(O\text{-}x_1, x_2, x_3)$. The field equations are given by

$$\sigma_{ji,j} = \rho u_{i,tt} \tag{4.237}$$

$$\varepsilon_{ijk} H_{k,j} = 0, \quad B_{i,i} = 0 \tag{4.238}$$

where σ_{ij}, u_i, H_i, and B_i are the components of stress tensor, displacement vector, magnetic field intensity vector, and magnetic induction vector, respectively, ρ is the mass density, ε_{ijk} is the permutation symbol, a comma followed by an index denotes partial differentiation with respect to the space coordinate x_i or the time t, and the summation convention over repeated indices is used.

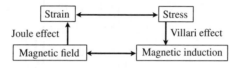

Figure 4.19   Magnetostriction

Constitutive relations can be written as

$$\varepsilon_{ij} = s^H_{ijkl}\sigma_{kl} + d'_{kij}H_k \tag{4.239}$$

$$B_i = d'_{ikl}\sigma_{kl} + \kappa^T_{ik}H_k \tag{4.240}$$

where ε_{ij} is the component of strain tensor, and s^H_{ijkl}, d'_{kij}, and κ^T_{ik} are the elastic compliance at constant magnetic field, magnetoelastic constant, and magnetic permeability at constant stress, respectively. Valid symmetry conditions for the material constants are

$$s^H_{ijkl} = s^H_{jikl} = s^H_{ijlk} = s^H_{klij}, \quad d'_{kij} = d'_{kji}, \quad \kappa_{ij} = \kappa_{ji} \tag{4.241}$$

Here and in the following, we have dropped the superscript T on the magnetic permeability.

The linearized boundary conditions are

$$[\![\sigma_{ji}]\!]n_j = 0 \tag{4.242}$$

$$\varepsilon_{ijk}n_j[\![H_k]\!] = 0$$
$$[\![B_i]\!]n_i = 0 \tag{4.243}$$

where n_i is the component of the outer unit vector **n** normal to the surface, and $[\![\ ]\!]$ means the jump of the quantity across the surface.

Terfenol-D ($Tb_{0.3}Dy_{0.7}Fe_{1.93}$) is a highly magnetostrictive alloy of iron and rare-earth elements (terbium and dysprosium) [50, 51]. The name Terfenol stems from the composition of the alloy $TbFe_2$ and its place of origin, the Naval Ordnance Laboratory (NOL). This material, which has outstanding elongation and energy density at room temperature, has shown great potential in applications. For Terfenol-D, the constitutive relations, Eqs. (4.239) and (4.240), are written in the following form:

$$
\begin{Bmatrix} \varepsilon_{11} \\ \varepsilon_{22} \\ \varepsilon_{33} \\ 2\varepsilon_{23} \\ 2\varepsilon_{31} \\ 2\varepsilon_{12} \end{Bmatrix} =
\begin{bmatrix}
s^H_{11} & s^H_{12} & s^H_{13} & 0 & 0 & 0 \\
s^H_{12} & s^H_{11} & s^H_{13} & 0 & 0 & 0 \\
s^H_{13} & s^H_{13} & s^H_{33} & 0 & 0 & 0 \\
0 & 0 & 0 & s^H_{44} & 0 & 0 \\
0 & 0 & 0 & 0 & s^H_{44} & 0 \\
0 & 0 & 0 & 0 & 0 & s^H_{66}
\end{bmatrix}
\begin{Bmatrix} \sigma_{11} \\ \sigma_{22} \\ \sigma_{33} \\ \sigma_{23} \\ \sigma_{31} \\ \sigma_{12} \end{Bmatrix}
$$
$$
+ \begin{bmatrix}
0 & 0 & d'_{31} \\
0 & 0 & d_{31} \\
0 & 0 & d'_{33} \\
0 & d'_{15} & 0 \\
d'_{15} & 0 & 0 \\
0 & 0 & 0
\end{bmatrix}
\begin{Bmatrix} H_1 \\ H_2 \\ H_3 \end{Bmatrix} \tag{4.244}
$$

$$\begin{Bmatrix} B_1 \\ B_2 \\ B_3 \end{Bmatrix} = \begin{bmatrix} 0 & 0 & 0 & 0 & d'_{15} & 0 \\ 0 & 0 & 0 & d'_{15} & 0 & 0 \\ d'_{31} & d'_{31} & d'_{33} & 0 & 0 & 0 \end{bmatrix} \begin{Bmatrix} \sigma_{11} \\ \sigma_{22} \\ \sigma_{33} \\ \sigma_{23} \\ \sigma_{31} \\ \sigma_{12} \end{Bmatrix}$$

$$+ \begin{bmatrix} \kappa_{11} & 0 & 0 \\ 0 & \kappa_{11} & 0 \\ 0 & 0 & \kappa_{33} \end{bmatrix} \begin{Bmatrix} H_1 \\ H_2 \\ H_3 \end{Bmatrix} \tag{4.245}$$

where

$$\varepsilon_{23} = \varepsilon_{32}, \ \varepsilon_{31} = \varepsilon_{13}, \ \varepsilon_{12} = \varepsilon_{21} \tag{4.246}$$

$$\sigma_{23} = \sigma_{32}, \ \sigma_{31} = \sigma_{13}, \ \sigma_{12} = \sigma_{21} \tag{4.247}$$

$$s_{11}^H = s_{1111}^H = s_{2222}^H, \ s_{12}^H = s_{1122}^H, \ s_{13}^H = s_{1133}^H = s_{2233}^H, \ s_{33}^H = s_{3333}^H$$

$$s_{44}^H = 4s_{2323}^H = 4s_{3131}^H, \ s_{66}^H = 4s_{1212}^H = 2(s_{11}^H - s_{12}^H) \tag{4.248}$$

$$d'_{15} = 2d'_{131} = 2d'_{223}, \ d'_{31} = d'_{311} = d'_{322}, \ d'_{33} = d'_{333} \tag{4.249}$$

4.8   Nonlinear Magneto-Mechanical Response

Terfenol-D has been studied due to its great potential as sensor, actuator, and switch elements in a wide variety of applications that can benefit from its remote operation, high energy density, and short response time. However, one limitation on the practical use of Terfenol-D is its non-linear behavior [52]. In this section, nonlinear magneto-mechanical response of the Terfenol-D laminated structures is discussed.

4.8.1   Terfenol-D/Metal Laminates

Here, we consider a Terfenol-D/metal laminate [53] as shown in Fig. 4.20. Let $x = x_1, y = x_2$, and $z = x_3$ denote the rectangular Cartesian coordinates. The origin of the coordinate system is located at the center of the bottom left side of a Terfenol-D layer, and the easy axis for magnetization of the Terfenol-D layer is the z-direction. The Terfenol-D layer of length a_1, width b_1, and thickness h_1 is perfectly bonded on the top surface of a metal layer of length a_2, width $b_2 = b_1$, and thickness h_2. The laminate is cantilevered with $z = 0$ denoting the clamped end.

When the magnetic field is applied along the z-direction (easy axis) of the Terfenol-D layer, both the longitudinal (33) and transverse (31) magnetostrictive deformation modes are excited as shown in Fig. 4.21. When the magnetic field is applied along the x-direction, the shear mode (15) is excited. Applying a stronger field leads to more definite reorientation of more

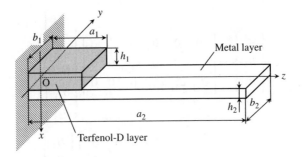

Figure 4.20   A Terfenol-D/metal laminate

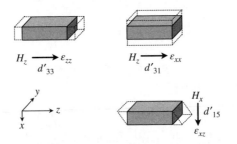

Figure 4.21   Various deformation modes

and more domains in the direction of magnetic field. When all the magnetic domains have become aligned with the magnetic field, the saturation point has been achieved as shown in Fig. 4.22.

As we know, nonlinearity of magnetostriction versus magnetic field curves arises from the movements of magnetic domain walls. A magnetic domain switching gives rise to the changes of the magnetoelastic constants. The constants d'_{15}, d'_{31}, and d'_{33} for the Terfenol-D layer in the x-direction magnetic field become

$$
\begin{aligned}
d'_{15} &= d^{m}_{15} + m_{15}H_x \\
d'_{31} &= d^{m}_{31} \\
d'_{33} &= d^{m}_{33}
\end{aligned}
\tag{4.250}
$$

where $d^{m}_{15}, d^{m}_{31}, d^{m}_{33}$ are the piezomagnetic constants, and m_{15} is the second-order magnetoelastic constant. The constants d'_{15}, d'_{31}, and d'_{33} for the Terfenol-D layer in the z-direction magnetic field are

$$
\begin{aligned}
d'_{15} &= d^{m}_{15} \\
d'_{31} &= d^{m}_{31} + m_{31}H_z \\
d'_{33} &= d^{m}_{33} + m_{33}H_z
\end{aligned}
\tag{4.251}
$$

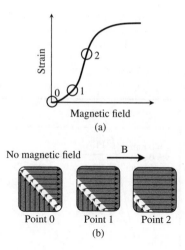

Figure 4.22 Schematic diagram of (a) strain versus magnetic field curve and (b) domain structure

where m_{31}, m_{33} are the second-order magnetoelastic constants. When the thickness of the Terfenol-D layer is much smaller than other two sizes, the longitudinal (33) magnetostrictive deformation mode is dominant. So we assume that only the constant d'_{33} varies with the magnetic field H_z, and the constant m_{31} equals zero. The constants m_{15} and m_{33} can predict the nonlinearity well, without complex computation and more parameters [54].

We perform finite element calculations to obtain the strain, deflection, and internal stress for the Terfenol-D/metal laminates using ANSYS, because the equations describing magnetostrictive materials are mathematically equivalent to those describing piezoelectric materials. From Eqs. (4.250) and (4.251), the magnetoelastic constants d'_{15} and d'_{33} vary with magnetic field because of domain wall movement. Making use of magnetic-field-dependent material properties, the model calculates the nonlinear behavior.

In order to evaluate the magnetostriction, we prepare the magnetostrictive laminates using Terfenol-D (ETREMA products, Inc., USA) of $a_1 = b_1 = 10$ mm and SUS316 of $a_2 = 40$ mm, $b_2 = 10$ mm, and $h_2 = 0.5$ mm. The thickness of the Terfenol-D layer is $h_1 = 1, 3$ and 5 mm. The material properties of Terfenol-D are

$$
\begin{aligned}
s_{11}^H &= 17.9 \times 10^{-12} \ \text{m}^2/\text{N} \\
s_{12}^H &= -5.88 \times 10^{-12} \ \text{m}^2/\text{N} \\
s_{13}^H &= -5.88 \times 10^{-12} \ \text{m}^2/\text{N} \\
s_{33}^H &= 17.9 \times 10^{-12} \ \text{m}^2/\text{N} \\
s_{44}^H &= 26.3 \times 10^{-12} \ \text{m}^2/\text{N} \\
d_{15}^m &= 28 \times 10^{-9} \ \text{m/A} \\
d_{31}^m &= -5.3 \times 10^{-9} \ \text{m/A} \\
d_{33}^m &= 11 \times 10^{-9} \ \text{m/A} \\
\kappa_{11} &= 6.29 \times 10^{-6} \ \text{H/m} \\
\kappa_{33} &= 6.29 \times 10^{-6} \ \text{H/m}
\end{aligned} \tag{4.252}
$$

Table 4.4   Second-order magnetoelastic constants

h_1 (mm)	1	3	5
m_{15} (m²/A²)	1.5×10^{-10}	9.3×10^{-11}	3.7×10^{-11}
m_{33} (m²/A²)	1.4×10^{-11}	4.4×10^{-12}	1.1×10^{-12}

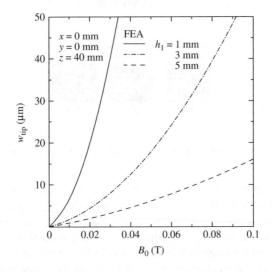

Figure 4.23   Tip deflection versus magnetic field

Young's modulus E and Poisson's ratio v of SUS316 are

$$E = 189 \times 10^9 \text{ N/m}^2$$
$$v = 0.3 \tag{4.253}$$

A strain gage is placed at the surface $(x = -h_1, y = 0, z = a_1/2)$ of the Terfenol-D layer. Magnetic field is then applied in the x- or z-direction, and the magnetostriction is evaluated.

Table 4.4 lists the second-order magnetoelastic constants obtained from the combined finite element analyses (FEAs) and tests. The second-order magnetoelastic constants decrease with an increase in the Terfenol-D layer thickness. This is due to the fact that the second-order magnetoelastic constants depend on the internal stress [52]. Figure 4.23 shows the tip deflection w_{tip} versus magnetic field B_0 in the z-direction at $x = 0$ mm, $y = 0$ mm, and $z = a_2 = 40$ mm for the Terfenol-D/SUS316 laminates obtained from the FEA. A nonlinear relationship between the tip deflection and magnetic field is observed as is expected. The tip deflection increases as the Terfenol-D layer thickness decreases.

4.8.2   Terfenol-D/PZT Laminates

Terfenol-D/PZT-laminated cantilevers are gaining increasing attention for applications in self-sensing/monitoring and energy-harvesting systems. Here, we review some theoretical

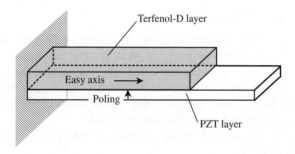

Figure 4.24   A Terfenol-D/PZT laminate

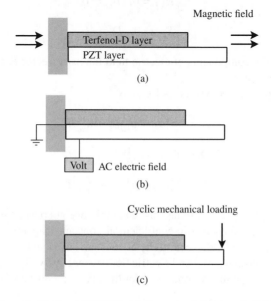

Figure 4.25   Illustration of a Terfenol-D/PZT laminate under (a) magnetic field, (b) AC electric field, and (c) concentrated mechanical loading

and experimental works on the detection and response characteristics of the Terfenol-D/PZT laminates as shown in Fig. 4.24.

The nonlinear electromagneto-mechanical behavior of the Terfenol-D/PZT laminates under magnetic field was studied theoretically and experimentally by Shindo et al. [55]. Figure 4.25(a) shows a schematic representation of the model and test setup. The effect of the magnetic field on the deflection of the laminate and the induced voltage of the PZT layer was discussed in detail. Mori et al. [56] investigated the dynamic bending of the Terfenol-D/PZT laminates under AC electric field as shown in Fig. 4.25(b). They discussed the effect of AC electric field on the dynamic deflection of the laminate and induced magnetic field of the Terfenol-D layer. Mori et al. [57] also considered the Terfenol-D/PZT laminates under cyclic concentrated load as shown in Fig. 4.25(c) and discussed the induced magnetic field of the Terfenol-D layer and output voltage of the PZT layer.

4.9 Magnetoelectric Response

Magnetoelectric (ME) effect has been observed in several single-phase materials including Cr_2O_3 and $BiFeO_3$, but the effect in general is weak at room temperature [58]. Later, it was shown that the piezoelectric–piezomagnetic composites may exhibit the ME effect [59], and the ME effect of $BaTiO_3$-$CoFe_2O_4$ composites was studied [60, 61]. In this section, the basic equations for modeling the electromagneto-mechanical response of magnetostrictive and piezoelectric particle-reinforced composites are presented.

Let us now take the rectangular Cartesian coordinates $x_i(O\text{-}x_1, x_2, x_3)$. The field equations are given by Alshits et al. [62]

$$\sigma_{ji,j} = \rho u_{i,tt} \tag{4.254}$$

$$\varepsilon_{ijk}H_{k,j} = 0, \quad B_{i,i} = 0 \tag{4.255}$$

$$\varepsilon_{ijk}E_{k,j} = 0, \quad D_{i,i} = 0 \tag{4.256}$$

where E_i and D_i are the components of electric field intensity vector $\mathbf{E}$ and electric displacement vector $\mathbf{D}$, respectively.

Constitutive relations can be written as [63]

$$\varepsilon_{ij} = s_{ijkl}^{EH}\sigma_{kl} + d_{kij}^{H}E_k + d_{kij}'^{E}H_k \tag{4.257}$$

$$D_i = d_{ikl}^{H}\sigma_{kl} + \epsilon_{ik}^{TH}E_k + \beta_{ik}^{T}H_k \tag{4.258}$$

$$B_i = d_{ikl}'^{E}\sigma_{kl} + \beta_{ik}^{T}E_k + \kappa_{ik}^{TE}H_k \tag{4.259}$$

where $s_{ijkl}^{EH}, d_{kij}^{H}, d_{kij}'^{E}, \epsilon_{ik}^{TH}$, and κ_{ik}^{TE} are the elastic compliance at constant electric and magnetic fields, direct (or inverse) piezoelectric coefficient at constant magnetic field, magnetoelastic constant at constant electric field, permittivity at constant stress and magnetic field, and magnetic permeability at constant stress and electric field, respectively. β_{ik}^{T} is the ME susceptibility at constant stress. Valid symmetry conditions for the material constants are

$$s_{ijkl}^{EH} = s_{jikl}^{EH} = s_{ijlk}^{EH} = s_{klij}^{EH}, \quad d_{kij}^{H} = d_{kji}^{H}, \quad d_{kij}'^{E} = d_{kji}'^{E}, \quad \epsilon_{ij}^{TH} = \epsilon_{ji}^{TH},$$

$$\kappa_{ij}^{TE} = \kappa_{ji}^{TE}, \quad \beta_{ij}^{T} = \beta_{ji}^{T} \tag{4.260}$$

The linearized boundary conditions are

$$[\![\sigma_{ji}]\!]n_j = 0 \tag{4.261}$$

$$\varepsilon_{ijk}n_j[\![H_k]\!] = 0$$
$$[\![B_i]\!]n_i = 0 \tag{4.262}$$

$$\varepsilon_{ijk}n_j[\![E_k]\!] = 0$$
$$[\![D_i]\!]n_i = 0 \tag{4.263}$$

If the magnetostrictive and piezoelectric particle-reinforced composites are fabricated, the aforementioned basic equations are useful to predict the mechanical and ME behavior of the composites. Further study on fabrication of the magnetostrictive and piezoelectric particle-reinforced composites is of great value.

4.10 Summary

We have discussed the basic macroscopic response of the magnetostrictive/metal laminates under magnetic field. We have also reviewed the recent works on the magnetostrictive/piezoelectric laminates under electromagneto-mechanical loading. In addition, we have given the basic equations of the magnetostrictive and piezoelectric particle-reinforced composites. This part offers useful guidelines for designing magnetostrictive materials and structures with self-sensing/monitoring and energy-harvesting capabilities.

Magnetostrictive materials such as Terfenol-D are very brittle and susceptible to fracture during service. Defects caused by manufacturing could also have an important influence on the performance of the magnetostrictive devices. It is therefore important to understand the crack behavior of the magnetostrictive materials. This is a challenging research area, and sooner or later, some progress will be made.

References

[1] W. F. Brown, Jr., *Magnetoelastic Interactions*, Springer-Verlag, Berlin, 1966.

[2] Y.-H. Pao and C.-S. Yeh, "A linear theory for soft ferromagnetic elastic solids," *Int. J. Eng. Sci.* **11**(4), 415 (1973).

[3] H. Parkus, "Thermoelastic equations for ferromagnetic bodies," *Arch. Mech. Stos.* **24**(5/6), 819 (1972).

[4] J. B. Alblas, "Electro-magneto-elasticity," *Topics in Applied Continuum Mechanics*, J. L. Zeman and F. Ziegler (eds.), Springer-Verlag, Wien, p. 71 (1974).

[5] A. A. F. van de Ven, "Magnetoelastic buckling of magnetically saturated bodies," *Acta Mech.* **47**(3/4), 229 (1983).

[6] K. Hutter and Y.-H. Pao, "A dynamic theory for magnetizable elastic solids with thermal and electrical conduction," *J. Elast.* **4**(2), 89 (1974).

[7] K. Hutter, "Wave propagation and attenuation in paramagnetic and soft ferromagnetic materials," *Int. J. Eng. Sci.* **13**(12), 1067 (1975).

[8] K. Hutter, "Wave propagation and attenuation in para- or soft ferromagnetic materials-II: the influence of the direction of the magnetic field," *Int. J. Eng. Sci.* **14**(10), 883 (1976).

[9] Y. Ersoy and E. Kiral, "A dynamic theory for polarizable and magnetizable magneto-electro thermoviscoelastic, electrically and thermally conductive anisotropic solids having magnetic symmetry," *Int. J. Eng. Sci.* **16**(7), 483 (1978).

[10] Y. Ersoy, "Propagation of waves in magneto-thermo-viscoelastic solids subjected to a uniform magnetic field," *Int. J. Eng. Sci.* **19**(1), 91 (1981).

[11] F. C. Moon and Y.-H. Pao, "Magnetoelastic buckling of a thin plate," *ASME J. Appl. Mech.* **35**(1), 53 (1968).

[12] D. V. Wallerstein and M. O. Peach, "Magnetoelastic buckling of beams and thin plates of magnetically soft material," *ASME J. Appl. Mech.* **39**(2), 451 (1972).

[13] C. H. Popelar, "Postbuckling analysis of a magnetoelastic beam," *ASME J. Appl. Mech.* **39**(1), 207 (1972).

[14] J. M. Dalrymple, M. O. Peach and G. L. Viegelahn, "Magnetoelastic buckling of thin magnetically soft plates in cylindrical mode," *ASME J. Appl. Mech.* **41**(1), 145 (1974).

[15] K. Miya, K. Hara and K. Someya, "Experimental and theoretical study on magnetoelastic buckling of a ferromagnetic cantilevered beam-plate," *ASME J. Appl. Mech.* **45**(2), 355 (1978).

[16] K. Miya, T. Takagi and Y. Ando, "Finite-element analysis of magnetoelastic buckling of ferromagnetic beam plate," *ASME J. Appl. Mech.* **47**(2), 377 (1980).

[17] A. A. F. van de Ven, "Magnetoelastic buckling of thin plates in a transverse magnetic field," *J. Elast.* **8**(3), 297 (1978).

[18] A. A. F. van de Ven, J. Tani, K. Otomo and Y. Shindo, "Magnetoelastic buckling of two nearby ferromagnetic rods in a magnetic field," *Acta Mech.* **75**(1/4), 191 (1988).

[19] K. Horiguchi and Y. Shindo, "Experimental and theoretical results for bending of a soft ferromagnetic plate in a transverse magnetic field," *Acta Mech.* **162**(1/4), 185 (2003).

[20] F. C. Moon and Y.-H. Pao, "Vibration and dynamic instability of a beam-plate in a transverse magnetic field," *ASME J. Appl. Mech.* **36**(1), 92 (1969).

[21] F. C. Moon, "The mechanics of ferroelastic plates in a uniform magnetic field," *ASME J. Appl. Mech.* **37**(1), 153 (1970).

[22] Y. C. Fung, *Foundations of Solid Mechanics*, Prentice-Hall, Inc., Englewood Cliffs, NJ, 1969.

[23] R. D. Mindlin, "Influence of rotatory inertia and shear on flexural motions of isotropic, elastic plates," *ASME J. Appl. Mech.* **18**(1), 31 (1951).

[24] G. C. Sih and E. P. Chen, "Dynamic analysis of cracked plates in bending and extension," *Mechanics of Fracture* Vol. 3, G. C. Sih (ed.), Noordhoff International Publishing, Leyden, p. 231 (1977).

[25] Y. Shindo, K. Horiguchi and T. Shindo, "Magneto-elastic analysis of a soft ferromagnetic plate with a through crack under bending," *Int. J. Eng. Sci.* **37**(6), 687 (1999).

[26] I. N. Sneddon, *Fourier Transforms*, McGraw-Hill, New York, 1951.

[27] Y. Shindo, K. Horiguchi and A. A. F. van de Ven, "Bending of a magnetically saturated plate with a crack in a uniform magnetic field," *Int. J. Appl. Electromagnet. Mater.* **1**(2/4), 135 (1990).

[28] K. Horiguchi and Y. Shindo, "A strain gage method for determination of magnetic moment intensity factors in through-cracked soft ferromagnetic plates," *J. Appl. Phys.* **96**(10), 5860 (2004).

[29] J. W. Dally and R. J. Sanford "Strain-gage methods for measuring the opening-mode stress-intensity factor, K_I," *Exp. Mech.* **27**(4), 381 (1987).

[30] Y. Shindo and K. Horiguchi, "Dynamic bending of cracked soft ferromagnetic plate in uniform magnetic field," *Theor. Appl. Fract. Mech.* **15**(3), 207 (1991).

[31] Y. Shindo, T. Shindo and K. Horiguchi, "Scattering of flexural waves by a cracked Mindlin plate of soft ferromagnetic material in a uniform magnetic field," *Theor. Appl. Fract. Mech.* **34**(2), 167 (2000).

[32] A. Ogihara, Y. Shindo and K. Horiguchi, "Flexural waves scattering by a through crack in a magnetically saturated Mindlin plate under a uniform magnetic field," *Eur. J. Mech. A. Solids* **22**(1), 163 (2003).

[33] Y. Shindo, "The linear magnetoelastic problem for a soft ferromagnetic elastic solid with a finite crack," *ASME J. Appl. Mech.* **44**(1), 47 (1977).

[34] Y. Shindo, "Singular stresses in a soft ferromagnetic elastic solid with two coplanar Griffith cracks," *Int. J. Solids Struct.* **16**(6), 537 (1980).

[35] Y. Shindo, "Magnetoelastic interaction of a soft ferromagnetic elastic solid with a penny-shaped crack in a constant axial magnetic field," *ASME J. Appl. Mech.* **45**(2), 291 (1978).

[36] Y. Shindo, "Singular stresses in a soft ferromagnetic elastic solid with a flat annular crack," *Acta Mech.* **48**(34), 147 (1983).

[37] Y. Shindo, "The linear magnetoelastic problem of two coplanar Griffith cracks in a soft ferromagnetic elastic strip," *ASME J. Appl. Mech.* **49**(1), 69 (1982).

[38] Y. Shindo, "Dynamic singular stresses for a Griffith crack in a soft ferromagnetic elastic solid subjected to a uniform magnetic field," *ASME J. Appl. Mech.* **50**(1), 50 (1983).

[39] Y. Shindo, "Diffraction of waves and singular stresses in a soft ferromagnetic elastic solid with two coplanar Griffith cracks," *J. Acoust. Soc. Am.* **75**(1), 50 (1984).

[40] Y. Shindo, "Impact response of a cracked soft ferromagnetic medium," *Acta Mech.* **75**(1/2), 99 (1985).

[41] Y. Shindo, D. Sekiya, F. Narita and K. Horiguchi, "Tensile testing and analysis of ferromagnetic elastic strip with a central crack in a uniform magnetic field," *Acta Mater.* **52**(15), 4677 (2004).

[42] Y. Shindo, T. Komatsu, F. Narita and K. Horiguchi, "Magnetoelastic analysis and tensile testing of a soft ferromagnetic strip with a single-edge crack," *J. Appl. Phys.* **100**(3), 034513 (2006).

[43] K. Yoshimura, Y. Shindo, K. Horiguchi and F. Narita, "Theoretical and experimental determination of magnetic stress intensity factors of a crack in a double cantilever beam specimen," *Fatigue Fract. Eng. Mater. Struct.* **27**(3), 213 (2004).

[44] Y. Shindo, I. Shindo and F. Narita, "The influence of magnetic field on the fracture toughness of soft ferromagnetic materials," *Eng. Fract. Mech.* **75**(10), 3010 (2008).

[45] ASTM E 1820-05, *Standard Test Method for Measurement of Fracture Toughness*, American Society for Testing and Materials, USA, 2005.

[46] Y. Shindo, F. Narita, K. Horiguchi and T. Komatsu, "Mode I crack growth rate of a ferromagnetic elastic strip in a uniform magnetic field," *Acta Mater.* **54**(19), 5115 (2006).

[47] Y. Shindo, I. Shindo and F. Narita, "Effect of magnetic field on fatigue crack propagation of single-edge cracked soft ferromagnetic specimens under mode I loading," *ASTM J. Test. Eval.* **35**(2), 151 (2007).

[48] E. du Tremolet de Lacheisserie, *Magnetostriction: Theory and Applications of Magnetoelasticity*, CRC Press, Boca Raton, FL, 1993.

[49] G. Engdahl, *Handbook of Giant Magnetostrictive Materials*, Academic Press, San Diego, CA, 2000.

[50] A. E. Clark, J. P. Teter, M. Wun-Fogle, M. Moffett and J. Lindberg, "Magnetomechanical coupling in Bridgman-grown $Tb_{0.3}Dy_{0.7}Fe_{1.9}$ at high drive levels," *J. Appl. Phys.* **67**(9), 5007 (1990).

[51] M. B. Moffett, A. E. Clark, M. Wun-Fogle, J. Linberg, T. P. Teter and E. A. McLaughlin, "Characterization of Terfenol-D for magnetostrictive transducers," *J. Acoust. Soc. Am.* **89**(3), 1448 (1991).

[52] Y. Wan, D. Fang and K.-C. Hwang, "Non-linear constitutive relations for magnetostrictive materials," *Int. J. Non Linear Mech.* **38**(7), 1053 (2003).

[53] Y. Shindo, F. Narita, K. Mori and T. Nakamura, "Nonlinear bending response of giant magnetostrictive laminated actuators in magnetic fields," *J. Mech. Mater. Struct.* **4**(5), 941 (2009).

[54] Z. Jia, W. Liu, Y. Zhang, F. Wang and D. Guo, "A nonlinear magnetomechanical coupling model of giant magnetostrictive thin films at low magnetic fields," *Sens. Actuators, A* **128**(1), 158 (2006).

[55] Y. Shindo, K. Mori and F. Narita, "Electromagneto-mechanical fields of giant magnetostrictive/piezoelectric laminates," *Acta Mech.* **212**(3), 253 (2010).

[56] K. Mori, Y. Shindo and F. Narita, "Dynamic electromagneto-mechanical behavior of clamped-free giant magnetostrictive/piezoelectric laminates under AC electric fields," *Smart Mater. Struct.* **21**, 115003 (2010).

[57] K. Mori, Y. Shindo, F. Narita and S. Okura, "Detection and response characteristics of giant magnetostrictive/piezoelectric laminated cantilevers under cyclic bending," *Mech. Adv. Mater. Struct.*, in press.

[58] G. T. Rado and V. J. Folen, "Observation of the magnetically induced magnetoelectric effect and evidence for antiferromagnetic domains," *Phys. Rev. Lett.* **7**(8), 310 (1961).

[59] J. van Suchtelen, "Product properties: a new application of composite materials," *Philips Res. Rep.* **27**(1), 28 (1972).

[60] J. van den Boomgaard, D. R. Terrell, R. A. J. Born and H. F. J. I. Giller, "An in situ grown eutectic magnetoelectric composite material: Part 1. Composition and unidirectional solidification," *J. Mater. Sci.* **9**(10), 1705 (1974).

[61] A. M. J. G. van Run, D. R. Terrell and J. H. Scholing, "An in situ grown eutectic magnetoelectric composite material: Part 2. Physical properties," *J. Mater. Sci.* **9**(10), 1710 (1974).

[62] V. I. Alshits, A. N. Darinskii and J. Lothe, "On the existence of surface waves in half-infinite anisotropic elastic media with piezoelectric and piezomagnetic properties," *Wave Motion* **16**(3), 265 (1992).

[63] A. K. Soh and J. X. Liu, "On the constitutive equations of magnetoelectroelastic solids," *J. Intell. Mater. Syst. Struct.* **16**(7/8), 507 (2005).

Index

1-3 piezocomposites, 127, 128, 130, 131

AFC, 127
Amperian-current, 219, 228
Angular frequency, 17, 62, 80, 237
Attenuation, 19, 21, 31, 224

BaTiO$_3$, 212
Bending, 1, 2, 9–11, 13–15, 33, 36, 43, 47,
    88–90, 92, 97, 99, 100, 102, 110, 111,
    116, 172, 189, 199, 204, 206, 207,
    209–211, 219, 226, 231, 233,
    235–238, 245, 246, 250–253, 255,
    271
  plate, 9, 13, 235, 237, 245, 246, 255
  three-point, 199, 204, 206, 207, 209–211
Bimorph, 90, 93, 95, 99–110
Boundary condition, 7, 8, 10, 11, 14, 27, 30,
    38, 42, 43, 49–51, 58–60, 70, 74, 86,
    87, 92, 93, 96, 98, 112, 115, 118, 121,
    125, 132, 142, 150–152, 154, 156,
    160, 162, 166, 168, 170, 173, 174,
    181, 183, 190, 221, 222, 224–227,
    229–231, 234, 237, 239, 240,
    245–248, 251, 252, 255, 257, 259,
    266, 272
Brittle, 33, 149, 185, 194, 201, 205, 273
Buckling, 224, 228, 231
  magnetic induction, 228, 231

Cantilever, 92, 93, 95, 97–99, 103, 231,
    270

Cavity, 152, 153
Classical lamination, 84, 100
Classical plate, 9, 24, 30, 31, 33, 36, 39, 40,
    235, 243, 245, 255
  bending, 9, 235, 245, 255
Clip-gage, 262, 264
Coefficient of thermal expansion, 83, 86
Composite, 2, 47, 82, 86, 126–132, 170,
    219, 272, 273
  1-3, 127
  disk, 126
  materials, 82
  particle-reinforced, 2, 219, 272, 273
  piezoelectric, 47, 127
  piezoelectric fiber, 126, 127
Compression, 70, 200
Compressive, 85, 116, 121, 122
  load, 121
Concentrated load, 95, 96, 195, 199, 202,
    232, 271
  cyclic, 271
Conducting elastic plate, 7, 16, 33
Conducting materials, 1, 2, 5, 7, 9, 13, 33,
    40, 45, 222
Conducting plate, 1, 11, 26, 33, 40
  perfectly, 11
Constant load-rate, 207, 209
Constitutive equation, 6, 7, 27, 40, 48, 79,
    92, 97, 114, 124, 151, 152, 165, 166,
    168, 169, 220, 225, 229, 233, 240,
    244, 248, 259

Electromagneto-Mechanics of Material Systems and Structures, First Edition. Yasuhide Shindo.
© 2015 John Wiley & Sons Singapore Pte Ltd. Published 2015 by John Wiley & Sons Singapore Pte Ltd.

linearized, 6, 48, 220
mechanical, 7, 40, 225, 229, 233, 244
Crack, 33, 35, 36, 38–44, 49–53, 55–57, 59,
    68–70, 149–154, 156–165, 167, 170,
    172–184, 186–189, 191, 194–205,
    207, 209–212, 219, 250–257, 260,
    262–265, 273
    central, 57, 163, 164, 174, 181, 184,
        257
    discharging, 173, 177–179, 183, 184,
        197, 198
    edge, 33, 182, 184
    Griffith, 49, 68
    growth, 149, 164, 187, 189, 207, 211,
        264
    growth rate, 164, 210, 211, 219, 264,
        265
    half-penny-shaped, 188
    impermeable, 160, 161, 164, 173–176,
        179, 180, 182–184, 188, 189, 197,
        198, 212
    open, 154, 160, 162, 173–177, 179, 180,
        183, 184, 197, 198, 212
    partial, 33
    penny-shaped, 40, 165, 167
    permeable, 159, 160, 163, 164, 167, 170,
        173–179, 181–184, 188,
        189, 195, 197, 198, 200–203, 207,
        212
    propagation velocity, 208, 209
    surface, 33
    through, 33, 35, 40, 41, 172, 173,
        250–252, 255, 256
    tip, 42, 49, 52, 53, 55, 56, 157, 159, 175,
        184, 201, 250, 253, 254, 260
Cracking, 47, 113, 149, 205, 211, 212
Cryogenic electromechanical response, 1
Cryogenic response, 140, 142
Current, 1, 5, 6, 8, 27, 29, 40–42, 44, 45, 47,
    128, 129, 223
    alternating, 128, 129
    density, 6, 27, 29, 40–42, 223
    electric, 1, 5, 40–42, 44, 45
    flow, 40, 41
    induced, 5
Cyclic load, 262, 264

DCB, 182–184
Deflection, 9, 86, 92, 93, 95–99, 104–107,
    109, 110, 113, 226, 227, 230–233,
    235, 251, 253, 269, 271
    load-point, 96, 97
    maximum, 96
    tip, 93, 95, 99, 104–107, 270
Deformation, 45, 47, 48, 51, 55, 80, 121,
    130, 220, 228, 267, 269
Depolarization, 140, 147, 149
Dielectric breakdown, 146, 149, 177, 184
Dielectric materials, 1, 47, 49, 55, 60, 68, 72
Dielectrics, 1, 47, 49
    elastic, 47, 49
Dipole, 73, 147, 219, 220
Dispersion relation, 18, 21, 237, 239, 246,
    256
Displacement, 5, 6, 9, 13, 27, 35, 42, 48, 52,
    53, 61, 69, 73, 86, 90, 100, 112, 115,
    117, 118, 121, 124, 130, 131, 154,
    157, 164, 168, 169, 174, 176, 178,
    179, 181, 189–191, 220, 224, 226,
    228, 229, 235, 237, 240, 243, 245,
    246, 248, 256, 259, 262, 264, 265
    crack opening, 176, 262, 264
    plastic accumulated, 164
Domain, 47, 77, 80, 81, 93, 99, 100, 102,
    104, 105, 107, 109, 123, 143, 145,
    206, 207, 209, 211, 212, 265, 268,
    269
    magnetic, 265, 268
    wall, 80, 81, 143, 268, 269
    wall motion, 47, 81, 99, 100, 102, 104,
        105, 107, 109, 145, 206, 207, 209,
        211, 212
Double Fourier series, 112
DT, 201, 202
Dual integral equations, 36, 39, 42, 43, 52,
    59, 71, 115, 125, 156, 167, 170, 173,
    252, 253
Ductility, 189

Easy axis, 267
Elastic compliance, 74, 80, 82, 83, 92, 95,
    99, 200, 266, 272
Elastic stiffness, 74, 79, 83, 223

Electric conductivity, 6, 9, 223
Electric displacement, 6, 40, 41, 48, 73, 82, 151, 156, 157, 160, 161, 176, 272
intensity factor, 156, 160
Electric field, 1, 5, 9, 13, 41, 47–49, 51, 52, 54–56, 58, 60, 67, 69, 71–73, 77, 80–83, 90–93, 95–97, 99, 100, 102, 104, 109, 111, 113, 115, 119–124, 126, 130–133, 136–138, 140, 142, 144, 145, 149, 150, 152, 154, 157, 160, 163–165, 167, 168, 170, 172–179, 181–184, 186, 187, 189–191, 193–207, 209–211, 271, 272
AC, 80–82, 97, 99, 100, 102, 104, 109, 131, 138, 144, 150, 205–207, 209, 211, 271
coercive, 77, 82, 104, 122, 131, 142, 143, 145, 179, 198, 207, 211
concentration, 113
DC, 77, 92, 93, 95, 96, 100, 113, 119, 121, 126, 130, 136, 138, 144, 149, 150, 204, 207, 210, 211
intensity, 5, 40, 48, 73, 151, 157, 223, 272
intensity factor, 164
Electric potential, 42, 51, 115, 118, 120, 128, 132, 137, 154, 156, 157, 165, 168, 169, 174, 183, 188, 190, 195, 199, 202
Electric susceptibility, 49
Electrical conductivity, 224
Electrical discharge, 176
Electrode, 47, 90, 94, 99, 103–106, 108, 109, 113–120, 123, 125, 126, 128, 129, 132, 135–139, 143–146, 149, 190, 199
film, 103, 105, 107
fully, 143–145, 149
partially, 113, 144–146, 149
tip, 113, 116, 118, 119, 126, 139, 145, 146, 149
Electroelastic crack mechanics, 1, 47, 72
Electroelasticity, 1
Electromagnetic body force, 42, 223

Electromagnetic field, 10, 13, 18, 19, 21, 25, 27, 33
quasistatic, 18, 19, 21, 25, 27, 33
Electromagnetic force, 1, 5, 40, 45
Electromagnetic fracture mechanics, 33
Electromagneto-mechanics, 1
Electromechanical field, 1, 47, 93, 102, 113, 116, 118, 123, 127, 136, 137, 149, 151–153, 176
concentrations, 1, 47, 113, 123, 127, 136, 149
Electromechanical interaction, 1
Electromechanical loading, 80, 149, 186, 190, 191, 212
Electrostriction, 72
Elliptic hole, 152
Energy density, 80, 81, 159–164, 167, 190, 191, 266, 267
strain, 190, 191
Energy harvesting, 270, 273
Energy release rate, 49, 56, 150, 158–160, 162–164, 167, 170, 173, 175–180, 182–184, 188, 189, 195, 197–211
Equation of motion, 8, 11, 26, 27, 61, 69, 70, 81, 83, 98, 234, 236, 239, 240, 244, 246, 248
displacement, 27, 61, 240, 248
stress, 8, 26, 27, 83, 234, 239, 240, 244, 248

Failure, 2, 47, 49, 113, 146, 149, 150, 191, 194, 204–207, 209, 211, 262
cycle to, 211
time-to-, 204–207
Fatigue, 1, 2, 150, 203–207, 209–211, 219, 256, 260, 264
cyclic, 209, 211
dynamic, 207, 209
life, 204
static, 204–206
transgranular, 211
FEA, 93, 95, 96, 102, 104, 107, 109, 110, 118, 119, 122, 126, 128, 130–132, 136–139, 175, 177, 180, 184, 186, 188, 190, 193, 195, 197, 199–202, 209, 270

Ferroelectric materials, 1, 47, 73, 77
Ferromagnetic materials, 1, 2, 219–221,
        224, 225, 228, 231, 233, 251,
        255–257, 261, 263–265
    soft, 1, 2, 219, 220, 225, 231, 251,
        255–257, 261, 263, 264
Ferromagnetics, 1
FGPM, 100–105, 107, 109, 110
Finite element, 93, 98, 102, 118, 127, 138,
        139, 175, 176, 183, 188, 195, 199,
        202, 204, 206, 207, 209–211, 269
First-principles free energy, 142
Five-element strip gage, 253, 260
Flexural rigidity, 11, 227
Fluctuating field, 5
Fourier series, 231
Fourier transform, 36, 39, 42, 52, 59, 71,
        109, 115, 152, 156, 170, 173, 252,
        253, 256
Fracture, 1, 2, 33, 40, 45, 47, 72, 113, 139,
        149, 150, 153, 159, 163, 164, 173,
        175, 179, 180, 185–187, 189, 190,
        193–197, 199–211, 219, 256,
        261–263, 273
    brittle, 149
    cryogenic, 198
    intergranular, 190, 204
    mechanics, 72, 153, 173, 175, 180, 185
    tensile, 2, 256, 261
    toughness, 185–187, 193, 194, 219, 262,
        263
    transgranular, 205, 211
    unstable, 33, 150
Fredholm integral equation, 36, 39, 42, 43,
        59, 71, 252, 253, 260
    of the second kind, 36, 39, 42, 59, 71,
        252, 253, 260
Free electric charge density, 6, 40, 223
Frequency equation, 28, 30, 34, 38, 66, 242,
        250
FRP, 82, 83, 172

Grain, 93, 119, 120, 190, 212

Hankel transform, 125, 167
Hooke's law, 7, 26, 225, 233, 239, 248,
        258

IDE, 132
IF, 185, 186
    toughness, 186
Impedance, 128–132
Indentation, 163, 186, 187, 189
Instability, 1, 219, 233
    magnetoelastic, 1, 219
Insulating materials, 49
Integral transform, 42, 260
Integrity, 32, 150

J-integral, 55, 56, 159, 175, 195, 200, 204,
        206, 207, 209–211

Lamé constant, 6, 49, 221
Laminate, 47, 83, 86, 89, 90, 92, 111–113,
        116, 170, 172, 219, 267, 269–271,
        273
    hybrid, 83, 111–113, 172
    piezoelectric, 47, 90, 113
    thermopiezoelectric, 89, 90, 111
Laminated beam, 90–93, 95–98
Laminated plate, 83, 89, 111, 112, 172,
        173
Landau-Devonshire potential, 140
Lead-free ferroelectric material, 212
Linear elastic fracture mechanics, 33, 150

Magnetic field, 1, 2, 5, 7, 9, 11, 16, 18, 19,
        21, 23, 24, 26, 28, 30, 32, 33, 36,
        39–41, 45, 219, 224–226, 228, 229,
        231–237, 239, 243, 245, 246, 248,
        250–252, 254–258, 260–273
    intensity, 5, 12, 15, 28, 220, 265
Magnetic force, 256
Magnetic induction, 6, 16, 18, 20, 23, 26,
        33, 36, 40–42, 220, 224, 225, 228,
        232, 233, 235, 237, 239, 243, 245,
        246, 248, 251, 252, 254, 255, 257,
        260, 262, 264, 265
Magnetic permeability, 6, 7, 40, 221, 224,
        266, 272
    specific, 221
Magnetic susceptibility, 252
Magnetically saturated materials, 1, 219,
        221, 228

Magnetization, 219–221, 267
   saturation, 221
Magneto-mechanics, 1
Magneto-solid mechanics, 2
Magnetoelastic constant, 266, 268–270, 272
   second-order, 268–270
Magnetoelastic interactions, 5, 9, 13, 222
   dynamic, 5
Magnetoelastic thin plate, 9
Magnetoelasticity, 1, 256
   dynamic, 1
Magnetostriction, 2, 265, 268–270
Magnetostrictive alloy, 266
Magnetostrictive constant, 222, 223
Magnetostrictive materials, 1, 2, 219, 269,
   273
Mass density, 6, 48, 73, 99, 129, 144, 220,
   265
ME susceptibility, 272
Mechanical behavior, 1, 33, 45, 265
Mechanical force, 256
Mechanical loading, 47, 51, 160, 204, 219,
   273
   electromagneto-, 219, 273
MEMS, 1, 136, 139
   mirror, 136, 139
MFC, 127, 132, 135, 136
Micro-electromechanics, 136
Microelectromechanics, 133, 136
Mindlin plate, 15, 23, 24, 30, 31, 33, 36, 37,
   40–42, 237, 238, 243, 246, 255,
   256
Mole fraction, 75, 141
Moment, 2, 10, 11, 14, 33, 35, 36, 39–44,
   73, 86, 88, 90, 92, 97, 172, 173, 219,
   226, 235–237, 250–256
   bending, 10, 14, 36, 43, 90, 92, 97, 172,
      226, 235, 237, 251–253
   resultant, 86
   twisting, 10, 14, 41–43, 235, 237
Moment intensity factor, 2, 33, 36, 39, 40,
   43, 44, 173, 219, 250, 252–256
   bending, 36
   twisting, 40, 43, 44
Monomorph, 90, 92, 93, 95–97, 99
MPB, 75, 140, 141, 144, 201

MSP, 189–191
Multilayer, 113, 116, 117, 119, 140

Nonlinear magneto-mechanical response, 2,
   219, 267

Ohm's law, 8, 41

Partially poled, 119, 120, 133, 137, 199
Path-independent integral, 49, 55, 56
Perfect conductivity, 20, 21, 24, 28, 31, 33,
   37, 39
Permittivity, 6, 7, 40, 49, 74, 80, 83, 150,
   152, 200, 223, 272
   speciific, 49
Perovskite oxide, 73
Phase, 18, 19, 21, 24, 30, 31, 67, 75, 129,
   130, 140, 141, 212, 243
   diagram, 75, 140, 141
   field, 212
   velocity, 18, 19, 21, 24, 30, 31, 67, 243
Piezo-magnetic constant, 268
Piezoelectric ceramics, 47, 114, 115,
   123–125, 150, 163, 186, 189, 193,
   198, 201, 203–205, 207, 209–211
Piezoelectric coefficient, 74, 79, 80, 82, 83,
   85, 93, 133, 141, 142, 147, 200, 213,
   272
   direct, 74, 79, 80, 82, 83, 85, 93, 147,
      272
   inverse, 74, 272
Piezoelectric constant, 74, 79, 83, 85, 104
Piezoelectric materials, 1, 47, 72, 80, 82,
   101, 111, 113, 119, 123, 149, 150,
   152, 154, 159, 160, 163–165, 167,
   173–176, 179–182, 184, 188, 203,
   212, 269
Piezoelectric property, 75, 100, 110, 140,
   147, 175
Piezoelectricity, 1, 72, 100, 170
Plane strain, 26, 27, 30, 31, 33, 49, 58, 61,
   69, 113–115, 151, 154, 167, 174, 181,
   184, 195, 199, 204, 206, 207,
   209–211, 225, 239, 240, 243, 248,
   258
   plate, 27, 30, 31, 239, 240, 243, 248

Poisson's ratio, 6, 18, 51, 85, 95, 99, 126, 129, 135, 144, 221, 270

Polarization, 47, 48, 55, 72–74, 77, 79–81, 93, 95, 96, 99, 115, 119, 120, 123, 124, 126, 130, 131, 136, 138, 140, 147, 154, 164, 169, 175, 179, 190, 191, 195, 198, 212, 222

  remanent, 73, 77, 115, 124, 169

  spontaneous, 73, 77, 80, 93, 99

  switching, 47, 74, 77, 79, 80, 93, 95, 96, 119, 126, 130, 131, 136, 138, 175, 179, 190, 191, 198, 212

Pole, 219

Poling, 73, 77, 79, 91, 99, 102, 113, 116, 117, 123, 125, 127, 137, 144, 147, 148, 151, 153, 154, 163, 165, 167, 168, 174, 181–184, 186–190, 194, 195, 198, 199, 201

PVDF, 83, 84, 111–113

Pyroelectric constant, 83

PZT, 75, 82, 84, 90, 93, 94, 99–104, 107, 111, 116–118, 121, 126, 128, 129, 132, 133, 135–143, 147, 149, 152, 176, 182–184, 186, 189, 190, 195, 196, 198, 201, 204, 205, 207–211, 270, 271

  rod, 128, 129

Resonance frequency, 81, 132, 139

Resonant frequency, 130, 137–139

Resultant force, 86

Scattering, 33, 40, 68, 127, 164, 170, 173

Self-monitoring, 82, 270, 273

Self-sensing, 105, 109, 270, 273

SEM, 199, 201, 204, 206, 210, 211

SEPB, 185, 193–195, 197–199, 203–207, 209–211

Shear, 6, 10, 13, 14, 35, 40, 43, 44, 68, 70, 71, 79, 85, 86, 91, 105, 127, 168–170, 172, 221, 226, 235, 237, 251, 256, 267

  antiplane, 127, 169, 170

  modulus, 6, 85, 221

Shear force, 10, 14, 40, 43, 44, 86, 226, 235, 237, 251

  intensity factor, 40, 43, 44

  vertical, 10, 14, 43, 86, 226, 235, 237

Simply supported, 95, 96, 111, 251

Sound pressure, 109, 110, 138

SP, 189

Strain, 72–74, 77, 80–82, 85, 86, 91, 93, 99, 115, 118, 119, 124, 126, 133, 135, 136, 144, 148, 149, 152, 154, 156, 157, 160, 165, 168, 169, 174, 181, 190, 191, 224, 254, 256, 260, 265, 266, 269, 270

  compressive, 119, 126, 136

  gage, 118, 135, 144, 148, 149, 270

  normal, 119, 144, 149, 152

  radial, 126, 254

  remanent, 73, 77, 85, 115, 124, 169

  spontaneous, 73, 93, 99

  tensile, 126, 136, 154, 165

Stress, 1, 5, 6, 8, 10, 26, 27, 29, 45, 47–50, 52, 54, 56, 58–60, 72–74, 77, 83–86, 91, 104, 105, 107, 109, 111, 113, 116, 121–123, 139, 145, 146, 149, 151, 152, 154, 156, 157, 160, 167, 170, 172, 175, 179, 180, 183, 207, 219–222, 225, 227–229, 233, 234, 239, 240, 244, 248, 256–258, 260, 262, 264–266, 269, 270, 272

  boundary condition, 10, 27, 29, 86

  compressive, 85, 121–123

  concentration, 113

  first Piola-Kirchhoff-, 222

  local, 47, 48, 52, 54

  magnetoelastic, 220, 225, 229, 233, 244, 258

  Maxwell, 6, 8, 29, 49, 50, 52, 54, 219, 221, 225, 228, 233, 258

  normal, 105, 107, 116, 139, 145, 149, 151, 152, 154, 156, 167, 175, 257

  plane, 84, 258, 260, 262, 264

  shear, 91, 105, 170

  singular, 1, 5, 45

  tensile, 49, 58, 85

Stress intensity factor, 33, 40, 49, 52, 55, 56, 59, 60, 71, 72, 150, 156, 159, 160, 163, 164, 167, 170, 187, 201, 219, 256, 260, 262–264

electric, 52, 55, 56, 59, 60, 71, 72
  magnetic, 260, 262, 264
Stress-temperature coefficient, 83, 85
Strip, 40, 57, 58, 60, 64, 65, 67, 163, 164,
      170, 253, 256, 260, 262, 264
Surface charge density, 7
Surface current density, 7

Temperature, 47, 75, 83, 111, 112,
      140–144, 146–149, 172, 182, 200,
      201, 212, 219, 232, 266, 272
  cryogenic, 146, 149, 182, 201
  Curie, 141, 147, 148
  high, 47, 140, 149, 212
Tensile, 2, 49, 58, 85, 116, 126, 136, 154,
      164, 165, 219, 256, 261, 263, 264
Tension, 116, 262, 264
Terfenol-D, 266–271, 273
Thermopiezoelectric layers, 82
Thermopiezoelectric materials, 82, 83, 172
Twisting, 10, 14, 41–43, 235, 237

Unimorph, 90, 100, 136, 137
Unpoled, 119–122, 132, 198

Vibration, 1, 5, 7, 45, 47, 60, 72, 81, 219,
      233
  electroelastic, 1, 47, 60, 72
  lattice, 81
  magnetoelastic, 1, 5, 7, 45, 219, 233
Viscoelastic damping, 83, 111, 172
Voltage, 73, 90, 95, 96, 99, 104–107, 109,
      110, 114, 116, 118, 120, 121, 123,
      125, 128–130, 132, 135, 137, 138,
      144, 148, 271
  AC, 99, 104–107, 109, 118, 128, 137,
      144

DC, 118, 120, 128, 137, 144, 148
driving, 105, 107, 110
induced, 271
output, 106, 107, 109, 110, 271

Wave, 1, 5, 7, 16–19, 21, 23, 24, 26–28, 30,
      31, 33–36, 38–40, 45, 47, 60–63,
      66–68, 70–72, 127, 164, 170, 173,
      219, 224, 233, 235, 237, 239–243,
      245, 246, 248, 250, 255, 256
  antiplane shear, 127, 170
  elastic, 18, 28, 40
  electroelastic, 1, 47, 60, 72
  flexural, 1, 16, 18, 21, 23, 26, 33, 34, 36,
      38, 40, 173, 235, 237, 239, 243,
      245, 246, 248, 255
  horizontally polarized shear, 170
  incident, 34, 38, 39, 70, 255
  incident flexural, 34, 255
  longitudinal, 28, 63, 241
  magnetic flexural, 16, 23, 235, 237, 239,
      245, 246, 248
  magnetoelastic, 1, 5, 7, 26, 45, 219, 224,
      233
  number, 17–19, 21, 24, 30, 31, 62, 67,
      71, 237, 243
  P, 68, 70, 71
  plane, 224
  Rayleigh, 45
  scattered, 35, 256
  shear, 18, 61, 170, 237
  SV, 68, 70, 71
Wavelength, 67, 224, 228

Yield strip model, 164
Young's modulus, 6, 92, 126, 129, 135, 144,
      221, 270